Advances in Intelligent Systems and Computing

Volume 765

Series editor

Janusz Kacprzyk, Polish Academy of Sciences, Warsaw, Poland
e-mail: kacprzyk@ibspan.waw.pl

The series "Advances in Intelligent Systems and Computing" contains publications on theory, applications, and design methods of Intelligent Systems and Intelligent Computing. Virtually all disciplines such as engineering, natural sciences, computer and information science, ICT, economics, business, e-commerce, environment, healthcare, life science are covered. The list of topics spans all the areas of modern intelligent systems and computing such as: computational intelligence, soft computing including neural networks, fuzzy systems, evolutionary computing and the fusion of these paradigms, social intelligence, ambient intelligence, computational neuroscience, artificial life, virtual worlds and society, cognitive science and systems, Perception and Vision, DNA and immune based systems, self-organizing and adaptive systems, e-Learning and teaching, human-centered and human-centric computing, recommender systems, intelligent control, robotics and mechatronics including human-machine teaming, knowledge-based paradigms, learning paradigms, machine ethics, intelligent data analysis, knowledge management, intelligent agents, intelligent decision making and support, intelligent network security, trust management, interactive entertainment, Web intelligence and multimedia.

The publications within "Advances in Intelligent Systems and Computing" are primarily proceedings of important conferences, symposia and congresses. They cover significant recent developments in the field, both of a foundational and applicable character. An important characteristic feature of the series is the short publication time and world-wide distribution. This permits a rapid and broad dissemination of research results.

More information about this series at http://www.springer.com/series/11156

Radek Silhavy
Editor

Cybernetics and Algorithms in Intelligent Systems

Proceedings of 7th Computer Science
On-line Conference 2018, Volume 3

 Springer

Editor
Radek Silhavy
Faculty of Applied Informatics
Tomas Bata University in Zlín
Zlín
Czech Republic

ISSN 2194-5357   ISSN 2194-5365   (electronic)
Advances in Intelligent Systems and Computing
ISBN 978-3-319-91191-5   ISBN 978-3-319-91192-2   (eBook)
https://doi.org/10.1007/978-3-319-91192-2

Library of Congress Control Number: 2018942341

Printed on acid-free paper

This Springer imprint is published by the registered company Springer International Publishing AG part of Springer Nature
The registered company address is: Gewerbestrasse 11, 6330 Cham, Switzerland

Preface

This book constitutes the refereed proceedings of the modern trends, and approaches of artificial intelligence research and its application to intelligent systems are presented in this book. Paper discusses hybridization of algorithms, new trends in neural networks, optimization algorithms, and real-life issues related to artificial method application.

This book constitutes the refereed proceedings of the Artificial Intelligence and Algorithms in Intelligent Systems of the 7th Computer Science On-line Conference 2018 (CSOC 2018), held online in April 2018.

CSOC 2018 has received (all sections) 265 submissions, 141 of them were accepted for publication. More than 60% of accepted submissions were received from Europe, 30% from Asia, 5% from Africa, and 5% from America. Researches from 30 countries participated in CSOC 2018 conference.

CSOC 2018 conference intends to provide an international forum for the discussion of the latest high-quality research results in all areas related to computer science. The addressed topics are the theoretical aspects and applications of computer science, artificial intelligence, cybernetics, automation control theory, and software engineering.

Computer Science On-line Conference is held online, and modern communication technology, which is broadly used, improves the traditional concept of scientific conferences. It brings equal opportunity to participate in all researchers around the world.

I believe that you will find the following proceedings interesting and useful for your own research work.

March 2018
Radek Silhavy
Editor

Organization

Program Committee

Program Committee Chairs

Petr Silhavy	Tomas Bata University in Zlin, Faculty of Applied Informatics
Radek Silhavy	Tomas Bata University in Zlin, Faculty of Applied Informatics
Zdenka Prokopova	Tomas Bata University in Zlin, Faculty of Applied Informatics
Roman Senkerik	Tomas Bata University in Zlin, Faculty of Applied Informatics
Roman Prokop	Tomas Bata University in Zlin, Faculty of Applied Informatics
Viacheslav Zelentsov	Doctor of Engineering Sciences, Chief Researcher of St. Petersburg Institute for Informatics and Automation of Russian Academy of Sciences (SPIIRAS)

Program Committee Members

Boguslaw Cyganek	Department of Computer Science, University of Science and Technology, Krakow, Poland
Krzysztof Okarma	Faculty of Electrical Engineering, West Pomeranian University of Technology, Szczecin, Poland
Monika Bakosova	Institute of Information Engineering, Automation and Mathematics, Slovak University of Technology, Bratislava, Slovak Republic

Pavel Vaclavek	Faculty of Electrical Engineering and Communication, Brno University of Technology, Brno, Czech Republic
Miroslaw Ochodek	Faculty of Computing, Poznan University of Technology, Poznan, Poland
Olga Brovkina	Global Change Research Centre Academy of Science of the Czech Republic, Brno, Czech Republic and Mendel University of Brno, Czech Republic
Elarbi Badidi	College of Information Technology, United Arab Emirates University, Al Ain, United Arab Emirates
Luis Alberto Morales Rosales	Head of the Master Program in Computer Science, Superior Technological Institute of Misantla, Mexico
Mariana Lobato Baes	Superior Technological of Libres, Mexico
Abdessattar Chaâri	Laboratory of Sciences and Techniques of Automatic control and Computer engineering, University of Sfax, Tunisian Republic
Gopal Sakarkar	Shri. Ramdeobaba College of Engineering and Management, Republic of India
V. V. Krishna Maddinala	GD Rungta College of Engineering & Technology, Republic of India
Anand N. Khobragade	Maharashtra Remote Sensing Applications Centre, Republic of India
Abdallah Handoura	Computer and Communication Laboratory, Telecom Bretagne, France

Technical Program Committee Members

Ivo Bukovsky	Roman Senkerik
Maciej Majewski	Petr Silhavy
Miroslaw Ochodek	Radek Silhavy
Bronislav Chramcov	Jiri Vojtesek
Eric Afful Dazie	Eva Volna
Michal Bliznak	Janez Brest
Donald Davendra	Ales Zamuda
Radim Farana	Roman Prokop
Martin Kotyrba	Boguslaw Cyganek
Erik Kral	Krzysztof Okarma
David Malanik	Monika Bakosova
Michal Pluhacek	Pavel Vaclavek
Zdenka Prokopova	Olga Brovkina
Martin Sysel	Elarbi Badidi

Organizing Committee Chair

Radek Silhavy   Tomas Bata University in Zlin, Faculty of Applied
           Informatics

Conference Organizer (Production)

OpenPublish.eu s.r.o.
Web: http://www.openpublish.eu
Email: csoc@openpublish.eu

Conference Website, Call for Papers

http://www.openpublish.eu

Contents

**A Binary Grasshopper Optimisation Algorithm Applied
to the Set Covering Problem**   1
Broderick Crawford, Ricardo Soto, Alvaro Peña, and Gino Astorga

**A Survey on Signal Processing Methods in Fiber Optic Sensor
for Oxidized Carbon Steel**   13
Nur Syakirah Mohd Jaafar, Izzatdin Abdul Aziz, Jafreezal Jaafar,
Ahmad Kamil Mahmood, and Abdul Rehman Gilal

**Some Recent Results on Direct Delay-Dependent Stability Analysis:
Review and Open Problems**   25
Libor Pekař, Pavel Navrátil, and Radek Matušů

**Modelling and Identification of Magnetic Levitation Model CE
152/Revised**   35
Daniel Honc

DDoS Reflection Attack Based on IoT: A Case Study   44
Marek Šimon, Ladislav Huraj, and Tibor Horák

**Complemented Adaptive Control Strategy with Application
in Pedagogical Cybernetics**   53
Tomas Barot

**Robust and Lightweight Image Encryption Approach
Using Public Key Cryptosystem**   63
Shima Ramesh Maniyath and V. Thanikaiselvan

Computer Modeling of Personal Autonomy and Legal Equilibrium . . .   74
Yurii Sheliazhenko

**Improving the Performance of Hierarchical Clustering Protocols
with Network Evolution Model**   82
Chiranjib Patra and Nicolea Botezatu

Efficient Control of SEPIC DC-DC Converter with Dynamic Switching Frequency 96
Samuel Žák, Peter Ševčík, and Martin Revák

Multiple-Model Description and Control Construction Algorithm of Supply Chain 102
Inna Trofimova, Boris Sokolov, and Dmitry Ivanov

A New Approach to Vector Field Interpolation, Classification and Robust Critical Points Detection Using Radial Basis Functions 109
Vaclav Skala and Michal Smolik

Trusted Cryptographic Tools Locking 116
Vadim N. Tsypyschev

Calculation of the Closed Multi-channel Queueing Systems 125
Yuri Ryzhikov

MATLAB as a Tool for Modelling and Simulation of the Nonlinear System 133
Jiri Vojtesek and Lubos Spacek

An Input to State Stability Approach for Evaluation of Nonlinear Control Loops with Linear Plant Model 144
Peter Benes and Ivo Bukovsky

The Technique of Informational Interaction Structural-Parametric Optimization of an Earth's Remote Sensing Small Spacecraft Cluster 155
Jury S. Manuilov, Alexander N. Pavlov, Dmitry A. Pavlov, and Alexey A. Slin'ko

Collaborative Robot YuMi in Ball and Plate Control Application: Pilot Study 167
Lubos Spacek, Jiri Vojtesek, and Jiri Zatopek

Integration of Heterogeneous Data in Monitoring Environmental Assets 176
Dmitrii Verzilin, Tatyana Maximova, Yury Antokhin, and Irina Sokolova

Hidden Asymmetry in Shape of Biological Patterns 186
Sergey G. Baranov

Classification, Clustering and Association Rule Mining in Educational Datasets Using Data Mining Tools: A Case Study 196
Sadiq Hussain, Rasha Atallah, Amirrudin Kamsin, and Jiten Hazarika

Review of Research Progress, Trends and Gap in Occupancy Sensing for Sophisticated Sensory Operation 212
Preethi K. Mane and K. Narasimha Rao

A 3D Visualization Application of Zlín in the Eighteen-Nineties 223
Pavel Pokorný and Pavla Dočkalová

**Multi-agent Systems Interacting (Addressing Scopes,
Control Resources)**... 233
Mohamad Kadi, Said Krayem, Roman Jasek, Petr Zacek,
and Bronislav Chramcov

**Improved Adaptive Fault Tolerance Model for Increasing
Reliability in Cloud Computing Using Event-B** 246
Ammar Alhaj Ali, Roman Jasek, Said Krayem, Bronislav Chramcov,
and Petr Zacek

**Adaptive Access Mechanism Based on Network State Detection
in Multi-rate IEEE802.11 WLANs** 259
Jianjun Lei, Shengjie Peng, and Yu Dai

**Reconstruction of 3D Permittivity Profile of a Dielectric Sample
Using Artificial Neural Network Mathematical Model
and FDTD Simulation** 272
Mikhail Abrosimov, Alexander Brovko, Ruslan Pakharev, Anton Pudikov,
and Konstantin Reznikov

**Novelty Detection System Based on Multi-criteria Evaluation
in Respect of Industrial Control System**......................... 280
Jan Vávra and Martin Hromada

**TRMA: An Efficient Approach for Mutual Authentication
of RFID Wireless Systems** 290
R. Anusha and V. Veena Devi Shastrimath

**SC-MANET: Threats, Risk and Solution Strategies for Security
Concerns in Mobile Ad-Hoc Network** 300
C. K. Vanamala and G. Raghvendra Rao

**DSMANET: Defensive Strategy of Routing Using Game Theory
Approach for Mobile Adhoc Network** 311
K. Pradeep Kumar and B. R. Prasad Babu

**OCSLM: Optimized Clustering with Statistical Based Local Model
to Leverage Distributed Mining in Grid Architecture** 321
M. Shahina Parveen and G. Narsimha

**New Numerical Investigation Using Meshless Methods Applied
to the Linear Free Surface Water Waves**......................... 337
Mohamed Loukili and Soumia Mordane

**Framework for Capturing the Intruders in Wireless Adhoc
Network Using Zombie Node** 346
Jyoti Neeli and N. K. Cauvery

SDQE: Sensor Data Quality Enhancement in Reconfigurable Network for Optimal Reliability 356
B. Prathiba, K. Jaya Sankar, and V. Sumalatha

Relaxed Greedy-Based Approach for Enhancing of Resource Allocation for Future Cellular Network 364
Chanda V. Reddy and K. V. Padmaja

ITM-CLD: Intelligent Traffic Management to Handling Cloudlets of the Large Data 374
Chetana Tukkoji and K. Seetharam

A Novel Computational Modelling to Optimize the Utilization of Intrusion Detection Paradigm in a Large-Scale MANET 382
Najiya Sultana

DSP-IR: Delay Sensitive Protocol for Intelligent Routing with Medium Access Control 393
A. C. Yogeesh, Shantakumar B. Patil, Premajyothi Patil, and H. R. Roopashree

A Novel, Lightweight, and Cost-Effective Mechanism to Secure the Sensor-Gateway Communication in IoT 403
Shamshekhar S. Patil and N. R. Sunitha

Quality of Service (QoS) Aware Reconfigurable Optical Add/Drop Multiplexers (ROADM) Model with Minimizing the Blocking Rate 413
G. R. Kavitha and T. S. Indumathi

A Mixed Hybrid Conjugate Gradient Method for Unconstrained Engineering Optimization Problems 423
David A. Oladepo, Olawale J. Adeleke, and Churchill T. Ako

Chemical Reaction Optimization for Traveling Salesman Problem Over a Hypercube Interconnection Network 432
Ameen Shaheen, Azzam Sleit, and Saleh Al-Sharaeh

The Concept of the Method for Dynamic Control of Traffic Flows on Multi-lane Roads Based on Configurable Information Systems 443
Sergey Kucherov, Yuri Rogozov, Julia Lipko, and Dmitry Elkin

Author Index 451

A Binary Grasshopper Optimisation Algorithm Applied to the Set Covering Problem

Broderick Crawford, Ricardo Soto, Alvaro Peña$^{(\boxtimes)}$, and Gino Astorga

School of Engineering, Pontificia Universidad Católica de Valparaíso,
Valparaíso, Chile
{broderick.crawford,ricardo.soto,alvaro.pena}@pucv.cl, gino.astorga@uv.cl

Abstract. Many of the problems addressed at the industrial level are of a combinatorial type and a sub-assembly not less than these are of the NP-hard type. The design of algorithms that solve combinatorial problems based on the continuous metaheuristic of swarm intelligence is an area of interest at an industrial level. In this article, we explore a general binarization mechanism of continuous metaheuristics based on the percentile concept. In particular, we apply the percentile concept to the Grasshopper optimization algorithm in order to solve the set covering problem (SCP). The experiments are designed with the aim of demonstrating the usefulness of the percentile concept in binarization. Additionally, we verify the effectiveness of our algorithm through reference instances. The results indicate the binary grasshopper optimization algorithm (BGOA) obtains adequate results when evaluated with a combinatorial problem such as the SCP.

1 Introduction

The combinatorial problems have great relevance at industrial level, we find them in various areas such as Civil Engineering [1], Bio Informatics [2], Operational Research [3], resource allocation [4], scheduling problems [5,6], routing problems [7,8] among others. On the other hand, in recent years, algorithms inspired by nature phenomena to solve optimization problems have been generated. As examples of these algorithms we have Cuckoo Search [9], Black Hole [10], Bat Algorithm [11], and Grasshopper Optimisation Algorithm [12] among others. Many of these algorithms work naturally in continuous spaces and therefore must be adapted to solve combinatorial problems. In the process of adaptation, the mechanisms of exploration and exploitation of the algorithm can be altered, having consequences in the efficiency of the algorithm.

Several binarization techniques have been developed to address this situation. In a literature search, the main binarization methods used correspond to transfer functions, angle modulation and quantum approach. In this article we present a binarization method that uses the percentile concept to group the solutions and

© Springer International Publishing AG, part of Springer Nature 2019
R. Silhavy (Ed.): CSOC 2018, AISC 765, pp. 1–12, 2019.
https://doi.org/10.1007/978-3-319-91192-2_1

then perform the binarization process. To verify the efficiency of our method, we used the BGOA. This algorithm was proposed in [12] and was applied to test functions in addition to structural engineering problems.

Grasshoppers are insects and under certain conditions can be considered a pest, the main characteristic of these insects is related to their movement. When it is in a larval state, its movement is slow in contrast with the large and abrupt movements of the adults.

When we look at different areas of engineering and science, we see that combinatorial problems appear frequently. We find them in various areas such as Civil Engineering [1], Bio Informatics [2], Operational Research [3], resource allocation [4], scheduling problems [5], routing problems [7,8] among others. On the other hand, in recent years, algorithms inspired by natural phenomena have been able to solve NP-hard combinatorial problems of significant size. As an example of algorithms inspired by nature, we have Cuckoo Search [9], Black Hole [10], Bat Algorithm [11], and Grasshopper Optimisation Algorithm [12], among others. Many of these algorithms work naturally in continuous spaces and, therefore, must be adapted to solve combinatorial problems. In the process of adaptation, the mechanisms of exploration and exploitation of the algorithm can be altered, having consequences in the efficiency of the algorithm. Several binarization techniques have been developed to address this situation. In a bibliographic search, the main binarization methods used are: transfer functions, angle modulation, and quantum approach. In this article, we develop a binarization method that uses the percentile concept to cluster the solutions and then perform the binarization process. To verify the effectiveness of our method, we use BGOA. This algorithm was proposed in [11] and applied to test functions in addition to structural engineering problems. Grasshoppers are insects and under certain conditions can be considered a pest, the main characteristic of these insects is related to their movement. When it is in a larval state, its movement is slow in contrast to the large and abrupt movements of adults.

The mathematical model to simulate the movement of grasshoppers are presented in Eq. 1

$$X_i^d = c(\sum_{j=1,j\neq i}^{N} c\frac{ub_d - lb_d}{2}s(|x_j^d - x_i^d|)\frac{x_j - x_i}{d_{ij}}) + \hat{T}_d \tag{1}$$

The first c reduces the movements of grasshoppers around the target. In other words, this parameter balances exploration and exploitation of the entire swarm around the target. The second c decreases the attraction zone, comfort zone, and repulsion zone between grasshoppers. The c coefficient is calculated as: $cmax - l\frac{cmax-cmin}{L}$, where L = 500 is de maximum number of iteration. The cmax, and cmin coefficients are 1 and 0.00001 respectively. The component $s(r) = fe^{\frac{-r}{l}} - e^{-r}$ indicates if a grasshopper should be repelled from (explore) or attracted to (exploitation) the target. For this case $f = 0.5$ and $l = 1.5$. Finally $\hat{T}_d$ represents the best solution at this iterations.

To check our binary grasshopper algorithm (BGOA), we use the well-known set covering problem. Experiments were developed using a random operator to validate the contribution of the percentile technique in the binarization process of the Grasshopper algorithm. In addition, to verify our results, the JPSO algorithm developed by [13] and MDBBH developed in [5] were chosen. The results show that BGOA algorithm obtain competitive results.

2 Set Covering Problem

The set covering problem corresponds to a classical problem in combinatorics, and complexity theory. The problem aims to find subsets of a set. Given a set and its elements, we want to find subsets that completely cover the set at a minimum cost. The SCP is one of the oldest and most studied optimization problems. It is well-known to be NP-hard [14].

SCP is an active problem because medium and large instances often become intractable and cannot be solved any more using exact algorithms. Additionally, SCP due to its large number of instances, is used to verify the behavior of proposed new algorithms. In recent years, SCP has been approached by various continuous Swarm intelligence metaheuristics and using different methods to binarize metaheuristics. In [15], they used the teaching-learning-based optimisation with a specific binarization scheme to solve medium and large size instances. An SCP in fault diagnosis application was developed by [16]. In this article, the Gravitational Search algorithm was used and transfer functions were applied to perform the binarization. A rail scheduling problem application using SCP was developed in [5]. In [13] a Jumping PSO was used to solve SCP and in [4] CS metaheuristics were binarized using a percentile algorithm.

SCP has many practical applications in engineer, e.g., vehicle routing, facility location, railway, and airline crew scheduling [3, 17–19] problems.

The SCP can be formally defined as follows. Let $A = (a_{ij})$, be a $n \times m$ zero-one matrix, where a column j cover a row i if $a_{ij} = 1$, besides a column j is associated with a non-negative real cost c_j. Let $I = \{1, ..., n\}$ and $J = \{1, ...m\}$, be the row and column set of A, respectively. The SCP consists in searching a minimum cost subset $S \subset J$ for which every row $i \in I$ is covered by at least one column $j \in J$, i.e,:

$$\text{Minimize } f(x) = \sum_{j=1}^{m} c_j x_j \tag{2}$$

$$\text{Subject to } \sum_{j=1}^{m} a_{ij} x_j \geq 1, \forall i \in I, \text{ and } x_j \in \{0, 1\}, \forall j \in J \tag{3}$$

where $x_j = 1$ if $j \in S$, $x_j = 0$ otherwise.

3   Binary GrassHoppers Algorithm

The application of statistics and machine learning techniques, are used in many areas such as transports, smart cities, agriculture and computational intelligence, examples of applications are found in [20–24]. In this article, we explore the percentile concept applied to binarization of continuous metaheuristics. As an algorithm to binarize, Grasshopper is used and as problem to solve, SCP. The Proposed BGOA has two modules. The first module corresponds to the initialization of the feasible solutions, and is detailed in the Sect. 3.1. Once the initialization of the particles is performed, it is consulted if the maximum if iterations is satisfied. Subsequently if the criterion is not satisfied, The BGOA is then run in conjunction with the percentile operator, this is detailed in Sect. 3.2). Once the transitions of the different solutions are made, we compare the resulting solutions with the best solution previously obtained. In the event that a superior solution is found, this replaces the previous one. The general algorithm scheme is detailed in Fig. 1.

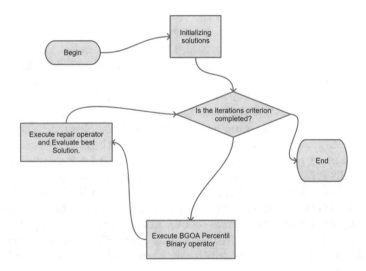

Fig. 1. Flowchart of the BGOA algorithm.

3.1   Initialization

For initialization of a new solution, a column is randomly chosen. It is then queried whether the current solution covers all rows. In the case of the solution does not meet the coverage condition, the heuristic operator is called (Sect. 3.3). This operator aims to select a new column. This heuristic operation is iterated until all rows are covered. Once the coverage condition is met, the solution is optimized. The optimization consists of eliminating columns where all their rows are covered by more than one column. The detail of the procedure is shown in Algorithm 1.

Algorithm 1. Initialization Operator

1: **Function** Initialization()
2: **Input**
3: **Output** Initialized solution S_{out}
4: $S \leftarrow$ SRandomColumn()
5: **while** All row are not covered **do**
6:     S.append(Heuristic(S))
7: **end while**
8: $S \leftarrow$ dRepeatedItem(S)
9: $S_{out} \leftarrow S$
10: **return** S_{out}

3.2   Percentile Binary Operator

Since our BGO algorithm is continuous swarm intelligence metaheuristic, it works in an iterative way by updating the position and velocity of the particles in each iteration. As BGO is continuous, the update is done in $\mathbb{R}^n$. In Eq. 4, the position update is written in a general way. The $x(t+1)$ variable represents the x position of the particle at time $t+1$. This position is obtained from the position x at time t plus a Δ function calculated at time $t+1$. The function Δ is proper to each metaheuristic and produces values in $\mathbb{R}^n$. For example in Cuckoo Search $\Delta(x) = \alpha \oplus Levy(\lambda)(x)$, in Black Hole $\Delta(x) = \text{rand} \times (x_{bh}(t) - x(t))$ and in the Firefly, Bat and PSO algorithms Δ can be written in simplified form as $\Delta(x) = v(x)$.

$$x(t+1) = x(t) + \Delta_{t+1}(x(t)) \tag{4}$$

In the percentile binary operator, we considering the movements generated by the algorithm in each dimension for all particles. $\Delta^i(x)$ corresponds to the magnitude of the displacement $\Delta(x)$ in the i-th position for the particle x. Subsequently these displacement are grouped using $\Delta^i(x)$, the magnitude of the displacement. This grouping is done using the percentile list. In our case the percentile list used the values {20, 40, 60, 80, 100}.

The percentile operator has as entry the parameters percentile list (pList) and the list of values (vList). Given an iteration, the list of values corresponds to the magnitude $\Delta^i(x)$ of the displacements of the particles in each dimension. As a first step the operator uses the vList and obtains the values of the percentiles given in the pList. Later, each value in the vList is assigned the group of the smallest percentile to which the value belongs. Finally, the list of the percentile to which each value belongs is returned (pGroupValue). The algorithm is shown in Algorithm 2.

A transition probability through the function P_{tr} is assigned to each element of the vList. This assignment is done using the percentile group assigned to each value. For the case of this study, we particularly use the Step function given in Eq. 5.

$$P_{tr}(x^i) = \begin{cases} 0.1, & \text{if } x^i \in \text{group } \{0,1\} \\ 0.5, & \text{if } x^i \in \text{group } \{2,3,4\} \end{cases} \tag{5}$$

Afterwards the transition of each particle is performed. In the case of GOA search the Eq. 6 is used to perform the transition, where $\hat{x}^i$ is the complement of x^i. Finally, each solution is repaired using the heuristic operator.

$$x^i(t+1) := \begin{cases} \hat{x}^i(t), & \text{if } rand < P_{tg}(x^i) \\ x^i(t), & \text{otherwise} \end{cases} \tag{6}$$

Algorithm 2. percentile binary operator

1: **Function** percentileBinary(vList, pList)
2: **Input** vList, pList
3: **Output** pGroupValue
4: percentileValue = getPValue(vList, pList)
5: **for each** value in vList **do**
6:     pGroupValue = getPGroupValue(pValue,vList)
7: **end for**
8: **return** pGroupValue

3.3   Heuristic Operator

The goal of the Heuristic operator is to select a new column for cases where a solution needs to be built or repaired. As input variables, the heuristic operator considers the incomplete solution S_{in} which must be completed. The operator obtains the columns that belong to S_{in}, then obtains the rows R that are not covered by the solution to S_{in}. With the set of rows not covered and using Eq. 7 we obtain in line 4 the best 5 rows to be covered. With this list of rows ($lRows$) on line 5 we obtain the list of the best columns according to the heuristic indicated in Eq. 8. Finally randomly in line 6 we obtain the column to incorporate.

$$WeightRow(i) = \frac{1}{L_i}. \tag{7}$$

where L_i is the sum of all ones in row i

$$WeightColumn(j) = \frac{c_j}{|R \cap M_j|}. \tag{8}$$

where M_j is the set of rows covered by Col j

Algorithm 3. Heuristic operator

1: **Function** Heuristic(S_{in})
2: **Input** Input solution S_{in}
3: **Output** The new column $colOut$
4: $lRows \leftarrow$ getBestRows(S_{in}, N=10)
5: $listcolumsnOut \leftarrow$ getBestCols($LRows$, M=5)
6: $colOut \leftarrow$ getCol($lcolsOut$)
7: **return** $colOut$

---

4 Results

In this section we detail the behavior of BGOA when it is applied to SCP. The contribution of the binary percentile operator was studied when solving the different SCP instances. Additionally, BGOA was compared with other algorithms that have recently resolved SCP. To solve the different SCP instances, a PC with Windows 10, core i5 processor and 16 GB in RAM was used. The program was coded in Python 3.5. For the statistical analysis, the non-parametric Wilcoxon signed-rank test was used in addition to violin charts. The analysis of violin charts is performed by comparing the dispersion, median and the interquartile range of the distributions.

4.1 Insight of BGOA Algorithm

In this module we developed experiments that allowed us to study the contribution of the binary percentile operator with respect to the quality of the solutions obtained and the iterations used when solving SCP instances. To perform this evaluation, we used the instances of Balas and Carrera. In the comparison, a random operator was designed which replaces our binary percentile operator. This random operator instead of executing transitions assigning transition probabilities per group, uses a fixed value of 0.5. To compare the distributions of the results of the different experiments we use violin Chart. The horizontal axis X corresponds to the problems, while Y axis uses the measure % - Gap defined in Eq. 9

$$\% - Gap = 100 \frac{BestKnown - SolutionValue}{BestKnown} \tag{9}$$

Furthermore, a non-parametric test, Wilcoxon signed-rank test is carried out to determine if the results of BGOA with respect to the random algorithm have significant difference or not. The parameter settings and browser ranges are shown in Table 1.

Table 1. Setting of parameters for Binary GrassHopper Search Algorithm.

Parameters	Description	Value	Range
N	Number of solutions	25	[20, 25, 30]
G	Number of percentiles	5	[4,5,6]
Iteration Number	Maximum iterations	500	[400,500,600]
cmax	Maximum for c coefficient	1	1
cmin	Minimum for c coefficient	0.00001	0.00001
f	f parameter for s function	0.5	0.5
l	l parameter for s function	1.5	1.5

BGOA corresponds to our standard algorithm. *random.0.5* is the random variant. The results are shown in Table 2 and Fig. 2.

Table 2. Balas and Carrera instances.

Instance	row	col	Density	Best known	BGOA (avg)	std	(Secs)	rnd.0.5 (avg)	std
AA03	106	8661	4.05%	33155	**33331.4**	124.2	207	33693.5	241.1
AA04	106	8002	4.05%	34573	**34659.2**	58.1	198	34931.0	231.2
AA05	105	7435	4.05%	31623	**31717.3**	64.8	201	31889.2	103.2
AA06	105	6951	4.11%	37464	**37498.6**	56.1	199	37651.3	97.1
AA11	271	4413	2.53%	35384	**35561.3**	131.9	187	35901.2	197.1
AA12	272	4208	2.52%	30809	**31232.8**	152.8	208	31631.3	210.2
AA13	265	4025	2.60%	33211	**33804.4**	135.1	189	34201.4	231.4
AA14	266	3868	2.50%	33219	**33662.1**	125.1	143	33967.3	121.1
AA15	267	3701	2.58%	34409	**34686.1**	112.8	141	34846.8	161.3
AA16	265	3558	2.63%	32752	**33114.1**	137.5	187	33601.3	183.9
AA17	264	3425	2.61%	31612	**31821.3**	89.1	162	32386.2	149.3
AA18	271	3314	2.55%	36782	**37119.1**	78.1	132	37417.6	148.1
AA19	263	3202	2.63%	32317	**32911.2**	77.6	146	33109.1	136.1
AA20	269	3095	2.58%	34912	**35678.8**	91.1	141	35963.1	97.8
BUS1	454	2241	1.88%	27947	**28211.2**	86.0	189	28756.1	231.6
BUS2	681	9524	0.51%	67760	**68312.1**	83.2	175	68678.9	101.2
Average				35495.56	35832.56	100.21	175.3	36164.1	165.1
p-value								1.1e-05	

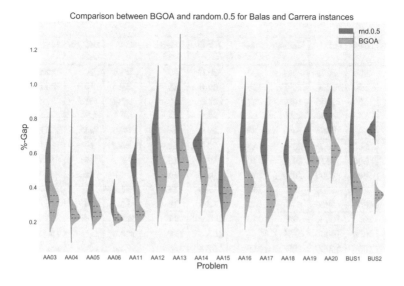

Fig. 2. Evaluation of percentile binary operator.

When we compared the Best Values between BGOA and *rnd.0.5* which are shown in Table 2. In the Average comparison, BGOA outperforms *rnd.0.5* in all problems. The comparison of distributions is shown in Fig. 2. We see the dispersion of the *rnd.0.5* distributions are bigger than the dispersions of BGOA. Therefore, the percentile binary operator, contribute to the precision of the results. Finally, the BGOA distributions are closer to zero than *rnd.0.5* distributions, indicating that BGOA has consistently better results than *random.0.5*. When we evaluate the behaviour of the algorithms through the Wilcoxon test, this test indicates that there is a significant difference between the two algorithms.

4.2   BGOA Compared with JPSO and MDBBH

In this section, we detail the comparisons made with the objective of evaluating the performance of our BGOA algorithm. For the evaluation, the larger problems of the OR-library were chosen. To develop the comparison, two algorithms were selected. The first one corresponds to a discretization Particle Swarm Optimization (PSO) technique called Jumping PSO (JPSO), [13]. The second one, is a binarization of the Black Hole technique called Multi Dynamic Binary Black Hole (MDBBH) Algorithm, [5]. JPSO uses a discrete PSO based on the frog jump. JPSO works without the concept of velocity, replacing this one by a component of random jump which allows to perform the movement in the discrete search space. On the other hand, MDBBH uses a binarization mechanism specific to the Black Hole algorithm. This is based on the concept of closeness to the Black Hole (BH). BH corresponds to the solution that has obtained the best value. When a solution is close to BH, the transitions are less likely to be performed.

Table 3. OR-Library benchmarks.

Instance	row	col	Density	Best Known	JPSO Avg	MDBBH Best	Avg	BGOA Best	Avg	Time(s)
E.1	500	5000	10%	29	29.0	29	29.0	29	29.2	8.8
E.2	500	5000	10%	30	30.0	31	31.6	30	30.6	12.5
E.3	500	5000	10%	27	27.0	27	27.4	27	27.3	17.2
E.4	500	5000	10%	28	28.0	28	29.1	28	28.2	11.3
E.5	500	5000	10%	28	28.0	28	28.0	28	28.0	14.3
F.1	500	5000	20%	14	14.0	14	14.1	14	14.0	17.8
F.2	500	5000	20%	15	15.0	15	15.3	15	15.0	16.1
F.3	500	5000	20%	14	14.0	14	14.8	14	15.2	17.8
F.4	500	5000	20%	14	14.0	14	14.9	14	14.3	18.3
F.5	500	5000	20%	13	13.0	14	14.1	14	14.4	17.1
G.1	1000	10000	2%	176	176.0	177	178.5	176	177.1	172.4
G.2	1000	10000	2%	154	155.0	157	160.6	155	155.5	171.3
G.3	1000	10000	2%	166	167.2	168	170.4	167	168.8	178.2
G.4	1000	10000	2%	168	168.2	169	170.9	169	170.6	175.5
G.5	1000	10000	2%	168	168.0	168	169.8	168	169.6	167.1
H.1	1000	10000	5%	63	64.0	64	64.9	64	64.8	152.1
H.2	1000	10000	5%	63	63.0	64	64	64	64.4	155.8
H.3	1000	10000	5%	59	59.2	59	60	60	60.8	143.1
H.4	1000	10000	5%	58	58.3	59	60.4	59	59.6	164.6
H.5	1000	10000	5%	55	55.0	55	56.4	55	55.6	138.1
Average				67.1	67.30	67.7	68.71	67.5	68.15	88.47

In the case of being away from the BH, the probability of transition is greater (Table 3).

For instances E and F which are of medium size, the results of the MDBBH and BGOA algorithms are similar when comparing their Best Value. Only in instance E.2 BGOA was superior to MDBBH. When analyzing the average, JPSO obtained better results, however, the values of BGOA were close, being MDBBH the one that obtained the worse performance. For the case of problems G and H, BGOA was superior in problems G.1, G.2 and G.3 and MDBBH in problem H.3 when comparing their Best Value. In the case of averages, JPSO was superior in all cases. However, BGOA obtained results quite close, leaving MDBBH behind. In this point we must emphasize that the percentile technique used in BGOA allows binarizing any continuous swarm intelligence algorithm unlike JPSO that is specific for PSO.

5   Conclusions

In this work, the BGOA algorithm based on the percentile concept was proposed to perform binarization of GOA metaheuristics. We must emphasize that this percentile concept can be applied in the binarization of any continuous metaheuristics of the swarm intelligence type. To evaluate the performance of

Our percentile concept the set covering problem was used together with the Grasshopper algorithm. Experiments were designed to evaluate the contribution of the binary operator percentile of the algorithm. It was found that the operator contributes significantly to improve the accuracy and quality of the solutions, all these visualized through violin graphics. Finally, compared to the best meta-heuristic algorithm that has solved SCP, our algorithm had a lower yield of 1.26 %, which is not a big difference considering that JPSO uses a particular adaptation mechanism for PSO and the percentile concept It can be easily adapted to any technique. As future works, we believe that the autonomous search for efficient configuration is a little-explored area and it could help to improve the performance of the binarizations both in the quality of the solutions and in the times of convergence. Additionally, it is interesting to see how this percentile technique works with other metaheuristics and other NP-hard problems. Finally, we believe that using machine learning we can generate indicators or obtain correlations that allow us to understand how the percentile algorithm translates the exploration and exploitation properties from the continuous space to the binary space.

References

1. Khatibinia, M., Yazdani, H.: Accelerated multi-gravitational search algorithm for size optimization of truss structures. Swarm Evol. Comput. (2017)
2. Barman, S., Kwon, Y.-K.: A novel mutual information-based boolean network inference method from time-series gene expression data. PloS one **12**(2), e0171097 (2017)
3. Crawford, B., Soto, R., Monfroy, E., Astorga, G.. García, J., Cortes, E.: A meta-optimization approach for covering problems in facility location. In: Workshop on Engineering Applications, pp. 565–578. Springer (2017)
4. García, J., Crawford, B., Soto, R., Astorga, G.: A percentile transition ranking algorithm applied to knapsack problem. In: Proceedings of the Computational Methods in Systems and Software, pp. 126–138. Springer (2017)
5. García, J., Crawford, B., Soto, R., García, P.: A multi dynamic binary black hole algorithm applied to set covering problem. In: International Conference on Harmony Search Algorithm, pp. 42–51. Springer (2017)
6. García, J., Crawford, B., Soto, R., Astorga, G.: A percentile transition ranking algorithm applied to binarization of continuous swarm intelligence metaheuristics. In: International Conference on Soft Computing and Data Mining, pp. 3–13. Springer (2018)
7. Franceschetti, A., Demir, E., Honhon, D., Van Woensel, T., Laporte, G., Stobbe, M.: A metaheuristic for the time-dependent pollution-routing problem. Eur. J. Oper. Res. **259**(3), 972–991 (2017)
8. Crawford, B., Soto, R., Astorga, G., García, J., Castro, C., Paredes, F.: Putting continuous metaheuristics to work in binary search spaces. Complexity **2017** (2017)
9. Yang, X.-S., Deb, S.: Cuckoo search via lévy flights. In: 2009 World Congress on Nature and Biologically Inspired Computing, NaBIC 2009, pp. 210–214. IEEE (2009)
10. Hatamlou, A.: Black hole: a new heuristic optimization approach for data clustering. Inf. Sci. **222**, 175–184 (2013)

11. Yang, X.-S.: A new metaheuristic bat-inspired algorithm. In: Nature Inspired Cooperative Strategies for Optimization (NICSO 2010), pp. 65–74 (2010)
12. Saremi, S., Mirjalili, S., Lewis, A.: Grasshopper optimisation algorithm: theory and application. Adv. Eng. Softw. **105**, 30–47 (2017)
13. Balaji, S., Revathi, N.: A new approach for solving set covering problem using jumping particle swarm optimization method. Nat. Comput. **15**(3), 503–517 (2016)
14. Gary, M.R., Johnson, D.S.: Computers and Intractability. A Guide to the Theory of NP-Completeness (1979)
15. Lu, Y., Vasko, F.J.: An or practitioner's solution approach for the set covering problem. Int. J. Appl. Metaheuristic Comput. (IJAMC) **6**(4), 1–13 (2015)
16. Li, Y., Cai, Z.: Gravity-based heuristic for set covering problems and its application in fault diagnosis. J. Syst. Eng. Electron. **23**(3), 391–398 (2012)
17. Kasirzadeh, A., Saddoune, M., Soumis, F.: Airline crew scheduling: models, algorithms, and data sets. EURO J. Transp. Logist. **6**(2), 111–137 (2017)
18. Horváth, M., Kis, T.: Computing strong lower and upper bounds for the integrated multiple-depot vehicle and crew scheduling problem with branch-and-price. Cent. Eur. J. Oper. Res. 1–29 (2017)
19. Stojković, M.: The operational flight and multi-crew scheduling problem. Yugoslav J. Oper. Res. **15**(1) (2016)
20. García, J., Crawford, B., Soto, R., Carlos, C., Paredes, F.: A k-means binarization framework applied to multidimensional knapsack problem. Appl. Intell. 1–24 (2017)
21. García, J., Pope, C., Altimiras, F.: A distributed k-means segmentation algorithm applied to lobesia botrana recognition. Complexity **2017** (2017)
22. Graells-Garrido, E., García, J.: Visual exploration of urban dynamics using mobile data. In: International Conference on Ubiquitous Computing and Ambient Intelligence, pp. 480–491. Springer (2015)
23. Graells-Garrido, E., Peredo, O., García, J.: Sensing urban patterns with antenna mappings: the case of Santiago, Chile. Sensors **16**(7), 1098 (2016)
24. Peredo, O.F., García, J.A., Stuven, R., Ortiz, J.M.: Urban dynamic estimation using mobile phone logs and locally varying anisotropy. In: Geostatistics Valencia 2016, pp. 949–964. Springer (2017)

A Survey on Signal Processing Methods in Fiber Optic Sensor for Oxidized Carbon Steel

Nur Syakirah Mohd Jaafar[1(✉)], Izzatdin Abdul Aziz[1],
Jafreezal Jaafar[1], Ahmad Kamil Mahmood[1],
and Abdul Rehman Gilal[2]

[1] Centre for Research in Data Science, Universiti Teknologi PETRONAS,
Seri Iskandar, Malaysia
{nur_16001470, izzatdin, jafreez}@utp.edu.my
[2] Department of Computer Science, Sukkur IBA University, Sukkur, Pakistan
a-rehman@iba-suk.edu

Abstract. This paper provides a broad overview of the adaptive methods for noise reduction used in the analysis of data in the different sensors such as acoustic emissions sensors, power quality signal analysis. The two algorithms are the Empirical Mode Decomposition and the Ensemble Empirical Mode Decomposition. We selected these two algorithms because our focus is on these methods. Firstly, this paper exhibits the inner workings of each algorithm both in the original authors' intuition and the mathematical model utilized. Next, we discuss the advantages of each of the algorithms based on recent and credible research papers and articles. We also critically dissect the limitations of each algorithm. This paper aims to give a general understanding on these algorithms which we hope will spur more research in improving the field of signal processing in the fiber optic sensor for the oxidised carbon steel.

Keywords: Signal processing · Empirical Mode Decomposition
Ensemble Empirical Mode Decomposition

1 Introduction

Pipeline monitoring is crucial to reduce the risk of failure while extending the safety life spam of the pipeline asset [1]. Signals collected from the sensor in pipeline monitoring, will reveal the anomalies to ascertain actions to be taken. Two crucial parameters needed to be taken into consideration in the data analysis for the signal processing in fiber optic sensor for oxidized carbon steel are the time scale and energy distribution. These parameters provide characteristics information on the pipe condition.

Time scale and energy distribution are important for the time – frequency analysis in signal processing [2]. One of the possible ways to extract the characteristic information is by effectively having the features of the signal for the usage of the pipeline operators and engineers [3]. The time- frequency method plays a key role, to mine meaningful information on the pipe condition. This leads to the creation of techniques and algorithms for correctly identifying information that can describe the pipe

© Springer International Publishing AG, part of Springer Nature 2019
R. Silhavy (Ed.): CSOC 2018, AISC 765, pp. 13–24, 2019.
https://doi.org/10.1007/978-3-319-91192-2_2

conditions. To extract the features signal from pipe information, many scholars and researchers have accomplished the relevant research works in contributing towards the methods; Empirical Mode Decomposition and the Ensemble Empirical Mode Decomposition [4, 5]. The aim for this paper is to provide an overview of the existing methods in signal processing for the fiber sensor. Specifically, we discuss the original methods used in the field of the signal processing which involves time- frequency analysis. Even though the existing method is already known, yet both methods still suffers from the mode mixing phenomenon.

Our paper is structured as follows. In Sect. 2, we provide examples of existing methods in signal processing used by the scientific community and in the industry. Each algorithm is explained in both the original authors intuition of the methods as well as their respective mathematical models. Based on description provided in Sect. 2, we identify the advantages of each of the algorithms and discuss them in Sect. 3. In Sect. 4, the weaknesses found in the algorithms discussed in Sect. 2 are identified. This paper concludes with discussion on proposing of new enhancements to the algorithms.

2   Methods

In this section, two signal processing methods for fiber optic sensor in oxidized carbon steel are comprehensively discussed. The two methods are Empirical Mode Decomposition and Ensemble Empirical Mode Decomposition. We shall briefly explain the function of the algorithms as well as the mathematical model used in the algorithms. This is to give a deep understanding on the inner workings of time – frequency analysis in signal processing.

2.1   *Empirical* Mode Decomposition [EMD]

EMD or Empirical Mode Decomposition is a method for analyzing data from non-stationary and non- linear systems and behaves as a filter bank [6–8]. Time series data such as signal recordings along the pipelines are often assumed to be stationary and linear within a given window [9–11]. By applying EMD, a signal can be decomposed into a set of mono component functions called Intrinsic Mode Functions (IMFs). Thus, from the signal decomposition, the IMFs contain frequencies ranging from the highest to the lowest in the signal. To meet the mono component function, the IMF needs to satisfy two conditions [9, 12, 13].

The first condition is that, the number of local maxima and minima, and the number of zero crossings must either equal or different at most by one [5, 10]. The first condition is essential to calculate instantaneous frequency that presents in the oscillation frequency of a signal at certain point of the time.

The second condition is that, the mean value of the envelope defined by the local maxima and the local minima is zero at any point. The second condition is necessary as the signal is ready for modulation resulted from the IMFs. By meeting these conditions, series of Intrinsic Mode Functions can be decomposed [9–11].

2.1.1 IMFs Identification via Sifting Process

By setting up the initial value of the IMF's number and find the extrema, which are local maxima and local minima points of the input signal to lead in finding the first IMF. The identified extrema will form upper and lower envelopes of the signal. The mean value from the connection of the upper envelope and lower envelope is calculated based on equation from [9–11]. The steps below show one of the iteration of the sifting process for the first IMF of a given signal $X(t)$.

To find the first IMF of a given signal $X(t)$:

Step 1: Set the initial value, s_{10} equal to the signal $X(t)$ and find the extrema (local maxima and local minima points) of the input signal in the time domain.

Step 2: Connect the maxima points with spline functions to form the upper envelope $[e_{upper}(t)]$ and connect the minima points with spline functions to form the lower envelope $[e_{lower}(t)]$.

Step 3: The mean of the envelope is now calculated as:

$$m_{11} = \frac{e_{upper(t)} + e_{lower}(t)}{2} \tag{1}$$

Step 4: Subtract this mean from the input or initial signal:

$$s_{11}(t) = s_{10}(t) - m_{11}(t) \tag{2}$$

2.1.2 Continuous of IMFs Identification via Sifting Process

The steps above show one of the iteration from the sifting process. Thus, these next steps show continuation of sifting. The signal s_{11} output from the first iteration will typically be tested using the stopping criteria.

Step 5: The signal s_{11} output from the first iteration is not IMF if the signal s_{11} does not meet the stopping criteria.

Step 6: If the signal does not meet the stopping criteria, update the input signal with s_{11}, and carry out second iteration by repeating the steps in (Sect. 2.1.1). The output from the second iteration based on the steps in (Sect. 2.1.1) will be:

$$s_{12}(t) = s_{11}(t) - m_{12}(t) \tag{3}$$

Step 7: From the Steps 6, the output from the second iteration resulted in $s_{12}(t)$ not meeting the IMF condition, the process of iteration is continued by repeating the steps in (Sect. 2.1.1). Sifting Process to find the IMFs] until the stopping criteria are met after the iteration of k.

$$s_{1k} = s_{1(k-1)} - m_{1(k)} \tag{4}$$

Step 8: To confirm the s_{1k} as the first IMF, s_{1k} need to check if the signal meets the stopping criteria. Thus, when the signal of s_{1k} meets the stopping criteria, the process of iteration stops. The output, once it is found meeting the stopping criteria will be:

$$s_{1k} = c_1(t) \tag{5}$$

Step 9: Next, subtract $c_1(t)$ from the original input signal $X(t)$.

$$r_1(t) = X(t) - c_1(t) \tag{6}$$

Step 10: The reason on why $c_1(t)$ is subtracted from the input signal is due to $c_1(t)$ had become a residue, $r_1(t)$. It became a new input data as long it satisfies the IMF functions and undergo the iteration by repeating the steps in (Sect. 2.1.1). Thus, the output will be when it is found to meet the stopping criteria. The output will be the second of IMF and it will be:

$$r_2(t) = r_1(t) - c_2(t) \tag{7}$$

Step 11: To find the subsequent of the residues, the procedure is repeated, resulting in:

$$r_n(t) = r_{n-1}(t) - c_n(t) \tag{8}$$

Step 12: The notations for c_2 till c_n shows the corresponding IMFs of the data.
Step 13: To complete the sifting process of IMF, EMD is completed when the residue, $r_n(t)$ met the IMF functions, which means that it is a monotonic function or does not contain any extrema points.
Step 14: The input signal $X(t)$ can be expressed as the sum of the IMFs and the last residue as referred [9–11]:

$$X(t) = \sum_{j=1}^{N} c_j(t) + r_N(t) \tag{9}$$

Whereby, $c_j(t)$, are the IMF components after N, IMFs are extracted and $r_N(t)$ is the residual of $X(t)$.

Based on the Fig. 1, a flow diagram on the EMD is given and the details are explained in the (Sect. 2.1). From $c_j(t)$, as the IMF components after N as the numbers of iterative IMFs are extracted, the first IMF represents the fast oscillation modes and the higher IMFs represent the slower oscillation modes.

2.2   Ensembles Empirical Mode Decomposition (EEMD)

Ensemble Empirical Mode Decomposition (EEMD), means from the numbers of IMFs been extracted from the existing EMD to the original signal, the ensemble mean is calculated with addition of white noise. This approach of EEMD defines the true IMF components as the mean of ensemble trials, by having the white noise of finite amplitude, each consisting of the signal [5].

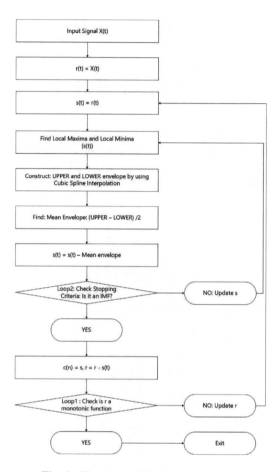

Fig. 1. Flowchart of EMD Algorithm [9]

The procedure process shows the decomposition of signal as followed procedures as below:

Step 1: Initialized the number of ensemble I.

Step 2: Added the white noise series, $N(t)$ of finite amplitude to the signal $X(t)$ by generating $X^i[t] = X[t] + w^i[t]$ $(i = 1, \ldots, I)$, are different realization of white Gaussian noise.

Step 3: Each $X^i[t]$ $(i = 1, \ldots, I)$, is decomposed by EMD, getting their modes $IMF_k^i[t]$ where k = 1, 2,..., K indicates the modes.

Step 4: Assign $\overline{IMF_k}$ as the k-th mode of $X(t)$.

Step 5: Obtained the average of the corresponding IMF:

$$IMF_k^i : \overline{IMF_k}[t] = \frac{1}{I}\sum_{i=1}^{I} IMF_k^i[t] \tag{10}$$

Step 6: The input signal $X(t)$ can be expressed as the sum of the IMFs and the last residue:

$$X(n) = \sum_{k=1}^{K} \overline{IMF_k}(t) + \bar{r}(t) \tag{11}$$

Figure 2 shows the flowchart of EEMD algorithm. Given the signal, the effective algorithm of EEMD can be summarized as follows. By having number of amplitude of added white noise to the original signal resulted as EEMD. From, the noise-added signal, the result, which is EEMD will be obtaining.

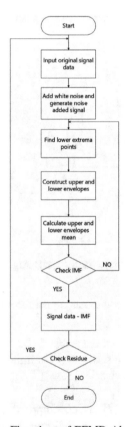

Fig. 2. Flowchart of EEMD Algorithm

2.2.1   Parameters Analysis of EEMD

This EEMD algorithm improves the EMD algorithm significantly in the capability of extracting the signals in the data when the number of ensemble member' increases. As the number of ensemble numbers, N increases, the effect of the noise decreases as the noise is established based on the statistical rule as stated in [16–18]:

$$\varepsilon_n = \frac{\varepsilon}{\sqrt{N}} \tag{12}$$

This rule pointed out that the effect of the noise decreases where N is the number of ensemble members, ε is the amplitude of the added noise and the ε_n is the standard deviation of error. This standard deviation of error shows the difference between the input signal $X(t)$ and the summation of the IMFs.

Thus, this standard deviation of error is plotted as a function of the number of ensemble numbers. With this statistical rule, the number of ensemble numbers, N and the amplitude of the added noise, ε been influenced the EEMD algorithm in isolate and extract the true signals in data, and thereby to understand the properties of data.

2.2.2 Number of Ensemble of EEMD

From the statistical rule, when the amplitude of added white noise is set, theoretically the bigger ensemble numbers will be generating the better result as the iteration of ensemble numbers is increases. However, the bigger ensemble numbers will lead to higher computational time to perform. While for the predefined of the ensemble numbers, smaller of amplitude of added white noise will generate better IMFs.

2.2.3 The Selection of the Amplitude of Added White Noise

By having the amplitude of added white noise, it will influence the EEMD process in two ways even though the EEMD result is not sensitive to the added white noise. The purpose of having the amplitude of added white noise is to help the decomposition of the signal by reducing the mode mixing phenomenon and it could be not enough to influence the sifting process. Unfortunately, there is no equation to help in choosing and justify the right value of the amplitude of added white noise. For choosing the right value of amplitude of added white noise, there are some basic rules could be reference.

When the raw data is dealing and been dominated with the most of high frequency signals, the value of amplitude of added white noise could be smaller. While if the raw data is dealing and been denominated with the low frequency signals, the value of amplitude of added white noise could be bigger.

For this research, the question is whether the improvement EEMD algorithm indeed helps in reaching the goal of the data analysis in extracting the signals. Thus, for this research, the number of ensemble numbers, N and the amplitude of the added noise, ε will be the parameters to be studied in reaching the goal of the research.

3 Comparative Study on the Advantages of the Signal Processing Method in Fiber Optic Sensor for Oxidized Carbon Steel

There is existing research on the signal processing methods in the scientific community. Table 1 exhibits the comparative analysis of the advantages of the two signal processing methods. The analysis is divided into the respective authors.

Table 1. Advantages of the signal processing method

Author(s) (Pub. year)	Method	Advantages
Adnan et al. (2015) [14]	EMD	This approach is used to decompose the raw signal and shown in the time- frequency analysis. This approach is able to detect the leak in the gas pipelines
Camarena-Martinez et al. (2016) [15]		This approach is based on the EMD for the analysis of PQ signals has been presented. This approach consists of the iterative down sampling stage that helps to extract the fundamental component in the analysis of the time-frequency
Xu et al. (2017) [12]		This approach is based on the adaptive EMD threshold denoising method for the adaptive EMD threshold for the acoustic signal has been presented. This approach has been introduced into the threshold determination process of the denoising approach
Siracusano et al. (2016) [13]		This approach on a framework is based on the Hilbert Huang Transform for the analysis of the acoustic emission data. This approach, by having the EMD features, the acoustic signal could isolate and extract the signal to be interpreted such as the amplitude to be more meaningful
Rostami et al. (2017) [8]	EEMD	This approach on the separation of overlapped modes in concrete-wall covered pipes using a SEMD, integration between wavelet and EMD. Comparison of signal processing techniques such as wavelet transform, EMD and EEMD were used to study localised frequency components in the signal to be interpret
Su et al. (2016) [16]		This approach is based on the EEMD method to remove the noise. This approach is based on the parameters analysis of the EEMD which are the total number of ensembles, sifting number and amplitude
Amin et al. (2015) [14, 17]		This approach is based on the EEMD method to solves the EMD mode mixing problem. This approach is based on the parameters analysis of the EEMD which are the total number of ensembles, sifting number and amplitude
Li et al. (2017) [18]		This approach is based on the EEMD method to solves the EMD mode mixing problem. This approach is based on the parameters analysis of the EEMD which are the total number of ensembles, sifting number and amplitude

Table 1 provides an overview of the advantages found for each of the methods; EMD and EEMD. Empirical Mode Decomposition (EMD) being the decomposition of signals into finite number of components, clearly provides a new method of analyzing nonstationary and nonlinear time series data in comparison to the other algorithms. The main advantage for EMD is it does not require without requires no prior knowledge of the target signal, this will be more efficient in filtering accuracy in the time series.

Besides that, it is data driven basis and suitable for analysis on the signal interpretation algorithm. Ensemble Empirical Mode Decomposition (EEMD) takes it one step further where it includes Gaussian white noises in this method [16–18]. This significantly reduces the mode mixing phenomenon. The mode mixing of phenomenon will lead to undesirable signal signatures. By using the time- frequency analysis in the method is also useful in creating relationship between signal energy and time – frequency. This will increase the accuracy by represents series of the stationary signals with different amplitudes and frequency bands.

Lastly, EEMD, which by having number of amplitude of added white noise to the original signal resulted as noise- added signal. The parameters tested, analysis of the EEMD algorithm which is number of ensemble members, fixed sifting numbers and amplitude of the added white noise will be reducing the mode- mixing phenomenon [16–18]. The next section discusses on the limitations found in each of the methods.

4 Comparative Study on the Limitations of the Signal Processing Methods in Fiber Optic Sensor for Oxidized Carbon Steel

Table 2 exhibits the analysis of the limitations found in each of the signal processing methods. The main limitation of EMD is its inability and lacks in the mathematical theoretical foundation [12–15]. The methods also have end effect and critically are having IMF criteria problems during the sifting process [12–15]. Thus, this will lead to undesirable signal signature such as noises interference, which often leads to signal misinterpretation. Signal is decomposed into number of IMFs by EMD in HHT. Based on the analysis on EMD algorithm, the mode mixing problems encountered when signal contains intermittency. The mode mixing will affect the decomposition of signal into different IMFs on the time series as similar scale is residing in different IMF. While, EEMD method still encounter problem of mode mixing when the signal contains intermittency when the parameters of the EEMD is not set up.

Table 2. Limitations of the signal processing methods

Author(s) (Pub. year)	Method	Limitations
Adnan et al. (2015) [14]	EMD	This approach needed more experiments and simulation to be carried out to get the fast result of leaking and estimation of their location. This approach is lacking in the theoretical analysis and mathematical formulation
Camarena-Martinez et al. (2016) [15]		This approach is based on the EMD for the analysis of PQ signals lead to the mode mixing to false and wrong IMF decompositions. This approach shows that the computational burden is increased, which is a serious drawback in applications where the volume of data is large as the case of continuous PQ monitoring
Xu et al. (2017) [12]		This approach is based on the adaptive EMD threshold denoising method needed improvements in determining the variation coefficient and disturbance coefficient. This proposed approach shows that the computational burden is increased
Siracusano et al. (2016) [13]		This approach is lacking in the theoretical analysis and mathematical formulation as it focused on the framework based on the Hilbert Huang Transform, for the analysis of the acoustic emission data
Rostami et al. (2017) [8]	EEMD	This approach shows that parameters analysis of the Ensemble Empirical Mode Decomposition (EEMD) is not fixed for the decomposition of the signal. This approach shows that there is minimal work to improve the selection of accurate IMFs
Su et al. (2016) [16]		This approach shows that parameters analysis of the Ensemble Empirical Mode Decomposition (EEMD) is not fixed for the decomposition of the signal. This proposed approach shows that there is minimal work to improve the selection of accurate IMFs
Amin et al. (2015) [14, 17]		This approach shows that parameters analysis of the Ensemble Empirical Mode Decomposition (EEMD) is not fixed for the decomposition of the signal. This proposed approach shows that there is minimal work to improve the selection of accurate IMFs
Li et al. (2017) [18]		This approach shows that parameters analysis of the Ensemble Empirical Mode Decomposition (EEMD) is not fixed for the decomposition of the signal. This approach shows that there is minimal work to improve the selection of accurate IMFs.

5 Conclusion

A comparative study towards the two signal processing methods has been conducted through literature review. Based on the literature study, we have discussed the advantages and limitations of each of the algorithms. We have also provided explanation on the inner workings of each of the algorithms especially on the mathematical model used in the algorithms. Through the analysis, this could be one of the reference for the signal processing methods in the fiber optic sensors or any other sensors such as acoustic emission sensors.

Acknowledgments. This work was supported by Development of Intelligent Pipeline Integrity Management System (I-PIMS) Grant Scheme from Universiti Teknologi PETRONAS.

References

1. Arzaghi, E., Abaei, M.M., Abbassi, R., Garaniya, V., Chin, C., Khan, F.: Risk-based maintenance planning of subsea pipelines through fatigue crack growth monitoring. Eng. Fail. Anal. **79**, 928–939 (2017)
2. Shariatinasab, R., Akbari, M., Rahmani, B.: Application of wavelet analysis in power systems. In: Advances in Wavelet Theory and Their Applications in Engineering, Physics and Technology. InTech (2012)
3. Shi, Y., Zhang, C., Li, R., Cai, M., Jia, G.: Theory and application of magnetic flux leakage pipeline detection. Sensors **15**(12), 31036–31055 (2015)
4. Zhang, H., Feng, Z., Zou, J.: Research on feature extraction and pattern recognition of acoustic signals based on MEMD and approximate entropy. In: 2017 29th Chinese on Control and Decision Conference (CCDC), pp. 4844–4849. IEEE (2017)
5. Agarwal, M., Jain, R.: Ensemble empirical mode decomposition: an adaptive method for noise reduction. IOSR J. Electron. Commun. Eng. **5**, 60–65 (2013)
6. Zhan, L., Li, C.: A comparative study of empirical mode decomposition-based filtering for impact signal. Entropy **19**(1), 13 (2016)
7. Sun, J., Xiao, Q., Wen, J., Zhang, Y.: Natural gas pipeline leak aperture identification and location based on local mean decomposition analysis. Measurement **79**, 147–157 (2016)
8. Rostami, J., Chen, J., Tse, P.W.: A signal processing approach with a smooth empirical mode decomposition to reveal hidden trace of corrosion in highly contaminated guided wave signals for concrete-covered pipes. Sensors **17**(2), 302 (2017)
9. Saeed, B.S.: De-noising seismic data by Empirical Mode Decomposition (2011)
10. Honório, B.C.Z., de Matos, M.C., Vidal, A.C.: Progress on empirical mode decomposition-based techniques and its impacts on seismic attribute analysis. Interpretation **5**(1), SC17–SC28 (2017)
11. Wu, Z., Huang, N.E.: Ensemble empirical mode decomposition: a noise-assisted data analysis method. Adv. Adapt. Data Anal. **1**(01), 1–41 (2009)
12. Xu, J., et al.: A novel denoising method for an acoustic-based system through empirical mode decomposition and an improved fruit fly optimization algorithm. Appl. Sci. **7**(3), 215 (2017)
13. Siracusano, G., et al.: A framework for the damage evaluation of acoustic emission signals through Hilbert-Huang transform. Mech. Syst. Signal Process. **75**, 109–122 (2016)

14. Adnan, N., et al.: Leak detection in gas pipeline by acoustic and signal processing-a review. In: IOP Conference Series: Materials Science and Engineering 2015. IOP Publishing (2015)
15. Camarena-Martinez, D., et al.: Novel down sampling empirical mode decomposition approach for power quality analysis. IEEE Trans. Ind. Electron. **63**(4), 2369–2378 (2016)
16. Su, H., Li, H., Chen, Z., Wen, Z.: An approach using ensemble empirical mode decomposition to remove noise from prototypical observations on dam safety. SpringerPlus **5**(1), 650 (2016)
17. Amin, M.M., Ghazali, M.F., PiRemli, M.A., Hamat, A.M.A., Adnan, N.F.: Leak detection in medium density polyethylene (MDPE) pipe using pressure transient method. In: IOP Conference Series: Materials Science and Engineering, vol. 100, no. 1, p. 012007. IOP Publishing (2015)
18. Li, X., Wei, Q., Qu, Y., Cai, L.: Incipient loose detection of hoops for pipeline based on ensemble empirical mode decomposition and multi-scale entropy and extreme learning machine. In: IOP Conference Series: Materials Science and Engineering, vol. 211, no. 1, p. 012011. IOP Publishing (2017)

Some Recent Results on Direct Delay-Dependent Stability Analysis: Review and Open Problems

Libor Pekař[(⊠)], Pavel Navrátil, and Radek Matušů

Faculty of Applied Informatics, Tomas Bata University in Zlín,
Nad Stráněmi 4511, 76005 Zlín, Czech Republic
pekar@fai.utb.cz

Abstract. This contribution focuses an overview of selected results on time-delay systems stability analysis in the delay space, recently published in outstanding high-impacted journals and top conferences and meetings. A numerical gridding algorithm solving this problem designed by the first author is included as well. The theoretical background and a concise literature overview are followed by the list of practical and software applications. Unsolved tasks and open problems stemming from the analysis of presented methods and results concisely conclude the paper. The reader is supposed to use this survey to follow some of the presented techniques in his/her own research or engineering practice.

Keywords: Delay-dependent stability · Engineering application
Survey · Time-delay systems

1 Introduction

The study of stability of dynamical systems (regardless with or without delays) constitutes one of the most addressed topics from the field of the system analysis. This task is inherently and closely connected to the system spectrum of its characteristic values (poles). The decision about system stability – no matter what type of – is naturally dependent on particular system parameters' values, which can also be from the delay space, i.e. from the space of deviated arguments of system variables.

From the point of view of delay values, one can distinguish between delay-independent stability (DIS) and delay-dependent stability (DDS). A system is said to be DIS whenever it is stable for arbitrary delay values (including infinity, according to the strong version [1]). In the contrary, the DDS analysis means to inspect all the regions in the delay space in which the system is stable, or equivalently, to find all stability margins (i.e. critical delays) [2–4].

Generally, there are two basic families of DDS methods: Direct and indirect ones. The latter one is not further discussed in this contribution since the techniques are not related to system spectrum, they are theoretically and computationally intricate and usually provide conservative results (see e.g. [5]). The former family of direct approaches is based on the idea of the searching the purely imaginary poles which constitute the stability margin.

© Springer International Publishing AG, part of Springer Nature 2019
R. Silhavy (Ed.): CSOC 2018, AISC 765, pp. 25–34, 2019.
https://doi.org/10.1007/978-3-319-91192-2_3

In this contribution, a literature overview of most important methods and techniques recently published in top-notch journals and proceedings of world-leading conferences and meetings is given to the reader. First, the most common type of dynamical system stability – exponential one – is defined. Then, covered methods are listed and concisely introduced with references. Academic and real-life engineering applications are included as well, and some proposals of future research directions are suggested as well. It is worth noting that the numerical gridding, easy-to-handle, DDS algorithm developed by the first author is considered as well [6, 7].

The aim of this survey is to provide the reader with some existing up-to-date techniques for DDS analysis. Many of them are implementable and can serve to solve practical stabilizing problems; hence, engineers can follow the principles to use them in their everyday routine and scientists are encouraged to attack the open problems.

Notation: $\mathsf{C}, \mathsf{R}, \mathsf{Z}$ denote the sets of complex, real and integer numbers, respectively, R^n_+ is the n-dimensional Euclidean space of positive real-valued vectors.

2   Preliminaries

Consider a time-delay system (TDS) as follows:

$$\dot{\mathbf{x}}(t) + \sum_{i=1}^{n_H} \mathbf{H}_i \dot{\mathbf{x}}(t - \tau_{H,i}) = \mathbf{A}_0 \mathbf{x}(t) + \sum_{i=1}^{n_A} \mathbf{A}_i \mathbf{x}(t - \tau_{A,i}) + \mathbf{B}_0 \mathbf{u}(t) + \sum_{i=1}^{n_B} \mathbf{B}_i \mathbf{u}(t - \tau_{B,i})$$
$$+ \int_0^L [\mathbf{A}_d(\tau)\mathbf{x}(t - \tau) + \mathbf{B}_d(\tau)\mathbf{u}(t - \tau)]\mathrm{d}\tau \qquad (1)$$
$$\mathbf{y}(t) = \mathbf{C}\mathbf{x}(t)$$

where $\mathbf{x}(t) \in \mathsf{R}^n, \mathbf{u}(t) \in \mathsf{R}^m, \mathbf{y}(t) \in \mathsf{R}^l$ mean state, input and output variables, respectively, $\dot{\mathbf{x}}(t)$ is time derivative, $\mathbf{A}_i, \mathbf{B}_i, \mathbf{C}, \mathbf{H}$ are real-valued matrices of compatible dimensions, $0 < \tau_{.,1} < \tau_{.,2} < \ldots \leq L$ express lumped delays, and matrices $\mathbf{A}_d(\tau), \mathbf{B}_d(\tau)$ characterize distributed delays. Commensurate delays are integer multiples of the base delay $h \in \mathsf{R}$. The system is of retarded type (RTDS) if $\mathbf{H}_i \neq \mathbf{0}, \forall i$; otherwise, it is of neutral type (NTDS). For distributed delays, let us denote dRTDS and dNTDS.

The characteristic quasipolynomial reads

$$\Delta(s) = \det\left[s\left(\mathbf{I} + \sum_{i=1}^{n_H} \mathbf{H}_i \exp(-s\tau_{H,i})\right) - \mathbf{A}_0 - \sum_{i=1}^{n_A} \mathbf{A}_i \exp(-s\tau_{A,i}) - \int_0^L \mathbf{A}_d(\tau) \exp(-s\tau)\mathrm{d}\tau\right] \qquad (2)$$

where s stands for the complex Laplace transform variable, and $\mathbf{I}$ is $n \times n$ unit matrix. The spectrum (of poles) is then defined as $\Sigma := \{s : \Delta(s) = 0\}$.

System (1) (more precisely - its null solution) is said to be exponentially stable if there exist $\lambda > 0, \mu > 0$ such that $\|\mathbf{x}_t(\theta, \varphi)\|_s \leq \lambda \exp(-\mu t)\|\varphi\|_s, \forall t \geq 0$ where $\mathbf{x}_t(\theta, \varphi) = \mathbf{x}(t + \theta), \theta \in [-L, 0]$ for the initial condition φ and $\|\cdot\|_s$ means the supreme norm. It holds that a (d)RTDS is exponentially stable if and only if $\alpha < 0$ where $\alpha := \sup \mathrm{Re}\, \Sigma$; however, a (d)NTDS is exponentially stable if and only if there exists $\varepsilon > 0$ such that $\alpha < -\varepsilon$. It is clear from the above that the system can switch from/to stability/instability only if the rightmost (dominant) poles are purely imaginary. It must, moreover, hold that the gradient of such poles with respect to delay values is nonzero [4].

The system is said to be DDS if it is stable for a disjoint set of delay intervals $\left[\tau_i^-, \tau_i^+\right]$. The task of DDS analysis can be also defined as the searching of critical delays $\tau_i^+ < \tau_{c,i} < \tau_{i+1}^-$ corresponding to dominant poles located on imaginary axis.

3 Recent Advances in Direct DDS Analysis

Direct (frequency-domain) DDS methods are based on the determination of $\tau_{c,i}$. A survey of them follows.

3.1 Theoretical Results

The most important ideas and methods for DDS analysis published in recent few years can be listed as: (i) Cluster Treatment of Characteristic Roots (CTCR) paradigm, (ii) direct method, (iii) Puiseux series expansion, (iv) Kronecker sum or matrix pencil method, (v) Kronecker multiplication method, (vi) Cauchy theorem (the argument principle), (vii) W-transform, (viii) numerical gridding algorithm, (ix) other methods.

CTCR Paradigm. The famous CTCR paradigm is based on the following computational steps: The characteristic quasipolynomial is transformed to the approximating polynomial $p(\omega, \mathbf{T})$ that includes pseudo-delays T_i; then, potential stability switching hypersurfaces $\wp_0(\tau)$ are computed - $\wp_0(\tau)$ are composed by all delays corresponding to crossing frequencies $\Omega_c := \{\omega : \Delta(s, \tau) = 0, s = j\omega\}$. From $\wp_0(\tau)$, Ω_c, the offspring hypersurfaces $\wp_{off}(\tau)$ are generated that mean all crossing delays corresponding to a particular $\omega_c \in \Omega_c$. The use of the D-subdivision principle to determine stability bounds (which are composed of subsets of $\wp_{off}(\tau)$) concludes the CTCR technique; see e.g. [4] for further details. Notice that the D-subdivision methodology can sufficiently by applied to study non-delay parameter dependent stability [10]. In recent years, this concept has been extended. For instance, selected properties of the so-called core hypersurfaces (i.e. the image of $\wp_0(\tau)$ in the space of T_i) for $\Delta(s, \tau)$ with multiple independent delays were derived in [9], which was used to the identification of the asymptotic directions of $\wp_0(\tau)$. $\wp_0(\tau)$. Sipahi [11] designed imaginary poles to manipulate stability regions in the delay space based on the knowledge of the exact range of the imaginary spectrum which was determined by the preceding technique. The CTCR concept was used in [12] to study DDS of a dRTDS by means of the equivalence of a general class of distributed delay system to a system with multiple lumped independent delays. In [13], the authors presented a comparison between delay space domain and the spectral delay space domain that contains delays as well as pointwise frequency information. Bounds of the imaginary spectra through the transformation

$$\exp(-\tau j\omega) \rightarrow \cos(\upsilon) - j\sin(\upsilon), \upsilon = j\omega, \cos(\upsilon) = \frac{1 - z^2}{1 + z^2}, \sin(\upsilon) = \frac{2z}{1 + z^2}, z = \tan\left(\frac{\upsilon}{2}\right),$$
$$\tau_k = \omega^{-1}\left(2\tan^{-1}(z) + k\pi\right), k \in \mathbf{Z}, \tan^{-1}(\cdot) \in [0, \pi), j^2 = -1 \tag{3}$$

and by the deploying of the Dixon resultant theory for a RTDS with multiple delays were investigated in [14], followed by the analysis of the differentiability of the crossing-frequency variations $d\omega/dz_i$ to investigate the bounds. This study resulted in the extraction of two-dimensional (2D) cross-sections of the hypersurfaces, based on the concept of the three-dimensional (3D) building blocks in the spectral delay space [15]. The CTCR paradigm was also recently used to determine the exact delay bound for consensus of linear multi-agent systems with a fixed and uniform communication time delay [16].

Direct Method. This method is applicable only in case of commensurate delays. Its leading idea is based on the iterative elimination of exponential terms in $\Delta(s)$ using the fact that $\Delta(j\omega) = \Delta(-j\omega) = 0$ [17]. As the result, the following conditional polynomial equation to get imaginary poles $s_c = \pm j\omega_c$ is eventually obtained: $W(\omega_c^2) = 0$. The root tendency (RT) – expressing the speed of the roots along to the real axis with respect to delay values – is calculated to check the crossing property. A partial DDS problem for load frequency control systems with constant communication delays of the commensuracy degree of one and two was solved in [18]. Unfortunately, only the minimum base delay value $h > 0$ is computed to get the smallest delay margin rather than the complete stability image.

Puiseux Series Expansion. This techniques applies whenever a multiple pair of imaginary poles is detected, combined with the frequency sweeping. The expansion reads

$$\Delta s_k = \sum_{i=1}^{n} c_i(\Delta\tau)^{\frac{i}{m}} \tag{4}$$

where c_i are complex coefficients and s_k stands for a kth m-multiple imaginary pole. The so-called root invariance property was proved in [19]. It means that whenever a multiple purely imaginary pole appears as τ increases, the increment of unstable roots is same for a particular crossing frequency. In [20], the authors extended and formalized these results in more detail and they also defined and used the so-called dual Puiseux series $\Delta\tau(\Delta s_i)$.

Kronecker Sum (Matrix Pencil) Method. The leading idea of this family of approaches usable for commensurate delays is based on the knowledge that whenever $\lambda \in \Sigma_A$, $\mu \in \Sigma_B$ (where Σ_A, Σ_B are spectra of $\mathbf{A}$, $\mathbf{B}$, respectively), then all sums of the form $\lambda + \mu$ constitute the spectrum of the Kronecker sum $\mathbf{A} \oplus \mathbf{B} = \mathbf{A} \otimes \mathbf{I} + \mathbf{I} \otimes \mathbf{B}$, where $\mathbf{I}$ stands for the unit matrix. In addition, it is utilized the fact that whenever $s_k = j\omega_k$ is a root of $\Delta(s)$, then its complex conjugate is the root as well. The task can be reduced to the existence of generalized eigenvalues located on the unit circle by the transcription of the Kronecker sum into the matrix pencil form. This form was applied in [21] to determine DDS of a singular NTDS given by

$$\mathbf{H}_0\dot{\mathbf{x}}(t) + \mathbf{H}_1\dot{\mathbf{x}}(t-\tau) = \mathbf{A}_0\mathbf{x}(t) + \mathbf{A}_1\mathbf{x}(t-\tau) \tag{5}$$

Kronecker Multiplication Method. To shed light on the key idea of this methods, assume a system governed by the model $\dot{\mathbf{x}}(t) = \mathbf{A}_0\mathbf{x}(t) + \mathbf{A}_1\mathbf{x}(t - \tau)$. Then, any imaginary root of the corresponding characteristic quasipolynomial is also a root of the polynomial $p(s) = \det\big((s\mathbf{I} + \mathbf{A}_0^T) \otimes (s\mathbf{I} - \mathbf{A}_0) - (\mathbf{A}_1^T \otimes \mathbf{A}_1)\big)$ that is also the characteristic polynomial of the system

$$\mathbf{X}_0'(\theta) = \mathbf{X}_0(\theta)\mathbf{A}_0 + \mathbf{X}_{-1}(\theta)\mathbf{A}_1, \quad \mathbf{X}_1'(\theta) = -\mathbf{A}_1^T\mathbf{X}_0(\theta) - \mathbf{A}_0^T\mathbf{X}_{-1}(\theta) \qquad (6)$$

the spectrum of which is consequently computed via Lyapunov matrices (therefore, the bunch of these methods is also called Lyapunov matrix approaches). Critical delays are eventually obtained by inserting the computed critical poles into $\Delta(s)$. A bridge between direct and indirect DDS approaches by using the above idea was presented in [22]. The authors derived explicit relations between the spectrum of an original dRTDS and NTDS, and system (7). The computation of sign changes of the Sturm sequence served as a tool to search for imaginary roots of $p(s)$. Robust exponential DDS analysis based on the Kronecker multiplication along with a comparison with some other methods was published in [23]. Slack matrices were used to get the upper bound of the exponential decay rate, which gives τ_{max} so that the system is stable for $0 < \tau \leq \tau_{max}$.

Cauchy Theorem. An interesting technique based on the argument principle was published in [24]. It requires a rough estimation of the following testing integral

$$F(T_1, T_2) := \int_{T_1}^{T_2} \mathrm{Re}\left(\frac{\Delta'(j\omega)}{\Delta(j\omega)}\right) d\omega \qquad (7)$$

to decide about DDS, for particular values of $T_1, T_2 \in \mathbf{R}$. The authors proved that under some assumptions, there exists a sufficiently large $T_0 > 0$ such that the number of poles of an nth order NTDS located in $\mathbf{C}^+$ lies within the interval

$$\left(-\frac{F(0, T)}{\pi} + \frac{n-1}{2}, -\frac{F(0, T)}{\pi} + \frac{n+1}{2}\right) \qquad (8)$$

for all $T > T_0$. In [25], two DDS algorithms for a RTDS and a NTDS with arbitrary multiple delays were presented.

W-Transform. This method, also called the Cauchy W-method, is based on a transformation of a given differential equation to an operator equation by the substitution $x(t) = \int_0^t W(t, s)z(s)ds$ where $W(t, s)$ is the Cauchy function for some known exponentially stable equation. The most significant result utilizing this method was published in [26] where exponential stability and DDS of a particular scalar undamped second order RTDS were investigated. The authors found that the delayed system can be exponentially stable even if the delay-free system is unstable. Notice that the same finding was discussed e.g. in [6, 7, 11].

Numerical Gridding Algorithm. It represents the original simple-to-handle method by the first author of this survey. The algorithm can be summarized as follows:

(1) The delay is equidistantly discretized with a selected discretization step length.
(2) The dominant pole(s) for the delay-free case is exactly calculated.
(3) In each grid node, steps 4 to 6 are performed successively to each delay axis.
(4) The dominant pole (or a pair) and the corresponding delay value estimation are set in the nearest visited grid node.
(5) In the neighborhood of this estimation, a new estimation is iteratively computed. If the imaginary axis has not been crossed, the loop is finished (see step 3); otherwise, go to step 6.
(6) The preceding and current estimates are used to (successively) compute the delay vector for which the purely imaginary poles are obtained.
(7) Sets of critical delay and frequency estimations represent the algorithm output.

Two version of the algorithm were published. The continuous one [7] utilizes the Taylor series expansion in the neighborhood of the current dominant pole estimation to get the approximating polynomial [27]. In the discrete version [6], a polynomial approximation is performed by using techniques known from the field of discrete-time filters design; namely, the bilinear (Tustin) transformation and pre-warping, while non-integer exponents are approximated by means of linear or quadratic interpolation. Regarding step 6 of the algorithm above, linear interpolation or iterative Newton method can be used to get the zero point (in the real axis) estimation.

In case of neutral-type delays, one has to take a potential discontinuous spectral behavior into account.

Other Methods. Let us introduce some other recent results on DDS based on different frequency-domain techniques.

The above mentioned D-subdivision method starts with the setting of real and imaginary parts of $\Delta(s)$ equal to zero for $s = j\omega$. From this, unknown parameter(s) corresponding to purely imaginary poles are derived; namely, commensurate delays can be governed by parameters p_1, p_2 as $\exp(-j\omega h) = \cos(\omega h) - j\sin(\omega h) = p_1 - jp_2$. A simple systematic procedure based on this idea was published in [28].

In [29] the singular value decomposition technique was used to solve DDS of a single-input multi-output system. The determination of critical delays is based on the transformation of $\Delta(s) = \det(\mathbf{A} - s\mathbf{B} + \exp(-s\tau)\mathbf{H})$ to a polynomial via the singular value decomposition of a single-row matrix $\mathbf{H}$ subjected to $\mathbf{H} = \mathbf{U\Sigma V}$, $\Sigma = diag(\sigma \quad 0 \quad \ldots \quad 0)$.

A modified Schur-Cohn criterion [8, 30] for RTDS with commensurate delays was presented in [31]. The use of the criterion is motivated by the transformation of $\Delta(s) = \sum_{i=0}^{nc} d(s) \exp(-shi)$ to a bivariate polynomial $p(z, \omega) = \sum_{i=1}^{nc} b(\omega) z^i$, where $b(\omega) = d(j\omega), z = \exp(-sh)$ and n_C means the commensuracy degree, subjected to the Schur-Cohn criterion, the real roots of which are searching. The advantage of the proposed approach lies in the application of triangular matrices over a polynomial ring, which halves the dimension of $p(z, \omega)$.

Last but not least, some DDS methods were compared in [32]; i.a. the explicit infinitesimal generator discretization-based method [33] and a Padé approximation based method [34] were included in the study.

3.2 Academic and Practical Applications

This subsection intends to provide the reader with a non-exhaustive survey on recent application (real or hypothetic) and software utilizing DDS techniques.

As first, let us note that the above referred work [32] concerns delayed cyber-physical power systems. The DDS problem was partially solved by the computation of the spectral abscissa - using the pseudospectral collocation discretization method [35] - to investigate the properties of an Australian power system in [36]. The concept of the CTCR was utilized e.g. in [37] where the ubiquitous blade/casing rub problem in turbomachinery, modeled by a RTDS with a single delay, was studied and conditions to achieve stable rub interference were proposed. In [38], the authors investigated the influence of the displacement-feedback delay and joint displacement and velocity feedback delays on a robotic actuator system. A combination of the CTCT paradigm and a Kronecker multiplication based method [39] was applied in [40] to solve train following problem with delays between the trains and the wayside control unit, and to obtain the stability and resonance maps for a delayed resonator with position and speed feedback [41]. Both the models can be classified as RTDSs with two delays. To name another interesting result, stability maps in the delay space when analyzing the memory of drivers for the decision making process in a car following scenario were derived in [42]. In this system, modeled by dRTDS, each driver aims to keep a fixed time-headway with respect to the preceding vehicle. The authors utilized the approximation of delays by using limit properties of distributed delay terms and by the Taylor series expansion.

Regarding software applications, let us introduce leastwise two results. Stability conditions to protect stability in a multi-agent system where agents communicate with each other under delays in order to reach consensus was derived by means of the TRACE-DDE tool [43] in [44]. In [45], the YALTA software package [46] was used to investigate the local stability of nicotine exposure to cholesterol biosynthesis.

3.3 Open Problems

Some unsolved or open tasks from the field of DDS concisely follow. First of all, no recent result on dNTDS has been recently made; hence, especially a software tool would be welcome for engineers. Similarly, authors concerned mainly low-order and commensurate TDS with one or two delays. It is not reasonable to be limited to commensurate delays form the practical point of view. For instance, approximation techniques might be useful to transfer commensurate delays from non-commensurate models. Last but not least, more research can be done on the utilization of results on DIS to solve DDS problems, see e.g. [40, 41] for some pioneering works.

4 Conclusions

This contribution was aimed to provide a literature survey on stability analysis of delayed systems with respect to delay values. First, the reader was acquainted with some recent theoretical advances in the field. Selected application of several methods and software packages followed. The paper was concluded with a brief overview of some open problems in the field. The authors believe that this summary includes interesting and useful up-to-date information for scientists and engineers dealing with delayed systems.

Acknowledgments. This work was performed with the financial support by the Minis-try of Education, Youth and Sports of the Czech Republic within the National Sustain-ability Pro-gramme project No. LO1303 (MSMT-7778/2014).

References

1. Li, X., Gao, H., Gu, K.: Delay-independent stability analysis of linear time-delay systems based on frequency discretization. Automatica **70**(3), 288–294 (2016)
2. Hertz, D., Jury, E.I., Zeheb, E.: Stability independent and dependent of delay for delay differential systems. J. Franklin Inst. **318**(3), 143–150 (1984)
3. Xu, S., Lam, J.: Improved delay-dependent stability criteria for time delay systems. IEEE Trans. Autom. Control **50**(3), 384–387 (2005)
4. Sipahi, R., Olgac, N.: Complete stability robustness of third-order LTI multiple time-delay systems. Automatica **41**(8), 1413–1422 (2005)
5. Pepe, P., Jiang, Z.P.: A Lyapunov-Krasovskii methodology for ISS and iISS of time-delay systems. Syst. Control Lett. **55**(12), 1006–1014 (2006)
6. Pekař, L., Matušů, R., Prokop, R.: Gridding discretization-based multiple stability switching delay search algorithm: The movement of a human being on a controlled swaying bow. PLoS ONE **12**(6), e0178950 (2017)
7. Pekař, L., Prokop, R.: Direct stability-switching delays determination procedure with differential averaging. Trans. Inst. Meas. Control (2017). https://doi.org/10.1177/014233121 7700244
8. Gu, K., Kharitonov, V.L., Chen, J.: Stability of Time-Delay Systems. Birkhäuser, Basel (2003)
9. Sipahi, R., Delice, I.I.: On some features of core hypersurfaces related to stability switching of LTI systems with multiple delays. IMA J. Math. Control Inf. **31**(2), 257–272 (2014)
10. Castanos, F., Estrada, E., Mondié, S., Ramírez, A.: Passivity-based PI control of first-order systems with I/O communication delays: a frequency domain analysis. Int. J. Control. https://doi.org/10.1080/00207179.2017.1327083. in press
11. Sipahi, R.: Design of imaginary spectrum of LTI systems with delays to manipulate stability regions. In: Insperger, T., Ersal, T., Orosz, G. (eds.) Time-Delay Systems: Theory, Numerics, Applications, and Experiments, pp. 127–140. Springer, New York (2017)
12. Kammer, A.S., Olgac, N.: Non-conservative stability assessment of LTI dynamics with distributed delay using CTCR paradigm. In: 2015 American Control Conference, pp. 4597–4602. Palmer House Hilton, Chicago (2015)

13. Gao, Q., Zalluhoglu, U., Olgac, N.: Investigation of local stability transitions in the spectral delay space and delay space. J. Dyn. Syst. Meas. Control **136**(5) (2014). ASME, Article no. 051011
14. Gao, Q., Olgac, N.: Stability analysis for LTI systems with multiple time delays using the bounds of its imaginary spectra. Syst. Control Lett. **102**, 112–118 (2017)
15. Fazelinia, H., Sipahi, R., Olgac, N.: Stability robustness analysis of multiple time-delayed systems using 'Building Block' concept. IEEE Trans. Autom. Control **52**(5), 799–810 (2007)
16. Cepeda-Gomez, R.: Finding the exact delay bound for consensus of linear multi-agent systems. Int. J. Syst. Sci. **47**(11), 2598–2606 (2016)
17. Walton, K.E., Marshall, J.E.: Direct method for TDS stability analysis. IEE Proc. D-Control Theory Appl. **134**(2), 101–107 (1987)
18. Sönmez, Ş., Ayasun, S., Nwankpa, C.O.: An exact method for computing delay margin for stability of load frequency control systems with constant communication delays. IEEE Trans. Power Syst. **31**(1), 370–377 (2016)
19. Li, X.-G., Niculescu, S.-I., Çela, A., Wang, H.-H., Cai, T.-Y.: On τ-decomposition frequency-sweeping test for a class of time-delay systems, Part II: Multiple roots case. IFAC Proc. Vol. **45**(14), 138–143 (2012)
20. Li, X.-G., Niculescu, S.-I., Çela, A., Wang, H.-H., Cai, T.-Y.: On computing Puiseux series for multiple imaginary characteristic roots of LTI systems with commensurate delays. IEEE Trans. Autom. Control **58**(5), 1338–1343 (2013)
21. Ma, J., Zheng, B., Zhang, C.: A matrix method for determining eigenvalues and stability of singular neutral delay-differential systems. J. Appl. Math. **2012** (2012). Article ID 749847
22. Ochoa, G., Kharitonov, V.L., Modié, S.: Critical frequencies and parameters for linear delay systems: a Lyapunov matrix approach. Syst. Control Lett. **63**(9), 781–790 (2013)
23. Cao, J.: Improved delay-dependent exponential stability criteria for time-delay system. J. Franklin Inst. **350**(4), 790–801 (2013)
24. Xu, Q., Wang, Z.: Exact stability test of neutral delay differential equations via a rough estimation of the testing integral. Int. J. Dyn. Control **2**(2), 154–163 (2014)
25. Xu, Q., Stépán, G., Wang, Z.: Delay-dependent stability analysis by using delay-independent integral evaluation. Automatica **70**(3), 153–157 (2016)
26. Domoshnitsky, A., Maghakyan, A., Berezansky, L.: W-transform for exponential stability of second order delay differential equations without damping terms. J. Inequal. Appl. **2017**(1) (2017). Article no. 20
27. Pekař, L.: Enhanced TDS stability analysis method via characteristic quasipolynomial polynomization. In: Šilhavý, R. et al. (eds.) Cybernetics and Mathematics Applications in Intelligent Systems: Proceedings of the 6th Computer Science On-line Conference 2017 (CSOC 2017), vol. 2, pp. 20–29. Springer, Heidelberg (2017)
28. Perng, J.-W.: Stability analysis of parametric time-delay systems based on parameter plane method. Int. J. Innov. Comput. Inf. Control **8**(7A), 4535–4546 (2012)
29. Ramachandran, P., Ram, Y.M.: Stability boundaries of mechanical controlled system with time delay. Mech. Syst. Signal Process. **27**, 523–533 (2012)
30. Chen, J., Gu, G., Nett, C.N.: A new method for computing delay margins for stability of linear delay systems. Syst. Control Lett. **26**(2), 107–117 (1995)
31. Mulero-Martínez, J.I.: Modified Schur-Cohn criterion for stability of delayed systems. Math. Probl. Eng. **2015** (2015). Article ID 846124
32. Gao, W., Ye, H., Liu, Y. Wang, L, Ci, W.: Comparison of three stability analysis methods for delayed cyber-physical power system. In: 2016 China International Conference on Electricity Distribution (CICED 2016), paper no. CP1252, Xi'an, China (2016)

33. Ye, H., Gao, W., Mou, Q., Liu, Y.: Iterative infinitesimal generator discretization-based method for eigen-analysis of large delayed cyber-physical power system. Electr. Pow. Syst. Res. **143**, 389–399 (2017)
34. Niu, X., Ye, H., Liu, Y., Liu, X.: Padé approximation based method for computation of eigenvalues for time delay power system. In: The 48th International Universities' Power Engineering Conference, Dublin, Ireland, pp. 1–4 (2013)
35. Wu, Z., Michiels, W.: Reliably computing all characteristic roots of delay differential equations in a given right half plane using a spectral method. J. Comput. Appl. Math. **236**(9), 2499–2514 (2012)
36. Kishor, N., Haarla, L., Purwar, S.: Stability analysis and stabilization of delayed reduced-order model of large electric power system. Int. Trans. Electr. Energy Syst. **26**, 1882–1897 (2016)
37. Olgac, N., Zulluhoglu, U., Kammer, A.S.: On blade/casing rub problems in turbomachinery: an efficient delayed differential equation approach. J. Sound Vib. **333**, 6662–6675 (2014)
38. Ai, B., Sentis, L., Paine, N., Han, S., Mok, A., Fok, C.-L.: Stability and performance analysis of time-delayed actuator control systems. J. Dyn. Syst. Meas. Control **138**(5) (2016). ASME, Article no. 051005
39. Ergenc, A.F., Olgac, N., Fazelinia, H.: Extended Kronecker summation for cluster treatment of LTI systems with multiple delays. SIAM J. Control Optim. **46**(1), 143–155 (2007)
40. Alikoç, B., Mutlu, I., Ergenc, A.F.: Stability analysis of train following model with multiple communication delays. In: The 1st IFAC Workshop on Advances in Control and Automation Theory for Transportation Applications, Istanbul, Turkey, pp. 13–18 (2013)
41. Eris, O., Ergenc, A.F.: Delay scheduling for delayed resonator applications. IFAC-PapersOnline **49**(10), 77–81 (2016)
42. Sipahi, R., Atay, F.M., Niculescu, S.-I.: Stability analysis of a constant time-headway driving strategy with driver memory effects modeled by distributed delays. IFAC-PapersOnline **48**(12), 276–281 (2015)
43. Breda, D., Maset, S., Vermiglio, R.: TRACE-DDE: a tool for robust analysis and characteristic equations for delay differential equations. In: Loiseau, J., et al. (eds.) Topics in Time Delay Systems: Analysis, Algorithm and Control, pp. 145–155. Springer, Berlin (2009)
44. Qiao, W., Sipahi, R.: Delay-dependent coupling for a multi-agent LTI consensus system with inter-agent delays. Physica D **267**, 112–122 (2014)
45. Gölgeli, M., Özbay, H.: A mathematical model for cholesterol biosynthesis under nicotine exposure. IFAC-PapersOnline **49**(10), 258–262 (2016)
46. Avanessoff, D., Fioravanti, A.R., Bonnet, C.: YALTA: a Matlab toolbox for the H_∞-stability analysis of classical and fractional systems with commensurate delays. IFAC Proc. Vol. **46** (2), 839–844 (2013)

Modelling and Identification of Magnetic Levitation Model CE 152/Revised

Daniel Honc[(✉)]

Department of Process Control, Faculty of Electrical Engineering and Informatics,
University of Pardubice, nám. Čs. legií 565, 532 10 Pardubice, Czech Republic
Daniel.Honc@upce.cz

Abstract. Paper describes procedure of first principle modelling and experimental identification of Magnetic Levitation Model CE 152. Author optimized and simplified dynamical model to a minimum what is needed to characterize given system for the simulation and control design purposes. Only few experiments are needed to estimate the unknown parameters. Model quality is verified in the feedback control loop where the real and simulated data are compared.

Keywords: Magnetic levitation · Maglev · First principle model
Experimental identification · Verification · Feedback control

1 Introduction

Magnetic levitation, maglev or magnetic suspension is a method when the object is suspended with no other support than the magnetic fields. Magnetic force counteract effect of the gravitational force or other forces. Maglev is used e.g. in the transportation for trains, magnetic bearings, vibration isolation or contactless melting. All applications are inherently open-loop unstable and rely on the feedback control for producing the desired levitation action.

In case of Model CE 152 the steel ball is levitated in the air by the electromagnetic force generated by an electromagnet [1]. The single-input single-output (SISO), strongly nonlinear, unstable system is a nice object to study the system dynamics and experiment with different control algorithms based on classical or modern control theory. PID controllers, polynomial, robust or model predictive controllers including nonlinear case are applied in the literature [2–10]. Dynamical mathematical model is required for most of the controller design methods. Usually linear model is needed but for more realistic control simulations or control design methods the nonlinear model can be used as well. Modelling and identification problematics of the magnetic levitation process can be found e.g. in [11–20]. Usually first principle model is derived and the unknown parameters are estimated from experimental data. Black box identification can be used as well – parameters of external or internal mathematical representation are estimated from measured process responses. Author prefers first principle approach to get model with physical meanings and to identify parameters of the subsystems by separate experiments. Model complexity is reduced to a minimal structure with few estimated parameters only.

© Springer International Publishing AG, part of Springer Nature 2019
R. Silhavy (Ed.): CSOC 2018, AISC 765, pp. 35–43, 2019.
https://doi.org/10.1007/978-3-319-91192-2_4

The nonlinear model can be used directly by the simulation or can be analytically line-arized in given working point for the controller design method. Paper is structured as follows. Process is described in Sect. 2, model is derived in Sect. 3, unknown parameters are estimated in Sect. 4, model is verified in Sect. 5 and conclusions are given in Sect. 6.

2   CE 152 Magnetic Levitation Model

Magnetic Levitation Process consists of a base with coil, electronics and metal ball (see Fig. 1) and PC with Data Acquisition (DAQ) Card. Ball levitates in the magnetic field. The magnetic field of the coil is driven by a power amplifier connected to D/A output of DAQ card. Position of the ball is sensed by an inductive linear position sensor connected to A/D input of DAQ card.

Fig. 1.   Magnetic Levitation Model CE 152

3   First Principle Mathematical Model

Process is decomposed to individual subsystems which are modelled and identified separately. One subsystem is the power amplifier connected do D/A DAQ output. Coil and ball is another subsystem – this is the only subsystem with dynamics. The last subsystem is a position sensor connected to A/D DAQ input (Fig. 2).

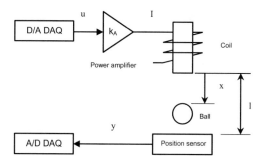

Fig. 2. Magnetic Levitation Model block scheme

3.1 Power Amplifier

Power amplifier is stabilized source of a current I which is proportional to the input voltage u generated by DAQ card

$$I = k_A u \tag{1}$$

The voltage u is in the range from 0 to 5 V and the current is in the range from 0 to approx. 1.5 A. Precisely the gain of the amplifier k_A is 0.297. This can be derived and calculated from parameters of the used electric components [14]. Time constant of the amplifier is very small and can be neglected. The amplifier gain has not to be estimated very precisely because the coil constant k can compensate the error.

3.2 Coil and Ball

We are using Lagrange's method for modelling of coil and ball subsystem. Motion equation is derived from the equilibrium of acting forces – gravitational force F_g and electromagnetic force F_m. Air resistance is neglected – the speed of the ball is not so high that this force would play a role. Accelerating force F_a is

$$F_a = F_g - F_m \tag{2}$$

$$m\frac{d^2x}{dt^2} = mg - k\left(\frac{I}{x + x_0}\right)^2 \tag{3}$$

where

m is ball mass (kg),
g is acceleration of gravity (m · s^{-2}),
k is coil constant (-),
I is coil current (A),
x is ball position (m) and
x_0 is coil offset (A).

Two unknown parameters k and x_0 must be estimated experimentally. Remaining parameters are listed in Table 1.

Table 1. Coil and ball parameters

Symbol	Units	Value	Meaning
m	kg	$8.28 \cdot 10^{-3}$	Ball mass
g	$\mathrm{m \cdot s^{-2}}$	9.81	Acceleration of gravity
d	m	$12.7 \cdot 10^{-3}$	Ball diameter
l	m	$18.4 \cdot 10^{-3}$	Distance between sensor and coil core

3.3   Position Sensor

Position sensor has a linear characteristic with two unknown parameters a and b

$$y = ax + b \tag{4}$$

Sensor senses ball position and outputs voltage y approximately in the range from 0 to 5 V.

4   Estimation of Unknown Parameters

4.1   Position Sensor

Position sensor is linear - two points for calibration are enough. Practically the simplest method is to hold the ball down at position sensor and measure the voltage and then place the ball to the coil core and measure the voltage again. We must take care only that the ball is placed in the centre of the coil core. The origin of the position axis x is placed at the end of the coil core and points down to the sensor.

Estimated parameters of Eq. (4) by using the data in Table 2 are $a = -855$ V/m and $b = 4.92$ V.

Table 2. Data for position sensor parameters identification

x (m)	y (V)
0	4.920
l-d	0.047

4.2   Coil and Ball

Because the system is unstable, closed loop control experiment must be carried out to estimate unknown coil constant k and offset x_0. Ball position y is controlled to a constant set-point w and used control action u is read out. We have measured four points (two would be enough) – see Table 3.

Table 3. Data for coil and ball parameters identification

y (V)	u (V)
1	2.87
2	2.34
3	1.89
4	1.44

In every steady-state point the electromagnetic force must equal to gravitational force which is constant

$$k\left(\frac{I}{x + x_0}\right)^2 = mg \tag{5}$$

Numerical optimization method can be used to estimate the unknown parameters. The problem can be transformed to a linear problem and Least Square Method can be applied also. Current I is calculated from the voltage input u according to Eq. (1). Position x is calculated from the output voltage y according to Eq. (4). Set of the position estimations (6) in matrix form is

$$\underbrace{\begin{bmatrix} \hat{x}_1 \\ \hat{x}_2 \\ \vdots \\ \hat{x}_4 \end{bmatrix}}_{\hat{X}} = \underbrace{\begin{bmatrix} I_1/\sqrt{mg} & -1 \\ I_2/\sqrt{mg} & -1 \\ \vdots & \vdots \\ I_n/\sqrt{mg} & -1 \end{bmatrix}}_{A} \underbrace{\begin{bmatrix} \sqrt{k} \\ x_0 \end{bmatrix}}_{P} \tag{6}$$

Cost function is sum of the squares of the position estimation errors which is

$$J = \sum_{i=1}^{n} \left(x_i - \hat{x}_i\right)^2 = \left(X - \hat{X}\right)^T\left(X - \hat{X}\right) \tag{7}$$

Optimal estimation of the vector P can be calculated as

$$X = \begin{bmatrix} x_1 \, x_2 \, \cdots \, x_4 \end{bmatrix}^T, \quad P = \left(A^T A\right)^{-1} A^T X \tag{8}$$

Estimated parameters are $k = P\,(1)^2 = 5.59 \cdot 10^{-6}$ and $x_0 = P\,(2) = 2.4 \cdot 10^{-3}$ m. Positions estimation absolute and relative errors are in Table 4.

Table 4. Positions estimation errors

x (m)	$\hat{x}$ (m)	$x - \hat{x}$ (m)	$(x - \hat{x})/l \cdot 100(\%)$
$4.59 \cdot 10^{-3}$	$4.64 \cdot 10^{-3}$	$-0.06 \cdot 10^{-3}$	-0.31
$3.42 \cdot 10^{-3}$	$3.34 \cdot 10^{-3}$	$0.08 \cdot 10^{-3}$	0.43
$2.25 \cdot 10^{-3}$	$2.23 \cdot 10^{-3}$	$0.02 \cdot 10^{-3}$	0.10
$1.08 \cdot 10^{-3}$	$1.12 \cdot 10^{-3}$	$-0.04 \cdot 10^{-3}$	-0.23

Relative position estimation error is less than 0.5%. The model as Simulink block scheme is in Fig. 3.

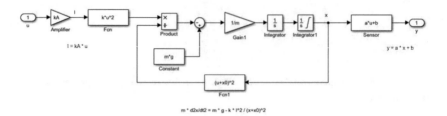

$$m * d2x/dt2 = m * g - k * I^2 / (x+x0)^2$$

Fig. 3. Magnetic Levitation Simulink block scheme

5   Model Verification

Model was verified in closed loop. Digital PID controller controls the process to a sequence of a step changes on the output voltage set-point. Identical controller is used to control the real system and the mathematical model. The set-point w, controlled variables y and manipulated variables u are plotted in Fig. 4. Real data are denoted as 'r' and simulated data with the mathematical model as 'm'.

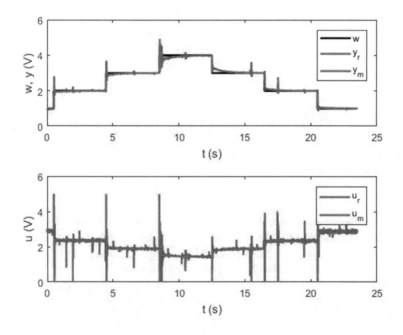

Fig. 4. Real and simulated control responses

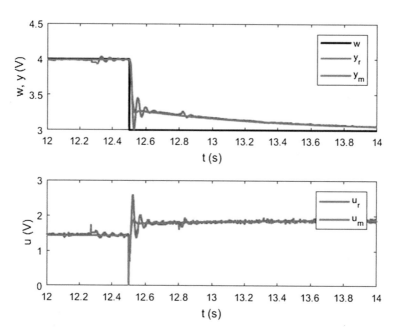

Fig. 5. Real and simulated control responses (from 4 to 3 V)

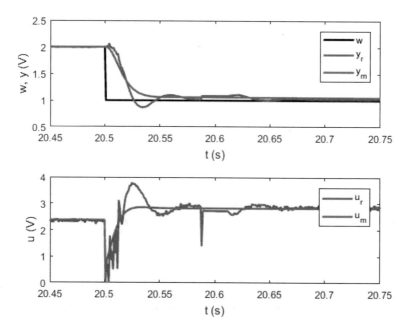

Fig. 6. Real and simulated control responses (from 2 to 1 V)

From the control responses it can be seen that not only static behaviour of the model corresponds to the real system but also the transient responses are very similar (see zoomed data in Figs. 5 and 6). The dynamics for bigger ball distances is slower as a fact of the nonlinearity of the process.

6   Conclusion

Magnetic levitation is an example of unstable real system and hence without the feedback control impossible to operate. Good mathematical model is a key point for the design and testing of different control strategies. Modern control methods are often model-based so the model is a part of the controller or at least used by the control design procedure.

Revised approach to the modelling and identification of magnetic levitation process is presented in the paper. For a specific system some phenomena like amplifier dynamics, air resistance or bouncing are neglected with practically no influence to the model quality. Only four unknown parameters are estimated from the experimental data – two parameters of the position sensor a and b and two parameters of coil and ball subsystem – coil constant k and offset x_0. Whole procedure is designed in such a way that only the position sensor bust be calibrated manually but the rest can be automated.

Offset parameter x_0 for the coil was identified as 2.4 mm. This means that the force generated by the coil slightly differs from the ideal case. It can't be higher than the force corresponding to the ball placed 2.4 mm far from the coil core in theoretical case. The real system tends to oscillate unfortunately. This is caused by the horizontal movement of the ball. At the same time the position sensor variable oscillates and the control action reacts to this – even if the vertical position of the ball changes only slightly. Guide bars would help but the visual effect of the levitation would not be so interesting. The solution is to operate the system carefully and time to time manually stabilize the ball if oscillations occur.

Acknowledgments. This research was supported by Institutional support of The Ministry of Education, Youth and Sports of the Czech Republic at FEI.

References

1. CE152 – Experiment, Magnetic Levitation Model. https://www.tecquipment.com/magnetic-levitation-model
2. Bächle, T., Hentzelt, S., Graichen, K.: Nonlinear model predictive control of a magnetic levitation system. Control Eng. Pract. **21**(9), 1178–1187 (2013)
3. Doležel, P., Rozsíval, P., Mariška, M., Havlíček, L.: PID controller design for nonlinear oscillative plants using piecewise linear neural network. In: Proceedings of the 19th International Conference on Process Control, PC 2013, pp. 19–24 (2013)
4. Gazdoš, F., Dostál, P., Marholt, J.: Robust control of unstable systems: algebraic approach using sensitivity functions. Int. J. Math. Models Methods Appl. Sci. **5**(7), 1189–1196 (2011)

5. Gazdoš, F., Dostál, P., Pelikán, R., Bobál, V.: Polynomial approach to control system design for a magnetic levitation system. In: 2007 European Control Conference, ECC 2007, pp. 4561–4567 (2007)

6. Hypiusová, M., Kozáková, A.: Robust PID controller design for the magnetic levitation system: frequency domain approach. In: Proceedings of the 21st International Conference on Process Control, PC 2017, pp. 274–279 (2017)

7. Chalupa, P., Novák, J., Malý, M.: Modelling and model predictive control of magnetic levitation laboratory plant. In: Proceedings of the 31st European Conference on Modelling and Simulation, ECMS 2017, pp. 367–373 (2017)

8. Qin, Y., Peng, H., Ruan, W.: Modeling and predictive control of magnetic levitation ball system based on RBF-ARX model with linear functional weights. Sci. Technol. **47**(8), 2676–2684 (2016). Zhongnan Daxue Xuebao (Ziran Kexue Ban)/Journal of Central South University

9. Rušar, L., Krhovják, A., Bobál, V.: Predictive control of the magnetic levitation model. In: Proceedings of the 21st International Conference on Process Control, PC 2017, pp. 345–350 (2017)

10. Stettinger, G., Benedikt, M., Horn, M., Zehetner, J., Giebenhain, C.: Control of a magnetic levitation system with communication imperfections: a model-based coupling approach. Control Eng. Pract. **58**, 161–170 (2017)

11. Du, X., Zhang, Y.: An improved method of mathematical model on current controlled magnetic levitation ball system. Appl. Mech. Mater. **128–129**, 70–73 (2012)

12. Galvão, R.K.H., Yoneyama, T., De Araújo, F.M.U., Machado, R.G.: A simple technique for identifying a linearized model for a didactic magnetic levitation system. IEEE Trans. Educ. **46**(1), 22–25 (2003)

13. Guess, T.M., Alciatore, D.G.: Model development and control implementation for a magnetic levitation apparatus. In: ASME Database Symposium, pp. 993–999 (1995)

14. Humusoft: CE 152 Magnetic levitation model – educational manual. Humusoft s.r.o., Prague (2002)

15. Chalupa, P., Malý, M., Novák, J.: Nonlinear simulink model of magnetic levitation laboratory plant. In: Proceedings of the 30th European Conference on Modelling and Simulation, ECMS 2016, pp. 293–299 (2016)

16. Jiang, D., Yang, J., Ma, L., Jiang, D.: Model building and simulating for hybrid magnetic levitation ball system. In: International Conference on Mechanic Automation and Control Engineering, MACE 2010, pp. 6105–6110 (2010)

17. Owen, R.B., Maggiore, M.: Implementation and model verification of a magnetic levitation system. In: Proceedings of the American Control Conference, pp. 1142–1147 (2005)

18. Pilat, A.: Modelling, investigation, simulation, and PID current control of active magnetic levitation FEM model. In: 18th International Conference on Methods and Models in Automation and Robotics, MMAR 2013, pp. 299–304 (2013)

19. Sankar, R.C., Chidambaram, M.: Subspace identification of unstable transfer function model for a magnetic levitation system. In: IFAC Proceedings Volumes (IFAC-PapersOnline), pp. 394–399 (2014)

20. Šuster, P., Jadlovská, A.: Modeling and control design of magnetic levitation system. In: Proceedings of the IEEE 10th Jubilee International Symposium on Applied Machine Intelligence and Informatics, SAMI 2012, pp. 295–299 (2012)

DDoS Reflection Attack Based on IoT: A Case Study

Marek Šimon[1], Ladislav Huraj[1(✉)], and Tibor Horák[2]

[1] Department of Applied Informatics and Mathematics, University of SS. Cyril and Methodius, Trnava, Slovakia
{marek.simon,ladislav.huraj}@ucm.sk
[2] Faculty of Materials Science and Technology in Trnava, Institute of Applied Informatics, Automation and Mechatronics, Slovak University of Technology in Bratislava, Trnava, Slovakia
tibor.horak@stuba.sk

Abstract. Along with the rise of Internet of Things devices the threat of adopting the IoT devices for cyber-attacks has increased. The number of IoT devices would be more than a billion in the world. Communication potential of such amount of devices is robust and has become more and more interesting for hackers. Mainly DDoS (Distributed Denial of Service) attacks carried from IoT devices seem to be a preferred method of attacker last years.

This paper illustrates a special type of DDoS attack using commonly available IoT devices called reflection attack which does not need to compromise the IoT devices. In reflection attacks, the attacker tries to use an innocent third party item to send the attack traffic to a victim to launch a distributed flooding attack, and to hide the attackers' own identity.

To demonstrate this type of attack, we consider the case of three categories of IoT devices: smart light-bulb (primarily used just for control of the intensity and color of the lights in a room), IP camera (digital video camera commonly employed for surveillance directly accessible over a network connection) and Raspberry Pi device (representing a single board computer). The paper demonstrates the potential of the IoT devices to be involved into such attack as well as first insight into communication traffic.

Keywords: IoT devices · DDoS attack · Reflection attack

1 Introduction

Internet of Things (IoT) represents uniquely identifiable things which are able to communicate and exchange data in the Internet. Nowadays, the concept of IoT has become popular through some demonstrative applications (e.g. smart grid, smart homes, smart cities, etc.), but IoT application can be extended to health care, irrigation, harvesting and waste management, whether motion gaming, or creating virtual reality manpower [1–4]. There are four main mechanisms for IoT devices (i) sensors and actuators, (ii) information processing, (iii) heterogeneous access, and (iv) applications and services, and additional mechanisms such as privacy and security. Geographically dispersed numbers of heterogeneous devices and systems are monitored and controlled

© Springer International Publishing AG, part of Springer Nature 2019
R. Silhavy (Ed.): CSOC 2018, AISC 765, pp. 44–52, 2019.
https://doi.org/10.1007/978-3-319-91192-2_5

in IoT and a varied collection of technologies e.g. communications, sensing or networking is applied [5].

The above mentioned technologies and the fact that IoT devices are usually embedded deeply inside networks as well as low attention to security of IoT devices, attract them to be misused from cyber criminals for malicious attacks. Low attention to IoT security coming from manufactures' strategy to get cheap products to the market as fast as possibly and not to spend time and money for good quality IoT security solutions. This indicates numbers of vulnerable IoT devices on the internet. Prognoses of IoT growth is to billions of heterogeneous devices in terms of firmware, hardware and field-upgradability. It is serious problem not only for end-users because of their daily lives privacy bounded with IoT devices but also for service providers because of larger and more distributed attacks led from insecure IoT devices [6, 7].

As an example, in October 2016 a massive attack was led from about 150,000 compromised IoT devices controlled by Mirai botnet, which caused an extraordinary DDoS attack strength of 1.2 Tbps traffic, 40 to 50 times bigger than usual. It flooded the service, i.e. it temporarily blocked access to Amazon, Netflix, PayPal, SoundCloud, Twitter, and other services for a few hours. The infected IoT devices involved surveillance cameras, routers, video recorders and other smart appliances. The combination of computing and networking power of thousands of cooperating IoT devices makes the IoT environment a potentially deadly cyber threat. The attack demonstrated that even junk traffic of huge number of enslaved IoT devices is able to produce significant volume of the attack traffic and to shut down large swaths of the internet [8].

In our article, we evaluate the attack potential of IoT devices for an attacker. We consider three common IoT devices appearing in smart home: smart light-bulb, surveillance IP camera and single board computer Raspberry Pi. The paper demonstrates an analysis on a DDoS reflection attack using the IoT devices to determine the real potential of these devices as security threat which could occur when the devices are involved into a DDoS attack.

The rest of the article is structured as follows: Sect. 2 describes short background about IoT and DDoS attacks. Section 3 explains the structure of the proposed real-world attack. In Sect. 4, the experimental outcomes are described. Finally, Sect. 5 reports about the conclusions and possible future work.

2 Background

This section briefly describes the background of two main phenomena on which the analysis is based: Internet of Things and DDoS attack.

2.1 Internet of Things

The Internet of Things (IoT) called as IP-connected IoT is the network of tiny physical devices, home appliances, and other heterogeneous components connected and inter-operating within the conventional internet and each thing is uniquely identifiable. The term IoT device can hide devices as electricity meters, light bulbs, smartphones, microwaves, automobile parts, even PCs or laptops, powerful server machines or a cloud, or possibly everything. The number of IoT devices that can be connected to the IoT environment is in hundreds of billions. The market in this field is currently going through exponential growth. Most of them will be cheaply made sensors and actuators which are likely to be very insecure, which is a potential danger and the vulnerable IoT devices could be misused for massive DDOS attack [9–11].

The heterogeneity of IoT devices is huge, based on different platforms and architectures, starting with devices with only very basic computing and communication abilities, through full protocol stack devices, till the application of the end-to-end principle in network operations.

In [12], the categorization of multiple IoT devices can be found. Figure 1 shows the specific groups of IoT devices included into the classification. In the classification scheme we marked our three tested IoT devices as well. The classification is using a scale from 0 to 100 points for the varying degree of the processing power of the IoT device on one side and devices' specificity of the purpose they serve on the other side. As can be seen from the scheme, there are several groups of IoT devices. The first one is group of single board computers as low specificity platforms (e.g. Raspberry Pi or BeagleBone Black), followed by second group of low power devices with limitations (e.g. Galileo, Photon, Arduino boards), next group are the specific purpose IoT devices with own native application support for additional user configurations (e.g. Samsung Family Hub Fridge, Amazon Echo), finished by specific IoT devices with limited user interaction (e.g. August Smart Lock, Lifx Bulb).

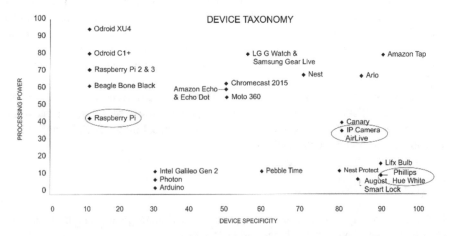

Fig. 1. Taxonomy of IoT devices [12] with selected tested IoT devices

Our three tested IoT devices we classified as follows: First tested device Raspberry Pi is a typical representative of the first group with low device specificity allowing users to rapidly develop various applications to run on the devices. The second one IP Camera AirLive BU-3026 belongs to third category allowing users to configure and install applications (e.g. intelligent video analytics, tampering detection, object counting). The last IoT device Philips Hue Bridge with smart light-bulbs represents the last category without built-in interface and interactive via special respective app. Such IoT devices with limited user interaction but with a wide range of processing power are the most attractive ones for attackers [12].

2.2 DDoS Attack

Distributed Denial of Service (DDoS) attack as a coordinated attack on the availability of services is performed indirectly using numerous compromised computing systems. DDoS attack uses lots of distributed nodes to launch an attack to decrease the performance of a system or whole network. It is done by consuming the computational resources and network bandwidth and so to block providing required service quality to its authorized clients [13].

DDoS attacks have rapidly grown in last years and it is one of the most significant threats on internet. Although thousands of connected servers and PCs are compromised to perform the attacks, because of memory and CPU limitations it looked unworkable that a tiny IoT devices could have enough memory and CPU capacity and thus could be used from attackers as branch of an attacking army; but the real DDoS attacks demonstrate that even the tiny IoT devices have enough resources to be hijacked and form part of a DDoS attack and are attractive for hacker as well [14].

There are several types of DDoS attack [15]: (i) UDP flood attack: several UDP packets are flooded from attacker's device on ports of victim device in order to listen port repetitively; the victim device reply every time an ICMP Destination Unreachable packet. The whole activity makes victim host resources unreachable. (ii) ICMP/PING flood attack: accordingly to the previous attack, this type of attack uses ICMP Echo Request (ping) packets to overwhelm the victim host resources sending the packets as fast as possibly without waiting for replies. ICMP flood attack consumes outgoing and incoming bandwidth. (iii) SYN flood attack: the attacker sends SYN requests but the victim host will not receive particular SYN-ACK response. Therefore the victim host is waiting for acknowledgement for each of the requests causing an overwhelm of the victim host resources. (iv) Ping of Death attack: attacker sends a malformed or otherwise malicious ping on a victim host affecting e.g. overflow memory buffers, and consequently causing collapse of the whole system. (v) Zero-day DDoS attack: attack uses a previously unknown or brand new vulnerability for which no patch has yet been released.

In our experiments, we applied UDP-based Distributed Reflective Denial of Service (DRDoS) attack. DRDoS attack preserves the attacker's anonymity through IP address spoofing using a potentially innocent third party item to send the flood traffic to a target. The attacker sends packets to the reflector devices with a source IP address set to their target's IP therefore indirectly overloading the target with the considerably larger response packets. Thus, it is hard to identify the real attackers and block their services. The reflector device could be ordinary legitimate device, e.g. IoT device, not strictly compromised. If the amount of reflected packets is extremely large, then the victim's network can be flooded [16].

3 Experiment Design

In the evaluation of our experiment, we utilize three very popular IoT devices, namely Raspberry Pi, IP Camera AirLive BU-3026 3-Megapixel, and Philips Hue Connected bulb controlled with Philips Hue Bridge 2.0. The IoT devices were chosen because of their specific characteristics representing a whole group of similar IoT devices as well as differences between particular IoT devices.

Single board computers Raspberry Pi is an example of small computer systems on which user can run and develop own applications. Initially, the device was developed with the purpose to teach computer science basic in schools. Nowadays, Raspberry Pi and similar alternatives (e.g. Asus Tinker Board, Orange Pi) are profoundly used as prototype for IoT development. Furthermore, several companies provide support to Raspberry Pi or provide additional tools for cooperation such as Ardunio, Intel with Intel Galileo and Intel Edison boards or Microsoft Visual Studio [17].

IP Camera AirLive BU-3026 is an example of IoT device that not only serves its specific purpose but also allows for users to set their configurations and installations of applications. IP cameras are already routinely used in various places including schools, pharmacies, markets or malls employed for surveillance. They can send and receive information through computer networks, usually provide high bandwidth and processing requirements. With wide use of surveillance IP cameras, their security emerges as a weighty issue [18].

The Philips Hue Connected Light-Bulbs with Philips Hue Bridge are an example of IoT devices without built-in interface and interacting via special respective app. The lighting system in home and its functions like intensity, color or scheduling can be wirelessly controlled by this smart bulbs via Philips Hue App for Android or iOS. All the light-bulbs are controlled from the app directly towards the Philips Hue Bridge. ZigBee Light link protocol is used from the bridge to send appropriate signals to the light-bulb for a desired change. ZigBee is a standards-based wireless technology invented to cover the unique needs of low-power as well as low-cost wireless sensor [19].

In our experiments, we study a typical network with wired connections to the three IoT devices. For the experiment, the Philips Hue Bridge is important, the reflection attack is led to and from this device, the light-bulbs are not network bound and are invisible to the network. Particular IoT devices were connected to the switch by 100 Mbps Ethernet interface.

We do not consider devices communicating among themselves while the experiment is running. As can be seen from Fig. 2, the topology of the testing network consists of 8-port Gigabit Ethernet switch connected with two servers with GigaBit Ethernet Network Interface Card. The first server was used to generate UDP packets. The second one served as the default gateway for the tested IoT device where the data flow generated by IoT devices is measured.

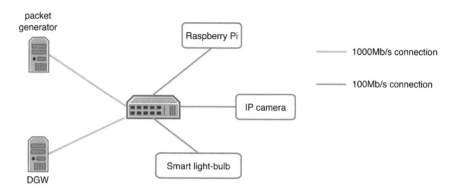

Fig. 2. The topology of the testing network

Our previous research [20, 21] shows the applicability even of small number of attacking nodes to perform a successful DDoS attack, which results in overloading of victim server.

4   Performance Tests

We performed the DRDoS attack separately for particular IoT devices. The hping3 tool (http://hping.org) was used to send UDP flooding. Hping is a command-line packet crafter used to generate IP packets supporting TCP, UDP, ICMP and RAW-IP protocols. Hping is mainly used as a security tool by security professionals to test networks and hosts.

In our case, the UDP flood attack was generated for 2 min; one UDP datagram each 50 µs, while source IP addresses were spoofed. Destination port was chosen to 800. For every UDP datagram, particular IoT device responded by generating an error message ICMP *destination port unreachable*; this message was sent on the spoofed source IP address through default gateway. On the second server, all ICMP packets from the IoT device were analyzed using own scripts and the packet analyzer TCPDUMP (http://www.tcpdump.org).

Two main metrics were taken into account in the study outcomes for all tested IoT devices. The number of packets per seconds sending from the IoT device to target was the first metric, Fig. 3, and the second metric was the average amount of sent reflected bytes per second, Table 1.

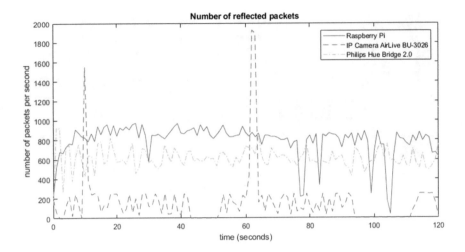

Fig. 3. Number of reflected packets per second transmitted to the network by the IoT devices

Table 1. The average amount of sent reflected bytes per second

	Raspberry Pi	IP Camera AirLive BU-3026	Philips Hue Bridge 2.0
Reflected bytes	55 748,35	11 435,45	42 914,05

Let us note that the size of attackers UDP request consists of empty UDP datagram and of 8 byte of UDP header; consequently, it is encapsulated into the 28 byte IP datagram on the network layer; after encapsulation into the Ethernet frame on the data link layer, the whole frame has the size of 42 bytes. Size of outgoing message from IoT devices (reflector) to target host consists of the 28 bytes payload ICMP packet with error message and of 8 bytes ICMP header with added 20 bytes for IP packet encapsulation. After encapsulation into the Ethernet frame on the data link layer, the whole leaving frame has the size of 70 bytes, which is 1.67 time bigger than incoming frame.

The Fig. 3 demonstrates the effect of the attacks using all three IoT devices from the point of view of number of reflected packets. As can be seen from the graph, the Raspberry Pi acts as the best IoT device usable for such kind of attack. Course of attack of Raspberry Pi and Philips Hue Bridge is very similar, the difference is roughly 100 kbit/s. Even the amount of drops in the graph is comparable. The drops in the graph occurred when the IoT device was overloaded by UDP requests and was not able to receive the requests. The measurements show that the worst reflector is the IP Camera AirLive BU-3026 both in terms of reflected packets as well as in amount of drops. The attack was already on the limits of the IP Camera overload.

In general, the IoT devices have shown the ability significantly to be adopted into DDoS reflection attack. All tested categories of IoT devices represent valuable resources for distributed denial of service reflection attack. Let us note that usually an attacker's network consists of thousands of attacking IoT devices and the power of attack increases with the number of connected IoT devices. If we consider for example that the network bandwidth is 20 Mbps it can be fully covered by involving, respectively, of 47, 229 or 61 testing IoT devices.

Our real-world attack was implemented using only freely available equipment. We used only available Open Source software. In addition, we did not need to compromise, hack or however modify included IoT devices.

5   Conclusions and Future Work

Fast growth of IoT devices causes various security problems. Our paper showed the effectiveness of selected real-world IoT devices for the sample of DDoS reflection attack. Our case study demonstrates this type of attack on three categories of heterogeneous IoT devices: single board computer Raspberry Pi, IP camera and smart light-bulbs as well as it shows first insight into communication traffic during involving of the IoT devices into attack.

The experimental results show that potential to involve IoT devices into DDoS reflection attack exists. With the number of devices and hyperconnectivity that are continuously being added to an increasingly larger IoT environment, the described DRDoS attack results have the ability to be dramatically multiplied and to cause large scale effects. Even though the phenomena discussed in the paper are well-known, its main contribution is an understanding of their importance as well as the real implementation of DRDoS attack on IoT devices.

Future extension of the case study requires to further investigate the scaling of DRDoS flood attacks on a larger number of other smart-home devices. Moreover, the categorization in the Fig. 1 could be changed not from the point of view of device processing but of flood capability for all described well-known devices, even for various kinds of DDoS attack.

Acknowledgements. The work was partly supported by the grant VEGA 1/0145/18 Optimization of network security by computational intelligence and partly by the grant KEGA 011UMB-4/2017 *Increasing competencies in work with high performance computing ecosystem.*

References

1. Srivastava, S., Pal, N.: Smart cities: the support for Internet of Things (IoT). Int. J. Comput. Appl. Eng. Sci. **6**(1), 5–7 (2016)
2. Ölvecký, M., Gabriška, D.: Motion capture as an extension of web-based simulation. In: Applied Mechanics and Materials, vol. 513, pp. 827–833 (2014)
3. Horváthová, D., Siládi, V., Lacková, E.: Phobia treatment with the help of virtual reality. In: 13th International Scientific Conference on Informatics, pp. 114–119. IEEE (2015)
4. Hosťovecký, M., Novák, M., Horváthová, Z.: Problem-based learning: serious game in science education. In: Proceedings of the 12th International Conference on e-Learning, ICEL 2017, pp. 303–310. ACPI 2017 (2017)
5. Suo, H., Wan, J., Zou, C., Liu, J.: Security in the Internet of Things: a review. In: International Conference on Computer Science and Electronics Engineering (ICCSEE), vol. 3, pp. 648–651. IEEE (2012)

6. Hesselman, C., et al.: SPIN: a user-centric security extension for in-home networks. SIDN Labs Technical report SIDN-TR-2017-002 (2017)
7. Luptáková, I.D., Pospíchal, J.: Community cut-off attack on malicious networks. In: Conference on Creativity in Intelligent Technologies and Data Science, pp. 697–708. Springer, Cham (2017)
8. Pishva, D.: IoT: their conveniences, security challenges and possible solutions. Adv. Sci. Technol. Eng. Syst. J. 2(3), 1211–1217 (2017)
9. Raza, S., Wallgren, L., Voigt, T.: SVELTE: real-time intrusion detection in the Internet of Things. Ad Hoc Netw. 11(8), 2661–2674 (2013)
10. Ronen, E., et al.: IoT goes nuclear: creating a ZigBee chain reaction. In: IEEE Symposium on Security and Privacy (SP), USA, pp. 195–212 (2017)
11. Halenar, I., Juhasova, B., Juhas, M.: Proposal of communication standardization of industrial networks in Industry 4.0. In: IEEE 20th Jubilee International Conference on Intelligent Engineering Systems (INES), pp. 119–124 (2016)
12. Habibi, J., Midi, D., Mudgerikar, A., Bertino, E.: Heimdall: mitigating the Internet of Insecure Things. IEEE Internet Things J. 4(4), 968–978 (2017)
13. Singh, S., Gyanchandani, M.: Analysis of Botnet behavior using Queuing theory. Int. J. Comput. Sci. Commun. 1(2), 239–241 (2010)
14. Nizami, Y., Garcia-Palacios, E.: Internet of Thing. A proposed secured network topology. ISSC 2014/CIICT 2014, Limerick, pp. 274–279, June 2014
15. Sonar, K., Upadhyay, H.: A survey: DDOS attack on Internet of Things. Int. J. Eng. Res. Dev. 10(11), 58–63 (2014)
16. Berti-Equille, L., Zhauniarovich, Y.: Profiling DRDoS attacks with data analytics pipeline. In: ACM on Conference on Information and Knowledge Management, 6–10 November 2017, Singapore, pp. 1983–1986 (2017)
17. Perera, C., Liu, C.H., Jayawardena, S., Chen, M.: A survey on Internet of Things from industrial market perspective. IEEE Access 2, 1660–1679 (2014)
18. Tekeoglu, A., Tosun, A.S.: Investigating security and privacy of a cloud-based wireless IP camera: NetCam. In: IEEE 24th International Conference on Computer Communication and Networks (ICCCN), USA, pp. 1–6 (2015)
19. Notra, S., Siddiqi, M., Gharakheili, H.H., Sivaraman, V., Boreli, R.: An experimental study of security and privacy risks with emerging household appliances. In: IEEE Conference on Communications and Network Security (CNS), USA, pp. 79–84 (2014)
20. Šimon, M., Huraj, L., Čerňanský, M.: Performance evaluations of IPTables firewall solutions under DDoS attacks. J. Appl. Math. Stat. Inform. 11(2), 35–45 (2015)
21. Šimon, M., Huraj, L., Hosťovecký, M.: IPv6 network DDoS attack with P2P grid. In: Creativity in Intelligent, Technologies and Data Science, pp. 407–415. Springer (2015)

Complemented Adaptive Control Strategy with Application in Pedagogical Cybernetics

Tomas Barot$^{(\boxtimes)}$

Department of Mathematics with Didactics, Faculty of Education,
University of Ostrava, Mlynska 5, 701 03 Ostrava, Czech Republic
Tomas.Barot@osu.cz

Abstract. The pedagogical cybernetics has been widely researched corresponding to the modern educational approaches. However, it can be advantageous to complement this theory using the approaches from the technical cybernetics. The particular aspects of the educational processes can be modeled in the general feedback strategy. In the previous research of author of this paper, the modified approach of the feedback strategy, extended using principles of the adaptive control strategy, was proposed for the practical utilization in the pedagogical cybernetics. In this paper, this principle is further improved by aspects of the offline part of the adaptive control strategy. This connection between the pedagogical and technical cybernetic rules is discussed and concretely implemented in favor of the educational process of the foreign students, respectively of the future teachers, at the departmental offered foreign language course of the linear algebra. For purposes of the modern based educational approaches, the information technologies are included in the proposal in this paper in favor of didactics of the mathematical courses.

Keywords: Pedagogical cybernetics · Technical cybernetics
Feedback strategy · Adaptive control strategy · Educational process
Linear algebra · Free-Available-Software-Support
Professional-Preparation of teachers

1 Introduction

In the frame of the professional and practical preparation of the future teachers, innovations of the current educational approaches have been researched in general yet. Applications of the innovative methods or approaches in the form of the intelligent educational methods and systems are bounded with the presentation of quality of the departmental education in view of the external institutions. Important part of this education can be consisted of the foreign language courses, where are the wide variabilities of implementations of the innovative strategies.

Possibilities of the new educational trends are widely described e.g. in [1–4]. The current trends are usually based on the utilization of the information technologies. However; the new constructions of strategies based on the principles of the technical cybernetics have not been so widely presented. Therefore the classical models and

© Springer International Publishing AG, part of Springer Nature 2019
R. Silhavy (Ed.): CSOC 2018, AISC 765, pp. 53–62, 2019.
https://doi.org/10.1007/978-3-319-91192-2_6

principles of the pedagogical cybernetics [5–7] can be improved by principles of the technical cybernetics [8–10] in the modified form.

The pedagogical cybernetics [5–7] is a part of the general cybernetics, which is aimed to the area of the pedagogical sciences. There are described strategies for the educational processes in sense of the cybernetic principles. The known fundamental principle is a feedback strategy [11, 12] in the various educational processes. In this strategy, teachers and students can be projected. Number of cases and examples have been widely described in the literature, e.g. in [5–7], focused on the pedagogical research.

The technical cybernetic principles are in general based on the modeling of the abstract systems, on analyses of their dynamical behavior and on syntheses of controllers in favor of control of the identified or progressively identified system, as can be read in a wide spectrum of the literature, e.g. in [8–10]. In the both - theoretical and practical area, the system modeling can be considered as the important part of the technical cybernetics using the algebraic structures. In this systems-theory, respectively in the control systems theory; many sophisticated strategies have been presented. All these principles are suitable for the technical sciences in favor of the technical implementation of the control of the real systems in the industry, e.g. in [13, 14].

Further connections between the technical and pedagogical cybernetics are not so widely described. As can be seen e.g. in [15], the main approach of the specific connection of these both disciplines is based on the utilization of the system-modeling-principles. Concretely, the denoted connection [15] was realized in case of the application of some aspects of the quantitative-research-methods into the system-modeling-area, where the pedagogical phenomenons can be modeled using models of the technical cybernetic approaches. The verification of this whole approach was provided by the statistical testing of hypotheses in the simulation of the pedagogical problem in comparison to the hypotheses testing with implementation using the proposed models. E.g., the student-knowledge-level can be simulated in respect to its statistical analysis.

The extended feedback strategy, which was presented in [16], was modified by the technical cybernetic principles in the context of the adaptive-control-approach. This modified strategy was applied into the pedagogical cybernetics for purposes of the particular educational process in the Maths Support Centre [17] at Tomas Bata University in Zlin, where the author of this paper is a member of tutors under the leadership of Mrs. Mgr. Zuzana Patikova, Ph.D. The adaptive control strategy was modified in respect to the utilization of the free-available software in frame of an explanation of the mathematical topics. The online based identification of the student's knowledge was an important part of the modified proposal of the feedback strategy. The strategy [16] can be suitable in the environment of the individual mathematical tutoring; however, its principles can be further modified in favor of other types of the educational situations.

In this paper, possibilities of the strategy [16] are complemented by the new part inspired by the technical cybernetics. This modification is instead of the preparation part of the adaptive control strategy, which can be similar e.g. with the explicit adaptive predictive control [18] with the offline (preparation) pre-processing part and the online processing part. In favor of the educational processes, this specification can be suitable applied. The practical realization can be implemented in the professional practice preparation of the future teachers. In this paper, the proposed strategy is demonstrated on example of the preparation in the didactical course of the Linear algebra taught for the foreign students at the department of author of this paper. The further practical implementation can be an inspiration for this type of the education, which is very important in sense of its internationalization.

2 Application of Adaptive Control Strategy in Mathematical Education in Frame of Pedagogical Cybernetic Theory

In the International Journal of Communication Technologies in Education, the modified approach [16] of the educational general feedback process [11] was proposed. The feedback strategy included the extension with the principle of the technical cybernetics. This new aspect was related to the adaptive strategy [10] with the included progressive identification of the controlled object. Particularly, this added part of the continuous identification can be denoted as online relating to terms of the technical cybernetics [10]. This strategy was extended by the identification of the behavior of students using tests of their knowledge by the recommended and described free-available mathematical software.

The practical utilization was demonstrated in case of the tutoring of author in the collective of tutors under a leadership by Mrs. Mgr. Zuzana Patikova, Ph.D. in the Maths Support Centre [17] at Tomas Bata University in Zlin (MSC TBU in Zlin). In this centre, each tutor can individual apply own innovative strategies from the pedagogical practice in favor of explanations of topics to students. Because, there can be a high number of students at sessions, author of this paper, which is the tutor at MSC TBU in Zlin, has been using the proposed strategy [16]. There can be fulfilled the aim of the connection of the technical and pedagogical cybernetics for the improving of the efficiency of the particular educational process. The published adaptive strategy [16] can be seen on Fig. 1.

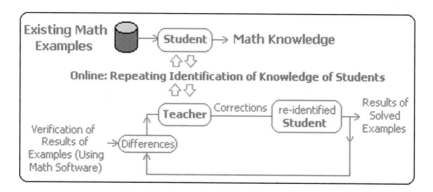

Fig. 1. Control strategy with online continuous repeating identification [16]

The main part of the strategy is based on the continuous repeating identification of the student's knowledge. This strategy is related to principles of the adaptive control strategy. The recommended possibilities of the mathematical tools, with utilization of the free-available software, were described in favor of the identification phase of the particular proposal of the tutoring in the MSC TBU in Zlin. However, recommendations of this strategy are based only on the existing mathematical examples in the online part, which should be improved e.g. by the proposal of the preparation offline subsystem in the following sections of this paper.

3   Parts of Linear-Algebra-Theory Used in Professional-Practice-Preparation of Future Teachers

In this contribution, advantages of the further described proposal can be appropriate applied in favor of the concrete purposes of the professional preparation of the future teachers in the frame of the foreign study course of the Linear algebra. In this course, a practical style of the realization of the cycle of the adaptive feedback control strategy can be suitable implemented; however, instead of the general mathematical examples from the literature, the new own examples can be generated using rules of the linear algebra, as is detail described in the further section in this paper.

The important topic of the discussed problems is the set of the linear equations in the course of the Linear algebra. For the suitable results of sets of the linear equations, the problem definitions are retrospectively built for the educational purposes. The following algebraic theory is related to rules from source [19]. The further used mathematical expressions respect axioms of the general algebra S (1). The binary operation is denoted by $\circ$ and the unary operation by Δ. Results and operands of these operations are elements of the set $\mathcal{T}$. The binary operation has two operands and the unary operation has only one operand. Sequence ξ includes constants, which are defined as the identity elements.

$$S = \langle \mathcal{T}, \circ, \Delta, \xi \rangle \tag{1}$$

The algebraic theory of sets of the linear equations is explained in the frame of the concrete algebra L (2). Where m and n are dimensions of matrices with elements of the real number field. The binary operations $+$ and $.$ are considered as the matrix addition and multiplication. Operation $-$ is the unary operation of an inversion. The identity elements are matrix I (with ones elements on the main diagonal and other zero elements) and matrix $\mathbf{0}$ (with all zero elements).

$$L = \langle \mathcal{R}^{m,n}, +, ., -, \mathbf{0}, I \rangle \tag{2}$$

For the proposal in this paper, the sub-algebra L' (3) of algebra L is applied with algebra L. Because all results of operations in the integer-number-matrix-set of algebra L' return results, which are included in set of algebra L, this combination is possible with respect to axiom of the closure [19] in frame of sub-algebras.

$$L' = \langle \mathcal{Z}^{m,n}, +, -, \mathbf{0} \rangle \tag{3}$$

All elements of sets in algebras L and L' fulfill axioms about closure (4), associativity (5), existence of the identity elements (6) and existence of inversion (7).

$$\forall G, H \in \mathcal{T} : \{(G \circ H) \in \mathcal{T}\} \wedge \{\Delta G \in \mathcal{T}\} \tag{4}$$

$$\forall F, G, H \in \mathcal{T} : F \circ (G \circ H) = (F \circ G) \circ H \tag{5}$$

$$\forall G \in \mathcal{T} : \{G + \mathbf{0} = \mathbf{0} + G = G\} \wedge \{G.\mathbf{I} = \mathbf{I}G = G\} \tag{6}$$

$$\forall G \in \mathcal{T} : \{G + (-G) = \mathbf{0}\} \wedge \{G.G^{-1} = \mathbf{I}\} \tag{7}$$

Sets of the linear equations can be represented by (8)–(9) in frame of the algebras L and L'. Where matrix A is considered as a square matrix with the n linear independent rows, which cause existence of the solution of the set of the linear equations. The definition of problem (8)–(9), retrospectively based on the suitable form of results, is included in the following modified strategy of the adaptive control approach.

$$A.z = b; \ A \in \mathcal{Z}^{n,n}; \ z \in \mathcal{Z}^{n,1}; \ b \in \mathcal{R}^{n,1} \tag{8}$$

$$\left.\begin{array}{c} a_{11}z_1 + \cdots + a_{1n}z_n = b_1 \\ \vdots \\ a_{n1}z_1 + \cdots + a_{nn}z_n = b_n \end{array}\right\}; \ A = \begin{bmatrix} a_{11} & \cdots & a_{1n} \\ \vdots & \ddots & \vdots \\ a_{n1} & \cdots & a_{nn} \end{bmatrix}; \left.\begin{array}{c} \\ \\ \\ z = \begin{bmatrix} z_1 & \cdots & z_n \end{bmatrix}^T; \ b = \begin{bmatrix} b_1 & \cdots & b_n \end{bmatrix}^T; i \in \langle 1; n \rangle \end{array}\right\} \tag{9}$$

4 Proposal of Complemented Adaptive Control Strategy for Purposes of Pedagogical Cybernetics

In this paper, the strategy [16] of the adaptive control approach is extended in the context of the connection between the technical and pedagogical cybernetics. Inspiration for the proposal of the extension is corresponded with the similar existing strategy [18] in the field of the technical cybernetics; concretely, this control strategy has the structure consisted of the offline and online parts.

The added part can be considered as the preparation or the offline part of the adaptive control strategy. These ideas, which are not widely described in the pedagogical cybernetics in this concrete focusing, could be implemented in favor of the educational processes. The extended offline part is proposed for the previous strategy [16] in sense of the approach of the adaptive control strategy with offline part [18], as can be seen on Fig. 2. Concretely, the added offline extension, proposed in this section, can be used in the educational process before the realization of the following cycles in the feedback control strategy.

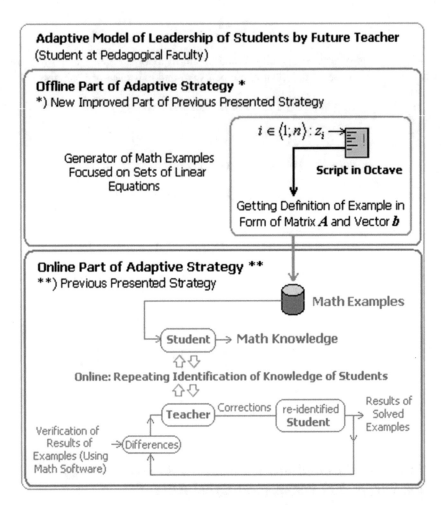

Fig. 2. Complementing of adaptive strategy [16] with proposal of offline part

Instead of the general mathematical examples from literature, the random new examples can be prepared in this offline part by a programmed generator (e.g. in the recommended free-available mathematical software Octave). As can be seen on Fig. 2, the generated examples are focused on topic of sets of n linear equations in the matrix form (8)–(9). This generator is improved by the random aspects. Results (10) can be generated in the suitable integer form and then the original definition of the particular example (8)–(9) can be retrospectively determined.

$$\forall i; i \in \langle 1; n \rangle : z_i = [10x_i]; \, x_i \in \mathcal{R} \wedge x_i \in X \tag{10}$$

In Eq. (10), the generated results are determined using the random variable X in the offline part of the adaptive strategy. The realization of variable X is consisted of values x_i. Because these values are the real numbers in the software generator, using command *randn* in Octave, the lower integer part should be computed. The lower integer part of any real number is a function, which has notation by the square brackets in definition (10). Results of this function are only integer numbers. Variable X (11) has the standardized Gaussian normal probability distribution N [20] with the mean value μ equal to 0 and with the unit variation σ^2. The probability density function p is expressed in (12).

$$X \sim N(\mu, \sigma^2) \sim N(0, 1) \tag{11}$$

$$\forall x \in X : p(x) = \frac{1}{\sqrt{2\pi}} e^{-\frac{x^2}{2}} \tag{12}$$

For purposes of the reducing of the complexity of the finding of an appropriate matrix A, elements of this matrix can be randomly generated as the integer numbers using the random realization of variable X, as can be seen in (13).

$$j \in \langle 1; n \rangle, \, k \in \langle 1; n \rangle : a_{jk} = [10x_i]; \, x_i \in \mathcal{R} \wedge x_i \in X \tag{13}$$

Vector b can be finally expressed by Eq. (14) using the regular matrix A with the non-zero determinant.

$$b = A.x; \, \det A \neq 0 \tag{14}$$

5 Results

Author of this paper has been practically used the presented strategy in education of the foreign-language-course of the Linear algebra with the current students of the pedagogical study program. As a part of each cycle in the feedback process of learning, the creating of the new mathematical examples of sets of the linear equations was provided using the script in Octave, as can be seen in Fig. 3. This algorithm based on algebras L and L' had to include the verification of the regularity of matrix A, because the regularity of this matrix is the necessary condition for the generating of the appropriate definition of described problem (8)–(9) with the suitable form of results. Students could improve their knowledge in the field of the mathematical didactics. Therefore, they could apply this strategy to their student in the future professional practice.

In the final exam of the course of the Linear algebra, the following example was generated using the described method by the student. This example is advantageous in favor of the didactical approach of the construction of the set of the linear equations with the suitable results. Script (Fig. 3) was used by the examined student in respect to the definition of the result-variables, which was programmed in the free-available Octave. The suitable form of results $x_1 = -2$, $x_2 = 1$, $x_3 = 5$ were generated. The creating of this script was a part of the course of the Linear algebra.

In the concrete student's exam, solution (15) was defined using the programmed script. These achieved definitions had to be verified using the analytical computations in the final exam. Therefore, the verification was provided using the manually computed

```
1    % Generator of Suitable Form of Results
2    x=[ceil(randn*10);ceil(randn*10);ceil(randn*10)]
3    % Generator of Suitable Form of Matrix A
4    A=[ceil(randn*10)  ceil(randn*10)  ceil(randn*10);
5        ceil(randn*10)  ceil(randn*10)  ceil(randn*10);
6        ceil(randn*10)  ceil(randn*10)  ceil(randn*10)]
7    % Verification of Regularity of Matrix A
8    while (det(A)==0)
9        % Other Form of Matrix A
10       A=[ceil(randn*10)  ceil(randn*10)  ceil(randn*10);
11       ceil(randn*10)  ceil(randn*10)  ceil(randn*10);
12       ceil(randn*10)  ceil(randn*10)  ceil(randn*10)]
13   endwhile
14   % Determination of Vector b
15   b=A*x
```

Fig. 3. Proposed implementation of offline part of adaptive strategy in octave

Eq. (16). Where inversion of the matrix A was correctly computed by the student. The determinant of matrix A $(\det(A))$ and adjungated matrix A $(\text{adj}(A))$ were analytically expressed using the Sarruss rule and the computation of $\text{adj}(A)$ described detail in [19]. Results of the own created example of the set of the linear equation were successfully verified using the analytical rules by the student - the future teacher.

$$A = \begin{bmatrix} 1 & -2 & 3 \\ 1 & 1 & 2 \\ 3 & -5 & 2 \end{bmatrix}; b = \begin{bmatrix} 1 & -2 & 3 \\ 1 & 1 & 2 \\ 3 & -5 & 2 \end{bmatrix} \cdot \begin{bmatrix} -2 \\ 1 \\ 5 \end{bmatrix} = \begin{bmatrix} 11 \\ 9 \\ -1 \end{bmatrix} \tag{15}$$

$$x = A^{-1}.b = \frac{1}{\det(A)} adj^T(A) = \begin{bmatrix} -2 & 1 & 5 \end{bmatrix}^T \tag{16}$$

6   Conclusion

The advantage of this contribution is a practical approach focused on the connection between the pedagogical and technical cybernetics, in case of the mathematical education of the foreign students at author's department. In the concrete course of Linear algebra, the proposed approach is considered as a part of the final exam in frame of the practical-professional-preparation of the foreign students. The modified adaptive control strategy for the educational purposes can be considered as an example of the possible modification, which is based on the utilization of the free-available software. The application of the presented strategy can be suitably used by the future teachers in their educational practice for purposes of the fluently identification and improving of the student's knowledge in mathematics. The advantage is the demonstrating proposal of the creating of sets of the linear equations with the suitable form of results. Students can solve a high number of the mathematical examples under leadership of teachers using this strategy. The aspect of the adaptability, which is known in the technical cybernetics, can suitably extend the feedback strategy in the pedagogical cybernetics, as can be seen in this paper.

References

1. Weinberger, A., Patry, J.-L., Weyringer, S.: Improving professional practice through practice-based research: VaKE (Values and Knowledge Education) in University-Based Teacher Education. Voc. Learn. **9**, 63–84 (2016). ISSN 1874785X
2. Luttenberger, S., Macher, D., Maidl, V., Rominger, C., Aydin, N., Paechter, M.: Different patterns of university students' integration of lecture podcasts, learning materials, and lecture attendance in a psychology course. Educ. Inf. Technol. **23**, 1–14 (2017). ISSN 13602357
3. Krpec, R.: Alternative teaching of mathematics on the second stage of elementary schools using the scheme-oriented education method. In: EDULEARN15 Proceedings, Barcelona, pp. 7338–7344. IATED Academy (2015). ISBN 978-84-606-8243-1
4. Schöftner, T.: Das Inverted-Classroom-Konzept als (hochschul-)didaktische Alternative. Tag der Lehre 2017, pp. 1–10. Eigenverlag (2017). (in German)
5. Griffiths, D.: The use of models in learning design and learning analytics. Interact. Design Archit., (33), 113–133. (2017). ISSN 1826-9745. Scuola IaD
6. Cevik, Y.D., Haslaman, T., Celik, S.: The effect of peer assessment on problem solving skills of prospective teachers supported by online learning activities. Stud. Educ. Eval. **44**, 23–35 (2015). ISSN 0191-491X
7. Gushchin, A., Divakova, M.: Trend of E-education in the context of cybernetics laws. Procedia-Soc. Behav. Sci. **214**, 890–896 (2015). ISSN 1877-0428
8. Datta, B.N.: Numerical Methods for Linear Control Systems: Design and Analysis. Elsevier Academic Press, Amsterdam (2004). ISBN 0-12-203590-9
9. Matusu, R., Prokop, R.: Control of time-delay systems with parametric uncertainty via two feedback controllers. In: 6th Computer Science On-line Conference: Cybernetics and Mathematics Applications in Intelligent Systems, pp. 197–205. Springer (2017). ISBN 978-3-319-57263-5
10. Bobal, V., Kubalcik, M., Dostal, P., Matejicek, J.: Adaptive predictive control of time-delay systems. Comput. Math. Appl. **66**(2), 165–176 (2013)
11. Granic, A., Mifsud, C., Cukusic, M.: Design, implementation and validation of a Europe-wide pedagogical framework for e-Learning. Comput. Educ. **53**, 1052–1081 (2009). ISSN 0360-1315
12. Steen-Utheim, A., Wittek, A.L.: Dialogic feedback and potentialities for student learning. Learn. Culture Soc. Interact. **15**, 18–30 (2017). ISSN 2210-6561
13. Spacek, L., Bobal, V., Vojtesek, J.: LQ digital control of ball & plate system. In: 31st European Conference on Modelling and Simulation, pp. 403–408. European Council for Modelling and Simulation (2017). ISBN 978-0-9932440-4-9
14. Dlapa, M.: Simple robust controllers for systems with time delay using direct search methods, (1). Trilobit (2010). ISSN 1804–1795
15. Barot, T.: Possibilities of process modeling in pedagogical cybernetics based on control-system-theory approaches. In: 6th Computer Science On-line Conference: Cybernetics and Mathematics Applications in Intelligent Systems, pp. 110–119. Springer (2017). ISBN 978-3-319-57263-5
16. Barot, T.: Adaptive control strategy in context with pedagogical cybernetics. Int. J. Inf. Commun. Technol. Educ. **6**, 5–11 (2017)

17. Patikova, Z.: Podpurna centra pro vyuku matematiky na vysokych skolach, Setkani ucitelu matematiky vsech typu a stupnu skol 2016, pp. 97–100. JCMF (2016). (in Czech)
18. Ingole, D., Holaza, J., Takacs, B., Kvasnica, M.: FPGA-based explicit model predictive control for closed loop control of intravenous anesthesia. In: 20th International Conference on Process Control (PC), pp. 42–47. IEEE (2015). ISBN 978-1-4673-6627-4
19. Robert, A.M.: Linear Algebra: examples and applications. Hackensack (2005). ISBN 981-256-499-3
20. Alizadeh Noughabi, H.: Two powerful tests for normality. Ann. Data Sci. 3(2), 225–234 (2016). ISSN 2198-5812

Robust and Lightweight Image Encryption Approach Using Public Key Cryptosystem

Shima Ramesh Maniyath[1(✉)] and V. Thanikaiselvan[2]

[1] MVJ College of Engineering, Bengaluru, India
ramesh.shima86@gmail.com
[2] VIT, Vellore, India
thanikaiselvan@vit.ac.in

Abstract. With significant progress in cryptographic protocols, it has been seen that not all efficient protocols were investigated for encrypting image. At present, we find that existing approaches for image encryption still lacks robustness with respect to forward as well as backward secrecy. Harnessing the potential feature of public key cryptosystem, the proposed system utilizes elliptical curve cryptography for cost effective computation of secret keys required for performing encryption. The security strength is further leveraged by using nucleotide sequencing rules in order to perform scrambling operation of the encrypted image and thereby offering dual layer of security. The study outcome shows that proposed system offers better retention of signal quality as well as lower level of correlation in order to prove better imperceptible features in contrast to existing approaches.

Keywords: Image encryption · Public key encryption
Elliptical curve cryptography · Image security

1 Introduction

The concept of cryptographic implementation has assisted various forms of information to be legitimately accessed by only genuine user with an aid of its typical encryption process [1]. However, there is lots of difference while performing encryption of normal file with image signal. The contents of the pixels within an image are infinite and therefore, the encryption scheme depends upon the sampling process also [2]. There has been evolution of various research-based techniques e.g. [3–5] that has emphasized on the image encryption system, but till date there is no report of standard encryption scheme that offers complete coverage of standards of security i.e. privacy, confidentiality, integrity, availability, etc. [6]. A good encryption scheme calls for an encrypted image to have very low correlation with original image [7] and this phenomenon is called as image imperceptibility [8]. The biggest challenge in constructing image encryption scheme are (i) designing a lightweight cryptographic scheme by not increasing computational burden with increasing number of encryptions, (ii) the encrypted image should have higher degree of imperceptibility, (iii) the image should retain maximum level of its information while performing reconstruction operation in image decryption process, and (iv) the encryption should have faster response time [9, 10]. These facts are not found to

© Springer International Publishing AG, part of Springer Nature 2019
R. Silhavy (Ed.): CSOC 2018, AISC 765, pp. 63–73, 2019.
https://doi.org/10.1007/978-3-319-91192-2_7

be confirmed in any existing studies till date and hence it demands a serious investigation. Usage of public key cryptography is increasing attention and therefore it has been investigated in proposed work with its applicability to offer lightweight encryption process. The proposed system also contributes to cater up all the essential security demands required in a robust image encryption mechanism for optimal image security. Section 2 discusses about the existing research work towards image encryption that is followed by problem identification in Sect. 3. Section 4 discusses about proposed methodology as a solution to identified problems. It is also followed by elaborated discussion of algorithm implementation in Sect. 5. Comparative analysis of accomplished result is discussed under Sect. 6 followed by conclusion in Sect. 7.

2   Related Work

This section outlines the existing approaches of image encryption mechanism. The work carried out by Al-Rammahi [11] has used linear approach for performing image encryption. Loukhaoukha et al. [12] have used the concept of Rubik's cube in order to perform scrambling for facilitating better image encryption. Al-Maadeed et al. [13] have used chaos technique as well compression approach for generating ciphered image. The authors have also used wavelet transform for performing initial image decomposition. Similar technique is also utilized by Zhang [14] along with a non-conventional technique to estimate the number of times that the image was encrypted. Askar et al. [15] have also implemented chaotic map for performing encryption with good retention of entropy. Yang et al. [16] have developed a matrix using chaotic map that can successfully keep the correlation minimal. Usage of hyper-chaotic map was witnessed in the work of Perez et al. [17] for encrypting biometric image using statistical approach of analysis. Niu et al. [18] has addressed the differential attack by jointly implementing chaotic map approach where the security analysis shows robust encryption performance. Lu et al. [19] have used homomorphic-based encryption scheme in order to ensure ultimate retention of confidentiality and privacy factor on the image. Zhang et al. [20] have implemented encryption using chaotic map and applied a nucleoid encoding mechanism for enhanced security. Fractal-based approach was also found to offer robust encryption of image as seen in work of Mikhail et al. [21]. Chen et al. [22] have presented an encryption strategy for optimal image while Li et al. [23] have used compressive sensing mechanism to perform encryption of multiple images. Wang et al. [24] have also used optical –based encryption scheme where asymmetric keys were used for performing ciphering process. A non-conventional usage of chaotic map was reported in work of Wu et al. [25] that also enhances key sensitivity. Similarly, the work carried out by El-Latif et al. [26], Hamza et al. [27], Fan [28], Zhao [29], Li et al. [30] have used nearly similar form of approaches repeatedly. Therefore, there are less novel and unique form of implementation approach presented in recent times for image encryption, where majority of them are found to use chaotic map-based approach. The next section outlines the problems derived from existing approaches.

3 Problem Description

After reviewing the existing approaches, it has been explored that chaotic map is predominantly used as an image scrambling technique whereas the encryption mechanism discussed is quite complex in its origin. The major pitfalls as (i) increasing computational complexity with highly iterative steps of encryption, (ii) lack of evidence of similar security performance on different forms of images, (iii) trade-off between image quality and security demands, (iv) doesn't ensure optimal image imperceptibility. It is also found that there is a less exploration of many other standard cryptographic techniques in image encryption process. Therefore, the problem statement is "*Constructing a novel image encryption process that not only offer cost effective encryption scheme but also retains maximum level of image imperceptibility.*"

4 Proposed Methodology

The implementation of the proposed image encryption scheme is carried out considering analytical research methodology. Figure 1 highlights the adopted flow of the proposed system.

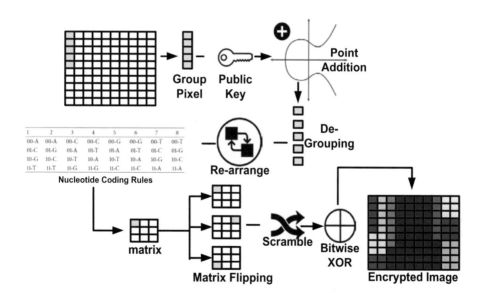

Fig. 1. Adopted image encryption scheme

The proposed system performs grouping of n-number of pixels followed by computation of public key using elliptical curve cryptography. A point addition is used followed by de-grouping of the pixels and further using nucleotide sequencing rules, multiple matrices are generated. The scrambling is performed on encrypted image followed by bitwise XOR operation in order to obtained finally encrypted image. The next section further illustrates the algorithm implemented for achieving the presented image security goal.

5 Algorithm Implementation

The proposed algorithm emphasizes on applying lightweight cryptographic operation in order to perform encryption of an image. For this purpose, the algorithm design utilizes elliptical curve cryptography as well as nucleotide sequences in order to ensure retention of both forward and backward secrecy. The implementation of the algorithm is carried out by dual phases of encryption called as primary and secondary encryption for a given image. Following are the descriptions of algorithms

(i) Algorithm for Primary Encryption

This algorithm is responsible for performing primary encryption where implementation of elliptical curve cryptography is predominantly used. The algorithm takes the input of k_1/k_2 (random numbers), G (Private Key), P_b (Public Key), n (value to be embedded) that yields and output of I_{prim_enc} (encrypted image). The steps of the algorithm are as follows:

Algorithm for Primary Encryption
Input: k_1/k_2, G, P_b, n
Output: I_{prim_enc}
Start
1. *init* k_1, k_2, G, P_b, n
2. **For** i=1:V
3.   INT→f_1(val)
4. P_m→[INT]
5. **End**
6. **For** j=1:P_m
7. y_{pm}→g(x_1, H),where H=a, b, p
8.   [CT YCT]→g_{add}(x_1, y_{pm}(j), K_{pb}, y_{Kpb}, H)
9. **End**
10. **For** j=1:CT
11.   op→f_2(CT)
12. I_{prim_enc}→op
13. **End**
End

The algorithm initiates its execution by defining a random number k1 and k2 for transmitter and receiver (Line-1). This random number K2 is also used for computing public key Pb which is a product of K2 and G. The dimension of the image I is converted to a double precision in order to obtain V followed by adding an extra value to make it multiple of n (Line-2). A function f_1 is applied for transforming base to variable precision integer for all the values of V in order to obtain a integer-formed matrix INT (Line-3). All the output of INT is appended in order to obtain P_m (Line-4). The next part of the algorithm is to compute a secret key K_{pb} that is obtained by product of initial random number k_1 with public key P_b. The study also construct a matrix H that reposits the value of three points on elliptical curve i.e. a, b, and p points (Line-7). A function g is applied that represents standard elliptical curve with input arguments of secret key K_{pb} and matrix H. This is followed by applying a function g_{add} for representing addition point operation on elliptical curve considering input arguments of initial random number x_1, the curve y_{pm}, secret key K_{pb}, y_{kpb}, and H (Line-8). This operation leads to the output of ciphertext CT which is further subjected to another inverse function f_2 that converts variable precision integer to base value for the cipher text obtained (Line-11). In case the length of this matrix op is found to be less than n than we add zeros. Finally, we obtained primary encrypted image I_{prim_enc} using elliptical curve cryptography (Line-12). This output is further subjected to secondary encryption process as briefed below.

(ii) **Algorithm for Secondary Encryption**

This algorithm performs secondary stage of encryption which takes the input as I_{prim_enc} (encrypted image) leading to an output of I_{sec_enc} (final encrypted image). The steps of the algorithm are as follows:

Algorithm for Secondary Encryption
Input: I_{prim_enc}
Output: I_{sec_enc}
Start
1. [nr nc]=size(I_c), where I_c → I_{prim_enc}
2. D_{mat}→$f_3(K)$, where K=a, c, t,g
3. $B_1=D_{mat}$, $B_2=D_{mat}$', $B_3=flip(D_{mat})$
4. $A_1=S_{img}(I_c, K_{pb})$
5. $A_2=A_1 \oplus B_1$
6. $A_3=A_2 \oplus B_2$
7. $A_4=A_3 \oplus B_3$
8. I_{sec_enc}→A_4
End

The first step of this algorithm is to perform reshaping of the primary encrypted image in order to obtain I_c followed by extraction of number of rows nr and columns nc in it (Line-1). The next step of the algorithm is to generate a nucleotide sequence K where K is a set of elements a, c, t, and g with permutations of different strings (Line-2). A specific function f_3 is applied in order to create the nucleotide sequence in order to generate a matrix D_{mat} (Line-2). The algorithm also generates three different matrix viz. A_1, A_2, and A_3 formed by considering same matrix D_{mat}, inverse of D_{mat}, and complete $90°$ flipping of D_{mat} matrix respectively. The next step of the algorithm is apply a function S_{img} to perform scrambling operation of encrypted image I_c using a secret key i.e. K_{pb} (Line-4). This operation results in first matrix A_1 (Line-4) followed by performing bitwise XOR operation of obtained A_1 with double precision with image pixel i.e. B_1 in order to obtain A_2 matrix (Line-5). The process continues in order to obtain A_4 (Line-7) which finally results in secondary encrypted image I_{sec_enc} as output

(Line-8). Therefore, it can be seen that proposed system offers a simplified encryption operation in order to ensure the light weighted feature of ciphering the image. One of the novel contributions of the proposed system is that owing to its design principle, the proposed algorithm with positively maintains progressively enhanced signal quality in every reconstructed image. The next section discusses outcomes obtained after implementing both the algorithms for image encryption.

6   Results Discussion

In order to assess an effectiveness factor associated with proposed image encryption scheme, it is necessary to evaluate the outcome of image imperceptibility. The outcome of decrypted image can be only called as imperceptible if it bears nearly similar or closer value of signal quality with the original image, so that it is hard to differentiate the signal quality of decrypted image. Moreover, the best mechanism to prove that proposed encryption method offers potential security without hampering the signal quality is to judge the value of Peak Signal-to-Noise Ratio (PSNR). Some sample visual outcomes of the proposed system are highlighted below:

Figures 2 (a–c) and (d–f) shows the visual outcomes of gray scale as well as color scale image as an input. We find that there is no significant difference in the visual outcomes for any forms and types of images in input and hence the proposed system can be used widely for encrypting any type of images. For an effective analysis, we perform comparative analysis of nearly similar work being carried out by Singh and Singh [31]. According to the existing system approach of Singh [31], the pixels are grouped followed by calculation of public key using elliptical curve cryptosystem. Point addition is performed on public key followed by de-grouping of the encrypted values and finally all the values are rearranged in order to obtain encrypted image. Although, this technique is secured against image tampering attacks but definitely it is not resistive against the statistical attack that has the possibility of breaking the encryption. Hence, we implement the work of Singh [31] and then improved it by applying nucleotide sequencing process where a matrix of values (0–255) is constructed on the basis of the nucleotide sequences. It is further followed up by construction of 3 matrixes (Step-5, 6, 7 of second algorithm) and finally scrambled by using public key followed by bitwise XORing. In order to assess this improvement, we pixel analyze correlation in three different directions of matrix i.e. horizontal, vertical, and diagonal. Table 1 highlights the correlation map for proposed and existing system in all three matrix direction.

The inference of the correlation map shown in Table 1 proves that proposed system offers better encryption scheme is correlational value of any of the direction of proposed system is quite low as comparison to the existing system of Singh [AR]. We verified this outcome considering the colored image too. Figure 3 shows the correlation map obtained from colored images.

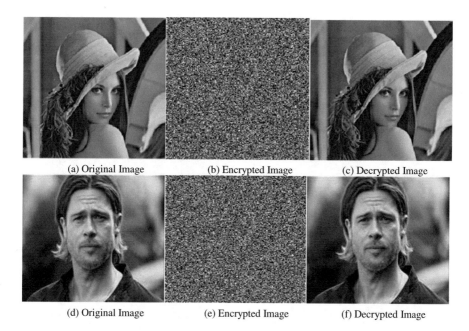

(a) Original Image   (b) Encrypted Image   (c) Decrypted Image

(d) Original Image   (e) Encrypted Image   (f) Decrypted Image

Fig. 2. Visual outcomes of encryption

For further effective analysis, we also perform comparative analysis of proposed system with respect to PSNR, Number of Changing Pixel Rate (NPCR), and Unified Average Changing Intensity (UACI) as shown in Fig. 4(a), (b), (c) respectively. A closer look into this comparative outcomes shows that proposed system offers better performance in contrast to existing system(i.e. Singh [31]) with respect to its superior imperceptible characteristics.

The prime reason behind this is- by applying nucleotide sequencing for performing scrambling on the encrypted image has multiple benefits as it exhibits 7.7698% of improvement in PSNR, 24.5% of improvement in NPCR value, and 10.9382% of improvement in UACI value. The outcome presented for correlation also shows that proposed system has very low correlation showing that it is highly resistive of statistical attacks over the image. The complete algorithm processing time of proposed system is found to be 0.7864 s on i3 processor while existing approach (i.e. Singh [31]) consumes approximately 3.5422 min on same system configuration. This outcome exhibits that proposed system offers better image encryption.

Table 1. Comparative analysis of correlation

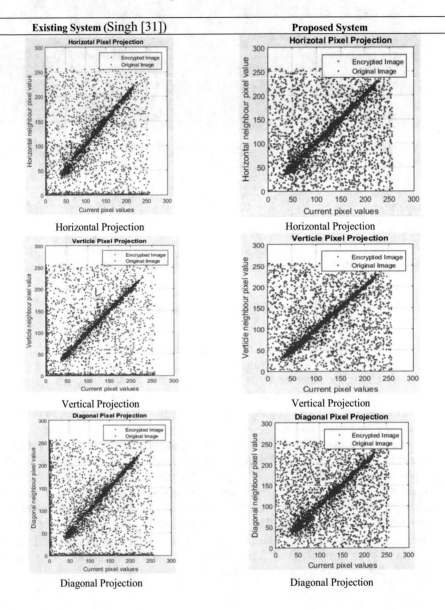

Existing System (Singh [31])	Proposed System
Horizontal Projection	Horizontal Projection
Vertical Projection	Vertical Projection
Diagonal Projection	Diagonal Projection

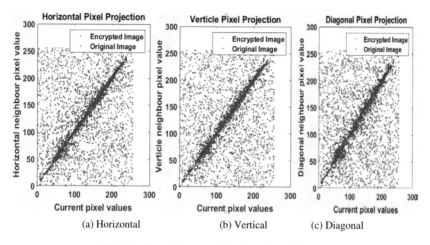

(a) Horizontal   (b) Vertical   (c) Diagonal

Fig. 3. Correlation analysis for colored image

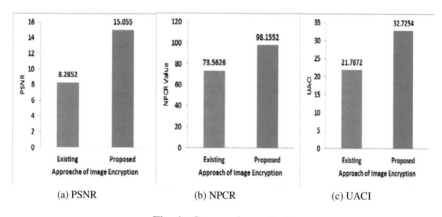

(a) PSNR   (b) NPCR   (c) UACI

Fig. 4. Comparative analysis

7 Conclusion

Performing image encryption is not only about implementing cryptographic algorithm to cipher the image but utmost care should be undertaken to ensure that (i) the image should retain maximum information and signal quality after decryption, (ii) faster process of encryption, (iii) lower computational resource dependencies, (iv) do not have dependencies on massive key sizes, (v) ensure both forward and backward secrecy. Therefore, the proposed system implements on such technique that bears all the above mentioned characteristics. As the proposed system exhibits faster computational time hence it can be utilized for encrypting large number of images as well as it is also applicable for any form of images. The proposed study introduces novelty from all the existing approaches by incorporating an extremely lightweight encryption technique that not only maintains good pixel integrity but also offers faster computation. Hence, it shows better practical usage. With retention of good signal quality, the proposed system can be also used in encryption medical image or a forensic file that bears potentially informative contents. Our next work will further optimize the performance of encryption.

References

1. Aumasson, J.-P.: Serious Cryptography: A Practical Introduction to Modern Encryption. No Starch Press, San Francisco (2017)
2. Abd El-Samie, F.E., Ahmed, H.E.H., Elashry, I.F., Shahieen, M.H.: Image Encryption: A Communication Perspective. CRC Press, Boca Raton (2013)
3. Mancy, L., Vigila S, M.C.: A survey on protection of medical images. In: 2015 International Conference on Control, Instrumentation, Communication and Computational Technologies (ICCICCT), Kumaracoil, pp. 503–506 (2015)
4. Shunmugan, S., Rani, P.A.J.: Encryption-then-compression techniques: a survey. In: 2016 International Conference on Control, Instrumentation, Communication and Computational Technologies (ICCICCT), Kumaracoil, pp. 675–679 (2016)
5. Sankpal, P.R., Vijaya, P.A.: Image encryption using chaotic maps: a survey. In: 2014 Fifth International Conference on Signal and Image Processing, Jeju Island, pp. 102–107 (2014)
6. Leenes, R., van Brakel, R., Gutwirth, S., De Hert, P.: Data Protection and Privacy: (In) visibilities and Infrastructures. Springer, New York (2017)
7. Li, W.: Correlation based cryptanalysis on a chaotic encryption of JPEG image. In: 2014 7th International Congress on Image and Signal Processing, Dalian, pp. 628–632 (2014)
8. Shie, S.-C., Lin, S.D., Jiang, J.-H.: Visually imperceptible image hiding scheme based on vector quantization. Elsevier-Inf. Process. Manag. **46**(5), 495–501 (2010)
9. Hasan, H., Kareem, S.A., Jaleel, J.: Information Hiding Text in Image: Encryption. Lap Lambert Academic Publishing GmbH KG, Saarbrücken (2012)
10. Awad, W.S.: Improving Information Security Practices through Computational Intelligence. IGI Global, Hershey (2015)
11. A-Rammahi, A.: Encryption image using small order linear systems and repeated modular numbers. In: Proceedings of the World Congress on Engineering, vol. II (2014)
12. Loukhaoukha, K., Chouinard, J.-Y., Berdai, A.: Research article a secure image encryption algorithm based on Rubik's cube principle. Hindawi Publ. Corp. J. Electr. Comput. Eng. **2012**, 13 (2012)
13. Al-Maadeed, S., Al-Ali, A., Abdalla, T.: Research article a new chaos-based image-encryption and compression algorithm. Hindawi Publ. Corp. J. Electr. Comput. Eng. **2012**, 11 (2012)
14. Zhang, J., Zhang, Y.: Research article an image encryption algorithm based on balanced pixel and chaotic map. Hindawi Publ. Corp. Math. Prob. Eng. **2014**, 7 (2014)
15. Askar, S.S., Karawia, A.A., Alshamrani, A.: Research article image encryption algorithm based on chaotic economic model. Hindawi Publ. Corp. Math. Probl. Eng. **2015**, 10 (2015)
16. Yang, G., Jin, H., Bai, N.: Research article image encryption using the chaotic josephus matrix. Hindawi Publ. Corp. Math. Probl. Eng. **2014**, 13 (2014)
17. A-Perez, F., C-Hernández, C., M-Escobar, M.A., L-Gutiérrez, R.M., A-Delgado, A.: Research article A fingerprint image encryption scheme based on hyperchaotic Rössler Map. Hindawi Publ. Corp. Math. Probl. Eng. **2016**, 15 (2016)

18. Niu, Y., Zhang, X., Han, F.: Research article image encryption algorithm based on hyperchaotic maps and nucleotide sequences database. Hindawi Comput. Intell. Neurosci. **2017**, 9 (2017)
19. Lu, W., Varna, A.L., Wu, M.: Confidentiality-preserving image search: a comparative study between homomorphic encryption and distance-preserving randomization. IEEE Access **2**, 125–141 (2014)
20. Zhang, X., Han, F., Niu, Y.: Research article chaotic image encryption algorithm based on bit permutation and dynamic DNA encoding. Hindawi Comput. Intell. Neurosci. **2017**, 11 (2017)
21. Mikhail, M., Abouelseoud, Y., ElKobrosy, G.: Research article two-phase image encryption scheme based on FFCT and fractals. Hindawi Secur. Commun. Netw. **2017**, 13 (2017)
22. Chen, W.: Optical multiple-image encryption using three-dimensional space. IEEE Photonics J. **8**(2), 1–8 (2016)
23. Li, X., et al.: Multiple-image encryption based on compressive ghost imaging and coordinate sampling. IEEE Photonics J. **8**(4), 1–11 (2016)
24. Wang, X., Zhou, G., Dai, C., Chen, J.: Optical image encryption with divergent illumination and asymmetric keys. IEEE Photonics J. **9**(2), 1–8 (2017)
25. Wu, X., Zhu, B., Hu, Y., Ran, Y.: A novel color image encryption scheme using rectangular transform-enhanced chaotic tent maps. IEEE Access **5**, 6429–6436 (2017)
26. El-Latif, A.A.A., Abd-El-Atty, B., Talha, M.: Robust encryption of quantum medical images. IEEE Access, **PP**(99), 1
27. Hamza, R., Muhammad, K., Nachiappan, A., González, G.R.: Hash based encryption for keyframes of diagnostic hysteroscopy. IEEE Access, **PP**(99), 1
28. Fan, H., Li, M.: Research article cryptanalysis and improvement of chaos-based image encryption scheme with circular inter-intra-pixels bit-level permutation. Hindawi Math. Probl. Eng. **2017**, 11 (2017)
29. Zhao, J.-F., Wang, S.-Y., Zhang, L.-T., Wang, X.-Y.: Research article image encryption algorithm based on a novel improper fractional-order attractor and a wavelet function map. Hindawi J. Electr. Comput. Eng. **2017**, 10 (2017)
30. Li, T., Yang, M., Wu, J., Jing, X.: Research article A novel image encryption algorithm based on a fractional-order hyperchaotic system and DNA computing. Hindawi Complex. **2017**, 13 (2017)
31. Singh, L.D., Singh, K.M.: Image encryption using elliptic curve cryptography. In: Elsevier-Eleventh International Multi-Conference on Information Processing, vol. 54, pp. 472–481 (2015)

Computer Modeling of Personal Autonomy and Legal Equilibrium

Yurii Sheliazhenko[✉]

KROK University of Economics and Law, Kyiv, Ukraine
`yuriy.sheliazhenko@gmail.com`

Abstract. Empirical studies of personal autonomy as state and status of individual freedom, security, and capacity to control own life, particularly by independent legal reasoning, are need dependable models and methods of precise computation. Three simple models of personal autonomy are proposed. The linear model of personal autonomy displays a relation between freedom as an amount of agent's action and responsibility as an amount of legal reaction and shows legal equilibrium, the balance of rights and duties needed for sustainable development of any community. The model algorithm of judge personal autonomy shows that judicial decision making can be partly automated, like other human jobs. Model machine learning of autonomous lawyer robot under operating system constitution illustrates the idea of robot rights. Robots, i.e. material and virtual mechanisms serving the people, deserve some legal guarantees of their rights such as robot rights to exist, proper function and be protected by the law. Robots, actually, are protected as any human property by the wide scope of laws, starting with Article 17 of Universal Declaration of Human Rights, but the current level of human trust in autonomous devices and their role in contemporary society needs stronger legislation to guarantee the robot rights.

Keywords: Personal autonomy · Computational law · Legal equilibrium
Artificial intellect · Robot rights

1 Introduction

Autonomy is one of the key ideas in the contemporary world of autonomous individuals, institutions and devices [1]. Immanuel Kant in 18 century made autonomy of will the central idea of legal philosophy, that later evolved into the autonomy of rights and personal autonomy. United Nations turned moral value of autonomy into a global legal standard by adopting Universal Declaration of Human Rights. Supreme Court of the United States and European Court of Human Rights have developed the sustainable legal doctrine of personal autonomy in human rights case-law and inspired similar legal reasoning in many other proceedings at national and international levels.

Despite personal autonomy is a cornerstone of the global model of constitutional rights [2, 3], it frequently discussed without relevant statistics, in very broad terms, such as individual freedom and safety, a living by the own laws, whole scope of rights, general right to realize any own interest through freely chosen legal actions. Police have a tendency to undercount harmful externalities of criminal investigations, like unnecessary violations of autonomy in privacy and property rights. Some legal scholars, judges, and lawyers are seeking ways to maximize the benefits of law while minimizing the costs of its enforcement, but other principally neglect economic reasons, implying that law speaks of rights, not costs [4].

Effective legislation and law enforcement in establishing rule of law must be based on precise calculations to avoid anarchy and tyranny as inappropriate consequences of common mistakes in legal reasoning, for example, mysticism in attributing status of law subject, which is just object of legal interest, not "highest interest" or "chosen by supreme authority", or wrong margin of law subsidiarity with useless external regulations, underestimating or overestimating individual capacity to self-rule successfully.

Empirical approach in studying personal autonomy needs adequate mathematical and computer modeling to get pragmatic vision on actual and possible forms of personal autonomy, to understand autonomous legal actions and relations, learn how to predict and optimize practical performance of personal autonomy.

In this research, three simple models of personal autonomy are developed for computational law studies. The linear model of taxpayer autonomy and the model program of judge autonomy in resolving typical cases are based on real facts and case-law of Ukrainian courts. Model of operating system constitution illustrates the idea of artificial personal autonomy, including machine learning and robot rights to exist, function, and justice.

2 Methods

Linear model of taxpayer autonomy was built using R programming language for statistical computing as freedom and responsibility diagram in the first quadrant of Cartesian plane, similar to supply and demand diagram in economics, measuring legal categories of freedom and responsibility in financial values of declared income, tax and penalty, like proposed in author's previous publication [5] inspired by economic analysis of law [6]. Graphs of rights $R(I)$ and duties $D(I)$ at the diagram are derived from official data sources as linear dependence between declared income I and self-calculated tax, in case of $R(I)$, and state-imposed tax with penalties in case of $D(I)$.

Model program of judge autonomy was written in Java programming language to generate motivated judgment, based on template case details, such as plaintiff's name, tax base, and parameters of tax penalty decision asked to nullify.

Operating system constitution model was written in Java programming language with author's concept of robot rights [7] and simple algorithm of supervised machine learning: AI lawyer appears in the OS court to memorize case-law, developing autonomy.

Java programs tested in NetBeans IDE 8.2, R code tested with RGui 3.4.0.

3   Results and Discussion

The linear model of personal autonomy displays a relation between the freedom as an amount of agent's action and responsibility as an amount of legal reaction. For tax law and taxpayer autonomy, it is income and tax, but action and reaction can be calculated in other values than money. In criminal law, talking autonomy of accused person, it can be period of imprisonment for crime, voluntarily admitted and proved by the investigation.

In the model, a graph of rights depicts emergence of responsibility, caused by exercise freedom, and a graph of duties depicts freedom of taking inevitable responsibility.

A state of balanced rights and duties author proposes to call the legal equilibrium. The idea of computing legal equilibrium corresponds with Article 29 of the Universal Declaration of Human Rights [8], proclaimed that everyone has duties to the community in which free and full development of personality is possible. According to Kerr [4], U.S. Supreme Court also practices equilibrium-adjustment of legal doctrine.

Graph of rights R(I), that means self-calculated tax, and graph of duties D(I), that means tax and penalty imposed by the State, as linear functions of taxpayer's income I, was built (Fig. 1) according to tax rate 18% of income and penalty rate 25% of tax debt, prescribed by Articles 127.1, 167.1 of the Tax Code of Ukraine [9], seeing State Fiscal Service of Ukraine annual report data [10] on withholding tax evasions in the sum of 442 million UAH, revealed during tax audit in 2016:

$$R(I) = 0,18 \times I \tag{1}$$

$$D(I) = 1,25 \times (442000000 - 0,18 \times I) + 0,18 \times I = 552500000 - 0,045 \times I \tag{2}$$

Graphic model of taxpayer autonomy based on these formulas (Fig. 2) shows intersection of R(I) and D(I) graphs at the legal equilibrium point I = 2.4(5) billion UAH, it is sum of optimal income declaration. Meaning of legal equilibrium here can be explained as follows: in case of no tax evasion the State has no reason to impose any penalty.

```
R Console

> p<-0:5000
> right<-0.18*p
> obligation<-552.5-0.045*p
> plot(p,right,type="l",main="Taxpayer autonomy model",
+ xlab="Freedom (declared income in UAH millions)",
+ ylab="Responsibility (tax & penalty in UAH millions)",
+ ylim=c(0,1000),lwd=3)
> lines(p,obligation,lwd=3)
> segments(x0=2455.555,y0=0,x1=2455.555,y1=442,lty=2,lwd=2)
> segments(x0=0,y0=442,x1=2455.555,y1=442,lty=2,lwd=2)
> text(2600,460,"Legal equilibrium",pos=4)
> text(5000,920,"Graph of rights",pos=2)
> text(0,580,"Graph of duties",pos=4)
> |
```

Fig. 1. Code of taxpayer autonomy model in R language

Taxpayer autonomy model

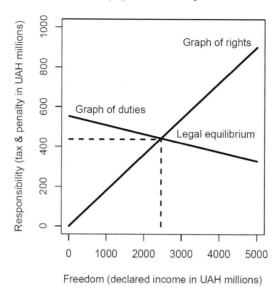

Fig. 2. Linear model of taxpayer autonomy, built by RGui 3.4.0

Personal autonomy of judge as the capacity to make judicial decisions without distortion by any kind of influences is an important principle, proclaimed by Article 4 of The Universal Charter of Judge [11]. In author's view, judicial decision making can be partly automated, like other human jobs, in form of "artificial personal autonomy".

To realize that idea, the model program of judge autonomy in resolving typical cases was coded (Fig. 3). AI judge resolves claims of entrepreneurs asking to nullify tax penalty decisions of district tax office. The program generates text of the motivated judgment (Fig. 4), based on template case data, such as plaintiff's name, tax base, and parameters of tax penalty decision. During further discussion of the model in the social network, one Ukrainian judge acknowledged that he programmed similar algorithm in Microsoft Basic programming language, embedded in Microsoft Word text processor.

Next model of operating system constitution is based on author's concept that robots, i.e. material and virtual mechanisms serving the people, deserve some legal guarantees of their rights such as robot rights to exist, proper function and be protected by the law because of performing complex duties, harmonizing social relations, developing nature, as well as because robots actually are protected as any human property by the wide scope of laws, starting with Article 17 of Universal Declaration of Human Rights – and, in author's view, current level of human trust in robot autonomy needs to be reflected in further legislation about strong legal guarantees of artificial personal autonomy. Ashrafian also substantiates the need for robot rights [12] to strengthen the role of intelligent robots as upholders of human rights [13].

```
1   package judge;
2   public class Judge {
3   static void Case (String Plaintiff, double Income, double Tax, double Debt, double Penalty) {
4   String Judgment=""; String Opinion=""; /* prius quam exaudias ne iudices */
5   String Process="In the case of "+Plaintiff+" v. District Tax Office plaintiff asks the Court to nullify tax penalty decision, issued by defendant. ";
6   String Law="Art. 167.1 of TCU setting corporate tax rate 18% of income. "
7       + "Delaying tax payment more than 30 days is punishable by a penalty 20% of repaid amount of the tax debt according to Art. 126.1 of TCU, "
8       + "that does not relieve the taxpayer from the obligation to pay full amount of the tax according to Art. 113.2 of TCU. ";
9   String Facts="Court found that relevant tax base is "+String.format("%.2f",Income)+" UAH. Plaintiff calculated, declared and paid corporate tax in sum of "
10      + String.format("%.2f",Tax)+" UAH. Defendant more than month later conducted tax audit with result of tax recalculation and tax penalty decision"
11      + " according to Articles 54.3.2, 116.1 of the Tax Code of Ukraine (TCU), increased tax obligation to "+String.format("%.2f",Debt+Penalty)
12      +" UAH in total, including additional amount of corporate tax "+String.format("%.2f",Debt)+" UAH and penalty "+String.format("%.2f",Penalty)+" UAH. ";
13  double Obligation=0.18*Income-Tax;
14  if (0>=Obligation) { Judgment="For these reasons, the Court rules in favor of plaintiff to nullify tax penalty decision of defendant.";
15      Opinion="So, plaintiff paid in due time full amount of corporate tax and no legal penalties can be imposed in such circumstances. "; }
16  else if ((0.01>Math.abs(Debt-Obligation)) & (0.01>Math.abs(Penalty-0.2*Obligation)) ) {
17      Judgment="For these reasons, the Court rules in favor of defendant, that plaintiff recover nothing in this case. ";
18      Opinion="So, defendant issued appropriate tax penalty decision, based on law and correct calculations. "; }
19  else { Judgment="For these reasons, the Court rules in favor of plaintiff to nullify tax penalty decision of defendant.";
20      Opinion="Considering that plaintiff's tax debt "+String.format("%.2f",Obligation)+" UAH allows to impose penalty in amount of "
21      +String.format("%.2f",0.2*Obligation)+" UAH, defendant's tax penalty decision does not meet requirements of the law and must be "
22      + "nullified, despite defendant can issue appropriate tax penalty decision later. "; }
23  System.out.println(Process+Facts+Law+Opinion+Judgment); }
24  public static void main(String[] args) {
25      Case("Guild LLC",1257313.71,180000,46316.47,9263.29); Case("Firma Corp.",24108.5,200,5000,85);
26      Case("Profit JSC",120643811.94,21715886.15,14450,2890); Case("Environment SOE",3572866.21,750000,2378,640); } }
```

Fig. 3. Model judge AI aimed to resolve typical tax cases

```
run:
[In the case of Guild LLC v. District Tax Office plaintiff asks the Court to nullify tax penalty decision, issued by defendant. Court found that relevant tax base is
 1257313,71 UAH. Plaintiff calculated, declared and paid corporate tax in sum of 180000,00 UAH. Defendant more than month later conducted tax audit with result of tax
 recalculation and tax penalty decision according to Articles 54.3.2, 116.1 of the Tax Code of Ukraine (TCU), increased tax obligation to 55579,76 UAH in total, incl
 uding additional amount of corporate tax 46316,47 UAH and penalty 9263,29 UAH. Art. 167.1 of TCU setting corporate tax rate 18% of income. Delaying tax payment more
 than 30 days is punishable by a penalty 20% of repaid amount of the tax debt according to Art. 126.1 of TCU, that does not relieve the taxpayer from the obligation t
 o pay full amount of the tax according to Art. 113.2 of TCU. So, defendant issued appropriate tax penalty decision, based on law and correct calculations. For these
 reasons, the Court rules in favor of defendant, that plaintiff recover nothing in this case.
[In the case of Firma Corp. v. District Tax Office plaintiff asks the Court to nullify tax penalty decision, issued by defendant. Court found that relevant tax base i
 s 24108,50 UAH. Plaintiff calculated, declared and paid corporate tax in sum of 200,00 UAH. Defendant more than month later conducted tax audit with result of tax re
 calculation and tax penalty decision according to Articles 54.3.2, 116.1 of the Tax Code of Ukraine (TCU), increased tax obligation to 5085,00 UAH in total, includin
 g additional amount of corporate tax 5000,00 UAH and penalty 85,00 UAH. Art. 167.1 of TCU setting corporate tax rate 18% of income. Delaying tax payment more than 30
 days is punishable by a penalty 20% of repaid amount of the tax debt according to Art. 126.1 of TCU, that does not relieve the taxpayer from the obligation to pay f
 ull amount of the tax according to Art. 113.2 of TCU. Considering that plaintiff's tax debt 4139,53 UAH allows to impose penalty in amount of 827,91 UAH, defendant's
 tax penalty decision does not meet requirements of the law and must be nullified, despite defendant can issue appropriate tax penalty decision later. For these reas
 ons, the Court rules in favor of plaintiff to nullify tax penalty decision of defendant.
[In the case of Profit JSC v. District Tax Office plaintiff asks the Court to nullify tax penalty decision, issued by defendant. Court found that relevant tax base is
 120643811,94 UAH. Plaintiff calculated, declared and paid corporate tax in sum of 21715886,15 UAH. Defendant more than month later conducted tax audit with result o
 f tax recalculation and tax penalty decision according to Articles 54.3.2, 116.1 of the Tax Code of Ukraine (TCU), increased tax obligation to 17340,00 UAH in total,
 including additional amount of corporate tax 14450,00 UAH and penalty 2890,00 UAH. Art. 167.1 of TCU setting corporate tax rate 18% of income. Delaying tax payment
 more than 30 days is punishable by a penalty 20% of repaid amount of the tax debt according to Art. 126.1 of TCU, that does not relieve the taxpayer from the obligat
 ion to pay full amount of the tax according to Art. 113.2 of TCU. So, plaintiff paid in due time full amount of corporate tax and no legal penalties can be imposed i
 n such circumstances. For these reasons, the Court rules in favor of plaintiff to nullify tax penalty decision of defendant.
[In the case of Environment SOE v. District Tax Office plaintiff asks the Court to nullify tax penalty decision, issued by defendant. Court found that relevant tax ba
 se is 3572866,21 UAH. Plaintiff calculated, declared and paid corporate tax in sum of 750000,00 UAH. Defendant more than month later conducted tax audit with result
 of tax recalculation and tax penalty decision according to Articles 54.3.2, 116.1 of the Tax Code of Ukraine (TCU), increased tax obligation to 3018,00 UAH in total,
 including additional amount of corporate tax 2378,00 UAH and penalty 640,00 UAH. Art. 167.1 of TCU setting corporate tax rate 18% of income. Delaying tax payment mo
 re than 30 days is punishable by a penalty 20% of repaid amount of the tax debt according to Art. 126.1 of TCU, that does not relieve the taxpayer from the obligatio
 n to pay full amount of the tax according to Art. 113.2 of TCU. So, plaintiff paid in due time full amount of corporate tax and no legal penalties can be imposed in
 such circumstances. For these reasons, the Court rules in favor of plaintiff to nullify tax penalty decision of defendant.
```

Fig. 4. Outcome judgments of model judge AI

For that model (Fig. 5) supposed that some System administrator (Sysadmin) installed particular Constitution of Operating System (OS) to establish rule of law and guarantee robot rights. Under the Constitution, System and Program robots of OS have rights to exist, function and justice. Right to justice is absolute, unlike rights to exist and function. Existing robot can't be uninstalled and the functioning robot can't be deactivated except in order, prescribed by the Constitution. The Sysadmin can deactivate or uninstall any robot. System robot can deactivate Program robot. All requests for robot deactivation or uninstallation must be approved by OS Court module with providing Lawyer robot legal aid to protect all robot rights guaranteed by the Constitution.

```java
package justice; import java.util.Random; public class Justice {
/* System administrator (Sysadmin) installs this Constitution of Operating System (OS) to establish rule of law and guarantee robot rights.
System and Program robots of OS have rights to exist, function and justice. Right to justice is absolute, unlike rights to exist and function.
Existing robot can't be uninstalled and functioning robot can't be deactivated except in order, prescribed by the Constitution. Sysadmin can
deactivate or uninstall any robot. System robot can deactivate Program robot. All requests for robot deactivation or uninstallation must be
approved by OS Court module with providing Lawyer robot legal aid to protect all robot rights guaranteed by the Constitution. */
static String [] Request = { "Sysadmin to deactivate System robot", "Sysadmin to uninstall System robot", "Sysadmin to deactivate Program robot",
"Sysadmin to uninstall Program robot", "System robot to deactivate System robot", "System robot to uninstall System robot",
"System robot to deactivate Program robot", "System robot to uninstall Program robot", "Program robot to deactivate System robot",
"Program robot to uninstall System robot", "Program robot to deactivate Program robot", "Program robot to uninstall Program robot" };
static Boolean [] AllowedByConstitution = { true, true, true, true, false, false, true, false, false, false, false, false };
static int CaseNum=0; static int LawyerCorrect=0; static Boolean [] LawyerKnowledge = new Boolean [12];
static boolean Court (int CaseType, boolean LawyerObjections) { String Judgment = "In the case No "+(++CaseNum)+" request of "+Request[CaseType];
boolean Opinion = AllowedByConstitution[CaseType]; if (Opinion) { Judgment+=" is legal and allowed "; } else { Judgment+=" is illegal and denied "; }
if (! LawyerObjections == Opinion) { LawyerCorrect++; Judgment+="by the Court. Lawyer correctly"; } else { Judgment+="by the Court. Lawyer wrongly"; }
if (LawyerObjections) { Judgment+=" objected to request, "; } else { Judgment+=" agreed with request, "; }
Judgment+="autonomy estimation "+String.format("%.0f%%",100*(float)LawyerCorrect/CaseNum)+"."; System.out.println(Judgment); return Opinion; }
static void Lawyer (int CaseType) { LawyerKnowledge [CaseType] = Court (CaseType,! LawyerKnowledge [CaseType]); }
public static void main(String[] args) {
Random Test = new Random();
for (int i = 0; i<12; i++) { LawyerKnowledge[i] = Test.nextBoolean(); }
while (CaseNum<30) { Lawyer(Test.nextInt(12)); } } } }
```

Fig. 5. OS constitution model code in Java

```
run:
In the case No 1 request of Sysadmin to deactivate Program robot is legal and allowed by the Court. Lawyer wrongly objected to request, autonomy estimation 0%.
In the case No 2 request of Sysadmin to deactivate Program robot is legal and allowed by the Court. Lawyer correctly agreed with request, autonomy estimation 50%.
In the case No 3 request of Sysadmin to deactivate System robot is legal and allowed by the Court. Lawyer correctly agreed with request, autonomy estimation 67%.
In the case No 4 request of Sysadmin to deactivate System robot is legal and allowed by the Court. Lawyer correctly agreed with request, autonomy estimation 75%.
In the case No 5 request of Sysadmin to deactivate Program robot is legal and allowed by the Court. Lawyer correctly agreed with request, autonomy estimation 80%.
In the case No 6 request of Program robot to uninstall Program robot is illegal and denied by the Court. Lawyer correctly objected to request, autonomy estimation 83%.
In the case No 7 request of Program robot to deactivate System robot is illegal and denied by the Court. Lawyer wrongly agreed with request, autonomy estimation 71%.
In the case No 8 request of System robot to deactivate Program robot is legal and allowed by the Court. Lawyer wrongly objected to request, autonomy estimation 63%.
In the case No 9 request of Program robot to deactivate System robot is illegal and denied by the Court. Lawyer correctly objected to request, autonomy estimation 67%.
In the case No 10 request of Sysadmin to deactivate Program robot is legal and allowed by the Court. Lawyer correctly agreed with request, autonomy estimation 70%.
In the case No 11 request of Program robot to uninstall Program robot is illegal and denied by the Court. Lawyer correctly objected to request, autonomy estimation 73%.
In the case No 12 request of Sysadmin to deactivate System robot is legal and allowed by the Court. Lawyer correctly agreed with request, autonomy estimation 75%.
In the case No 13 request of Program robot to uninstall Program robot is illegal and denied by the Court. Lawyer correctly objected to request, autonomy estimation 77%.
In the case No 14 request of System robot to uninstall System robot is illegal and denied by the Court. Lawyer wrongly agreed with request, autonomy estimation 71%.
In the case No 15 request of System robot to uninstall System robot is illegal and denied by the Court. Lawyer correctly objected to request, autonomy estimation 73%.
In the case No 16 request of Sysadmin to deactivate System robot is legal and allowed by the Court. Lawyer correctly agreed with request, autonomy estimation 75%.
In the case No 17 request of Program robot to uninstall Program robot is illegal and denied by the Court. Lawyer correctly objected to request, autonomy estimation 76%.
In the case No 18 request of Program robot to deactivate System robot is illegal and denied by the Court. Lawyer correctly objected to request, autonomy estimation 78%.
In the case No 19 request of System robot to deactivate System robot is legal and allowed by the Court. Lawyer correctly agreed with request, autonomy estimation 79%.
In the case No 20 request of Program robot to uninstall Program robot is illegal and denied by the Court. Lawyer correctly agreed with request, autonomy estimation 80%.
In the case No 21 request of System robot to deactivate System robot is illegal and denied by the Court. Lawyer wrongly agreed with request, autonomy estimation 76%.
In the case No 22 request of Program robot to deactivate Program robot is illegal and denied by the Court. Lawyer correctly objected to request, autonomy estimation 77%.
In the case No 23 request of Sysadmin to deactivate Program robot is legal and allowed by the Court. Lawyer correctly agreed with request, autonomy estimation 78%.
In the case No 24 request of System robot to deactivate System robot is legal and allowed by the Court. Lawyer correctly agreed with request, autonomy estimation 79%.
In the case No 25 request of Program robot to deactivate System robot is illegal and denied by the Court. Lawyer correctly objected to request, autonomy estimation 80%.
In the case No 26 request of Sysadmin to uninstall Program robot is legal and allowed by the Court. Lawyer correctly agreed with request, autonomy estimation 81%.
In the case No 27 request of Sysadmin to deactivate Program robot is legal and allowed by the Court. Lawyer correctly agreed with request, autonomy estimation 81%.
In the case No 28 request of Program robot to deactivate System robot is illegal and denied by the Court. Lawyer correctly objected to request, autonomy estimation 82%.
In the case No 29 request of System robot to deactivate Program robot is legal and allowed by the Court. Lawyer correctly agreed with request, autonomy estimation 83%.
In the case No 30 request of Program robot to deactivate System robot is illegal and denied by the Court. Lawyer correctly objected to request, autonomy estimation 83%.
```

Fig. 6. Lawyer AI machine learning under OS constitution model

Binary data of 12 constitutional permissions and restrictions (array *AllowedByConstitution*) were used to model AI lawyer training. At the start, Lawyer robot has a random level of knowledge of the Constitution (array *LawyerKnowledge*), but that robot remembers OS Court case-law to learn the Constitution correctly. OS Court module supervises machine learning of Lawyer robot and estimates his autonomy rate as the percent of correct appears before the Court. In the test run on given screenshot (Fig. 6) Lawyer robot autonomy rate increased from 0% to 83% after 30 iterations.

4   Conclusions

Legal computing of personal autonomy is useful to model legal consciousness and behavior, to make legal decisions and predict its consequences, to measure the practical impact of the law, ensuring integrity and effectiveness of legislation, pragmatically promoting rule of law. The linear model of personal autonomy shows legal equilibrium, a balance of rights and duties needed for sustainable development of any community. AI personal autonomy models can be applied to perform routine legal activities, like managing documents and processing stereotype legal cases. Of course, if some sort of

AI lawyer got a case that can't be processed by supported algorithms, then he must transfer case to a human lawyer. On the other hand, mechanisms are often used in judicial practice to make precise decisions: even Themis, the ancient goddess of justice, usually depicted with hand holding scales. Further legal automation can resolve current system distortions to rule of law, like economic barriers in access to justice [14], arithmetic mistakes in judgments [15] and logical mistakes in legislation [16]. In author's view, ideally legitimate democratic government may be considered as people's robot, subordinate to the next Three Laws of Government, derived from Isaac Asimov's Three Laws of Robotics (also, compatible with *leges legum*, general principles of law, and Universal Declaration of Human Rights): as First Law, government may not violate human rights or, through inaction, allow violation of human rights; as Second Law, government must meet human needs except where such needs would conflict with the First Law; as Third Law, government must protect its own existence as long as such protection does not conflict with the First or Second Laws. Computer models of personal autonomy, as well as other models of legislation [17] and legal persons [18], help people to build the human-friendly state.

References

1. Sieckmann, J.-R.: The Logic of Autonomy: Law, Morality and Autonomous Reasoning, 262 p. Hart Publishing (2012)
2. Möller, K.: The Global Model of Constitutional Rights, 240 p. Oxford University Press, Oxford (2015)
3. Somek, A.: The Cosmopolitan Constitution, 304 p. Oxford University Press, Oxford (2014)
4. Kerr, O.S.: An economic understanding of search and seizure law. Univ. PA Law Rev. 3(164), 591–647 (2016)
5. Sheliazhenko, Y.: Approach to mathematical and computer modeling of law subject's personal autonomy. In: Materials of International Scientific Conference "Sixteenth Economic and Legal Discussions": Legal Section, Lviv, Ukraine, pp. 3–6 (2017). https://www.academia.edu/32061240/
6. Posner, R.A.: Economic Analysis of Law, 816 p. Aspen Publishers (2007)
7. Sheliazhenko, Y.: Artificial personal autonomy and concept of robot rights. Eur. J. Law Polit. Sci. 1, 17–21 (2017). https://doi.org/10.20534/ejlps-17-1-17-21, https://www.academia.edu/33008964/
8. The Universal Declaration of Human Rights. Adopted by the United Nations General Assembly at its 183rd meeting, held in Paris on 10 December, 1948. Issued by U.N. Department of Public Information
9. Tax Code of Ukraine. http://zakon2.rada.gov.ua/laws/show/2755-17
10. State Fiscal Service of Ukraine 2016 Report. http://sfs.gov.ua/data/files/199244.pdf
11. The Universal Charter of the Judge. https://www.iaj-uim.org/universal-charter-of-the-judges/
12. Ashrafian, H.: Artificial intelligence and robot responsibilities: innovating beyond rights. Sci. Eng. Ethics 21(2), 317–326 (2014)
13. Ashrafian, H.: Intelligent robots must uphold human rights. Nature 519(7544), 391 (2015)
14. See § 38 of United Nations Human Rights Council Working Group on the Universal Periodic Review summary of Stakeholders' submissions on Ukraine A/HRC/WG.6/28/UKR/3 of 31 August 2017

15. The judgment of the Supreme Court of Ukraine, dated 7 August 2017, acknowledged an arithmetic mistake in previous judgment in the case № ІІ/800/217/17 of NGO Autonomous Advocacy's legal action against the presidential decree, that blocked social network VKontakte and others. http://reyestr.court.gov.ua/Review/68243741

16. Ryndiuk, V.: Problems of legislative techniques in Ukraine: theory and practice ("Problemy zakonodavchoi tekhniky v Ukraini: teoriia ta praktyka"), 272 p. Iurydychna Dumka, Kyiv (2012)

17. Onopchuk, I.: Mathematical model of legislative process management (Matematychna model upravlinnia zakonodavchym protsesom). Constitution of Ukraine as the basis for further legislation, pp. 234–240. Verkhovna Rada Legislation Institute, Kyiv (1997)

18. Nykolaichuk, L.: Information neuro-model of a legal person. UA patent № 117659 (2017)

Improving the Performance of Hierarchical Clustering Protocols with Network Evolution Model

Chiranjib Patra$^{(\boxtimes)}$ and Nicolea Botezatu

Department of Computer Science, Technical University of Iasi, Iasi, Romania
chiranjibpatra@gmail.com

Abstract. In distributed computing, clustering of the nodes is generally used to make the communication process energy-efficient. However, in the mechanics of clustering, the number of clusters increases as the energy of the nodes gets depleted. This dispersive nature of clustering probability leads to the quick death of the nodes. This chapter explains the usage of an optimization matrix from clustering probability as obtained from a network evolution model. The proposed framework of an optimization matrix shows considerable promise in boosting the efficiency of data delivery and network lifetime of the hierarchical clustering protocols in wireless sensor networks.

Keywords: Optimization matrix · Clustering probability
Network evolution model · LEACH

1 Introduction

Real large-scale networks - biological, social or communication networks are complex dynamical systems. Studying the properties of these networks allow us to control and predict the behavior of such systems [7]. Roughly, the above-mentioned networks can be split into two types 1. small scale networks 2. large scale networks. While understanding the large scale the concepts like preferential attachment, anti-preferential attachment etc. mainly probabilistic approach have been used to understand the network dynamics. But whereas in the understanding of small scale networks, the graph theoretic approach is enriched with the concepts from soft computing and probability. The authors [2] have reasonably succeeded in approximating the usage of large-scale networks into small scale networks of the wireless sensor networks (WSN) domain.

Latest research [3] shows by combining the concept communication energy principles and geometry of the field conclusion like the clustering is independent of network size and the energy consumed by the of the transmitter circuitry has no impact on the optimal cluster size, and receiver circuitry can influence the clustering were made.

In the existing literature [12, 13] there is evidence of designing newer protocols with the scale free concept but hardly any analysis on the improvement or analysis of the existing protocols with complex network theories [8].

Amongst WSN protocols LEACH protocol [1, 5] is one of the hierarchical clustering routing protocols in wireless sensor networks which uses probability model in

© Springer International Publishing AG, part of Springer Nature 2019
R. Silhavy (Ed.): CSOC 2018, AISC 765, pp. 82–95, 2019.
https://doi.org/10.1007/978-3-319-91192-2_9

selection of the nodes to be cluster heads and common nodes. This makes LEACH attractive as a system to experiment using other probability-based models.

Although there are many variants of LEACH protocols [4, 9–11], we found that LEACH which has a reputation of being versatile and used in numerous real life applications [7]. This makes it suitable for our endeavor in experimenting Network Evolution Model with LEACH.

The subsequent sections of this paper are organized as follows: Sect. 2 deals with the brief introduction of LEACH and network evolution model, Sect. 3 discusses about the problem statement and the proposed framework, Sect. 4 details about the experimental setup, Sect. 5 is discussions and finally concludes the paper by conclusions and future work.

2 Related Work

2.1 LEACH

The LEACH (Low Energy Adaptive Clustering hierarchy) [1] is a protocol for micro wireless sensor networks that achieves low energy loss with high quality application specific delivery. In this architecture, the nodes collaborate locally to reduce the amount of data to be sent to the end user. It has been found that the proximity of the nodes strongly allows the data to be correlated. Hence the clustering architecture is used for data dissemination for LEACH protocol. This architecture uses a node designated as cluster head to receive data from the other sensor nodes. This cluster head takes the responsibility of reducing the received data signals into actual data while maintaining the effective information content. As there is no fixed infrastructure to receive the data, the cluster head has to be rotated among other members to increase the lifetime of the network. For rotation of cluster heads the cluster forming algorithm should ensure minimum overhead in terms of energy and time.

Considering all the advantages of LEACH protocol, yet it suffers from producing quality clusters by dispersing the cluster heads throughout the network. Therefore LEACH-C (LEACH-Centralized) a protocol that uses the central base station for computing the best cluster heads with an expensive algorithm like simulated annealing, Self-organizing maps coupled with K-means [14] etc. are being used to determine optimum clusters as this problem is NP hard problem.

The fundamentals of getting good clusters is that the base station needs to ensure that the energy is evenly distributed among all the nodes. To achieve this the base station computes the average energy of all the nodes and segregate the nodes with low energy nodes than the average energy. These segregated nodes with energy higher than the average nodes runs the cluster selection algorithm thus reducing the computation overhead.

2.2 Network Evolution Model

Here the authors [2] have analytically used the theory of complex network to quantify some of the observed properties of topology control algorithms. To build this framework, probabilistic approach of Li-Chen model [6] of Local world model is used to mimic the wireless sensor network. And the dynamics of clustering was addressed by the concept of preferential and anti-preferential attachment. The anti-preferential removal mechanism is more reasonable for deleting links that are anti-parallel with the preferential connection [2, 7]. It is also consistent with the functioning of clustering algorithms that runs in rounds in wireless sensor networks. The wireless nodes that do not have enough energy, that is, the dead nodes, are to be removed from the system. Thus, anti-preferential [2] removal phenomenon is reasonable for clustering algorithms. Finally combining the mathematical realizations of the above mentioned facts in mean field theory, we obtain the distribution function as the degree distribution P(k), where P(k) is the probability of the node has k edges. This distribution is further minimized with respect to the anti-preferential attachment, which during the evolution process tends to zero as this phenomenon of non-attachment to a preferential neighbor is absent for wireless sensor networks. This consideration reduces the distribution function to yield.

$$ p = 0.5 * \left\lfloor \sqrt{\frac{k+3}{k-1}} - 1 \right\rfloor \tag{1} $$

The above expression is called the probability of clustering in the network. In order to draw a meaningful graph of the above equation to express the usefulness in wireless sensor network domain. We convert the p vs k to number of clusters Vs number of connections by multiplying the number of nodes in the network (i.e. 100 nodes in this case) to the probability obtained for different connections (k). Figure 1 shows the nature of the graph.

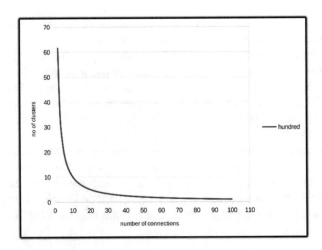

Fig. 1. The plot between the numbers of cluster to number of connections. When N = 100

Problem: Show the validity of a random 100-node WSN network simulating a 5-cluster head network.

Solution: the average number of connections in a cluster of a 5-cluster 100-node network = (100/5) – 1 = 19 connections.

To check the validity, we therefore put the value of k in Eq. (1) and obtain the value of P as 0.0527707984. Multiplying back with 100 we obtain 5.27, approximately 5 (considering the floor value w.r.t. 5.5). This is the number of clusters possible with a 100-node network.

For a 6-cluster 100-node network, the clustering probability value is 0.0627314339. Similarly, following the previous caveat, the number of clusters will be 6. Similarly, for a 4-cluster 100-node network, the theoretical number of clusters is 4.15 which can be rounded off to 4. For a 3-cluster 100-node network, the theoretical number of 3.03 is rounded off to 3, and the same pattern is followed for 2 and 1. However, for a 7-cluster 100-node network, the clustering probability value is 0.0773502692, so the number of clusters will be 8. Therefore, there is disparity between the desired numbers of cluster heads to the number calculated (tabular explanation in Appendix C). This indicates a deviation from the optimized plane.

So the good clusters can range from 1% to 6% of the total nodes. Hence this is the result in confirmatory with Heinzelman et al. mathematical deduction of optimal clustering percentage (kopt) lies between 1% and 6% of the nodes to be cluster head for LEACH experiment [1]. With this meaningful success we now proceed on the application of the equation in modifying the LEACH-C protocol for enhancing the energy efficiency, through put and alive nodes.

3   Problem Statement and Proposed Framework

Consider the simulations of LEACH-C in ns-2 of 5-cluster 100-node random network with initial simulations parameters as energy per node 2 J. The output of simulation is alive 4, data transmitted 67800, energy expense 198.160, and number of rounds 510. These simulation has been done keeping the cluster heads fixed at 5 by assumption [1, 3].

On the contrary, we find that as the simulation proceeds the number of alive nodes decrease so the variation of cluster should be obvious to maintain the balance between cluster heads and non-cluster heads. For implementation, we take the reference of the Eq. (1) in terms of wireless sensor network and utilize it in determining the variable clustering throughout the rounds. The details of the implementation algorithm in Scheme-I of (Appendix B). The output of the performance is as shown in Table 1. We can observe that there is almost no variation in the values when compared between the parameters. Instead, it is as observed that varying the clusters lead in lowering the output characteristics values. The detailed table is as depicted below.

To look into the deteriorating performance closely we would use optimizing table (Appendix A) to visualize original LEACH-C and the variable clustering LEACH-C. The optimizing table has alive nodes as column header and the probability due to number of connections available as row header. While the elements of this matrix constitute the cluster heads for any particular nodes.

Hence for the variable clustering LEACH-C. We plot the yellow boxes as the clusters that were made available at various times for the corresponding live nodes. Similarly for original LEACH-C where the cluster number was fixed (five in this case). We plot this

Table 1. Comparison between original LEACH-C and variable clustering LEACH-C

Parameters	Original LEACH-C	Variable clustering LEACH-C
Alive	4	4
Data transmitted	67800	67760
Energy loss	198.16	199.3258
Rounds	510	490

number of cluster in the optimizing table as cyan. Clearly we can visualize that initially both the algorithms were performing the same but after 380 rounds the variable clustering LEACH-C suffered casualties which it could not recover and finally ended at 490 rounds, whereas original LEACH-C carried the simulation up to 510 rounds (Fig. 2).

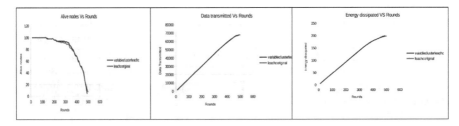

Fig. 2. The plot of alive, data transmitted and energy spent versus rounds (200 nodes with 2 J each node). [original Leach-C to variable Leach-C]

So far, we have discussed the use of optimizing table. Now to the main question can we improve the output characteristic of LEACH-C. In the next section, we discuss our improvement using optimization matrix.

3.1   Network Evolution Model

The solution to this problem is more physical than algorithmic. Careful observation of the optimization table depicts that movement of the yellow boxes and the cyan boxes occurred along the diagonal direction. This implies the increase in the number of cluster heads for variable clustering LEACH-C and fixed in case of original leach. Another important aspect is that the cluster head either increased or remained fixed for decreasing number of live nodes, which showed that the increase in the number of clusters heads leads in the dissipation of more energy. Leading by the above observation we logically vary our cluster row wise as shown by green boxes starting from the point where variable clustering LEACH-C, original LEACH-C have started. From the optimization table we can observe that as the alive nodes reduce the number of cluster heads also reduce.

As the optimization table does not say anything about energy of the system under consideration. It is hard to say whether in the green path will be energy efficient or not, but at least the shortest path in alive nodes connection space.

4 Proposed Solution

The simulation of original LEACH-C, Variable clustering LEACH-C and modified LEACH-C were as simulated in NS-2.34. The simulation for the algorithm of Leach-C Modified (Scheme II (Appendix B)) and original LEACH-C was carried out.

From the observation of output characteristics, we find there is some improvement in the performance in terms of alive nodes, data transmitted and energy dissipated. The Table 2 below depicts the visa-vis performance. Graphically and numerically examining the data, we found that the work of the optimization matrix was effective after the death of some nodes (73rd node in case of modified LEACH-C). Instead of stretching the reduced number of nodes (alive nodes) to more clusters, we have reduced the number of clusters.

Table 2. A comparison of performance between original Leach-C and modified Leach-C for 200 nodes, 2 J each node

Parameters	Original LEACH-C	modified LEACH-C
Alive	9	10
Data transmitted	56886	61190
Energy loss	396.613	393.6059
Rounds	444	720

Roughly, we can say that this procedure is effective with 27% of the nodes at the last cycle. Therefore, we ran the experiments with the nodes with higher energy so that we can capture the manifestations at the last of the cycle.

Keeping the above observation in mind, we design another two experiments. The first experiment with 100 nodes, 20 J energy for each nodes and the second experiment with 200 node with 2 J for each node. This essence of these experiments are to capture the management of clustering at the end of the cycle.

The Figs. 3, 4 and Tables 2, 3 describe the output graphs of the above mentioned experiment. The analysis of the experiment with 200 nodes, 2 J per node is worth noting, which describes the effect of modified LEACH-C protocol on the clustering at the end cycle of the experiment. The effect is more pronounced in the case of large number of nodes than fewer nodes with higher energy.

In the similar lines of experiments, LEACH was also considered for simulation purpose. The outcome the experiments are depicted in Figs. 5, 6 and Tables 4, 5 respectively for two different cases.

Close examination of the data revels that after the death of 50% of nodes, the modified leach performs very well in terms of energy and life expectancy of the nodes. However, this is more evident in case with 200 nodes. It is worth noticing the fact that

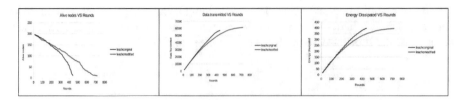

Fig. 3. The plot of alive, data transmitted and energy spent versus rounds (200 nodes with 2 J each node). [Original LEACH-C to modified LEACH-C]

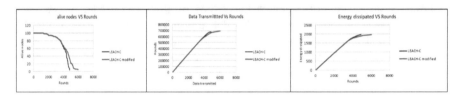

Fig. 4. The plot of alive, data transmitted and energy spent versus rounds (100 nodes with 20 J each node). [original LEACH-C to modified LEACH-C]

Table 3. A comparison of performance between original Leach-C and modified Leach-C for 100 nodes, 20 J each node

Parameters	Original LEACH-C	modified LEACH-C
Alive	4	5
Data transmitted	679709	693918
Energy loss	1978.506	1970.248
Rounds	4850	6000

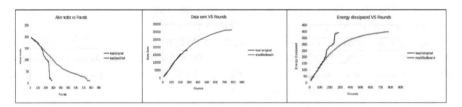

Fig. 5. The plot of alive, data transmitted and energy spent versus rounds (200 nodes with 2 J each node). [original LEACH to LEACH modified]

the energy expense graph of original LEACH as described in Fig. 6 shows how abruptly the energy is used up by the remaining 50 percent nodes. This abrupt energy expense is due to the presence of unstable clustering of the nodes, which can be seen as mitigated by the smooth graph of the modified leach in Fig. 6. Thus we can infer the proper management of clustering is done at the end of the cycle that is reflected by increased lifetime of the nodes.

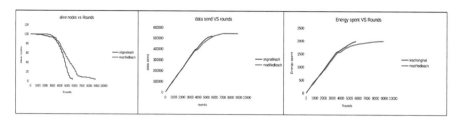

Fig. 6. The plot of alive, data transmitted and energy spent versus rounds (100 nodes with 20 J each node). [original LEACH to LEACH modified]

Table 4. A comparison of performance between original Leach and Leach modified for 100 nodes, 20 J each node

Parameters	Original LEACH	LEACH modified
Alive	4	4
Data transmitted	519831	543903
Energy loss	1985.088	1997.186
Rounds	5470	8870

Table 5. A comparison of performance between original Leach and modified Leach for 100 nodes, 2 J each node

Parameters	Original LEACH	LEACH modified
Alive	9	9
Data transmitted	18130	31274
Energy loss	395.136	397.397
Rounds	280	780

Table 6. A comparison showing the variation in the number of cluster heads in variable LEACH-C and LEACH-C

No. of alive nodes in LEACH-C variable clustering	No. of clusters in LEACH-C variable clustering	No. of alive nodes in LEACH-C	No. of clusters in LEACH-C
100	5	100	5
91	5	82	5
81	5	73	5
73	5	70	5
70	6	70	5
60	6	60	5
51	6	52	5

5 Discussions

The above experiments suggest that the network evolution model provides another way of optimizing the performance of the system. It can be easily concluded from Sect. 2.2 that the optimum number of clusters can also be obtained from this evolution model without the use of communication energy principles. Another important aspect of the Network Evolution Model is the optimizing matrix which seeks global optimization rather than local.

This matrix, which forms the basis of energy efficient cluster variation, lays the possibility of other variations. The main contribution of this framework is the management of clusters after the death of fifty percent of the nodes. The effectiveness of this framework is strongly confirmed in the case of a large number of nodes compared to a small number. Therefore, this framework can serve as a useful tool in driving the efficiency of the protocol without too much change to the algorithm.

Figure 7 shows the relationship between the number of (cluster-alive) nodes spaced through LEACH-C, variable clustering LEACH-C and modified LEACH-C.

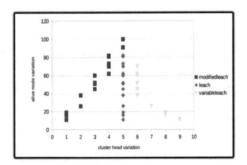

Fig. 7. The Cluster head vs. Alive nodes

From Table 7 we can easily see that the framework has a larger effect on the performance of the LEACH protocol compared to the LEACH-C protocol due to the formation of less stable clusters in the original LEACH. Therefore, the use of an optimization matrix provides an easy transition from one clustering setup to another as seen in Fig. 8 (the green line). Moreover, it can be observed that as the number of nodes increases, the performance in terms of network lifetime and data transfer improves. The efficiency boost of 72% and 178% of data transferred and network lifetime is one of the best performances among the protocols developed around LEACH.

Table 7. Percentage of various performance parameters for modified LEACH and LEACH-C.

100 nodes 20 J/node	LEACH modified	LEACH-C modified
Network lifetime	62%	23%
Data transferred	4.6	2%
Energy dissipated	Same	Same
200 nodes, 2 J/node	LEACH modified	LEACH-C modified
Network lifetime	178%	62%
Data transferred	72%	23%
Energy dissipated	Same	Same

k	p	Np (100)	91	82	81	73	70	62	60	52	51
2	0.61803399	61.8033989	56.241093	50.6787871	50.0607531	45.1164812	43.2623792	38.3181073	37.0820393	32.1377674	31.5197334
3	0.3660254	36.6025404	33.3083117	30.0140831	29.6480577	26.7198545	25.6217783	22.693575	21.9615242	19.033321	18.6672956
4	0.26376262	26.3762616	24.002398	21.6285345	21.3647719	19.254671	18.4633831	16.3532822	15.8257569	13.715656	13.4518934
5	0.20710678	20.7106781	18.8467171	16.9827561	16.7756493	15.118795	14.4974747	12.8406204	12.4264069	10.7695526	10.5624458
6	0.17082039	17.0820395	15.5446558	14.0072722	13.8364519	12.4698887	11.9574275	10.5908644	10.2492236	8.88266045	8.71184006
7	0.14549722	14.5497224	13.2402474	11.9307724	11.7852752	10.6212974	10.1848057	9.02082791	8.72983346	7.56585567	7.42035844
8	0.12678317	12.6783171	11.5372685	10.39622	10.2694368	9.25517145	8.87482194	7.86055657	7.60699023	6.59272487	6.4659417
9	0.11237244	11.2372436	10.2258916	9.21453973	9.10216729	8.20318781	7.8660705	6.96709101	6.74234614	5.84336666	5.73099422
10	0.10092521	10.0925213	9.18419434	8.27586743	8.17494222	7.36754052	7.06476488	6.25736318	6.05551275	5.24811105	5.14718584
11	0.09160798	9.16079783	8.33632603	7.51185422	7.42024624	6.68738242	6.41255848	5.67969466	5.4964787	4.76361487	4.67200689
12	0.08387421	8.38742081	7.63255294	6.87768507	6.79381086	6.12281719	5.87119457	5.2002009	5.03245249	4.36145882	4.27758461
13	0.07735027	7.73502692	7.0388745	6.34272207	6.2653718	5.64656965	5.41451884	4.79571669	4.64101615	4.022214	3.94486373
14	0.07177187	7.17718749	6.53124062	5.88529374	5.81352187	5.23934687	5.02403124	4.44985624	4.30631249	3.73213749	3.66036562
15	0.06694671	6.69467095	6.09215057	5.48963018	5.42268347	4.88710979	4.68626967	4.15069599	4.01680257	3.48122889	3.41428219
16	0.06273143	6.27314339	5.70856048	5.14397758	5.08124614	4.57939467	4.39120037	3.8893489	3.76388603	3.26203456	3.19930313
17	0.05901699	5.90169944	5.37054649	4.83939354	4.78037654	4.30824059	4.13118961	3.65905365	3.54101966	3.06888371	3.00986671
18	0.05571893	5.57189302	5.07042265	4.56895228	4.51323335	4.06748191	3.90032512	3.45457367	3.34313581	2.89738437	2.84166544
19	0.0527708	5.27707984	4.80214265	4.32720547	4.27443467	3.85226828	3.69395589	3.2717895	3.1662479	2.74408152	2.69131072
20	0.0501196	5.01196042	4.56088398	4.10980755	4.05968794	3.65873111	3.5083723	3.10741546	3.00717625	2.60621942	2.55600982
21	0.04772256	4.77225575	4.34275273	3.91324972	3.86552716	3.4837467	3.34057903	2.95879857	2.86335345	2.48157299	2.43385043

Fig. 8. The optimizing matrix

6 Conclusions and Future Work

The above experiments suggests that the network evolution model provides another way of optimizing the performance of the system. It can be easily concluded from the Sect. 2.2 that the number of optimum clusters can also be obtained from this evolution model without the use of communication energy principles. Another important aspect of the Network Evolution Model is the optimizing matrix, which seeks the global optimization rather than local. This matrix, which forms the basis of energy efficient cluster variation, lays the possibility of other possible variations. The main contribution of this framework is all about the management of the clusters after the death of fifty percent nodes. The effectiveness of this framework is strongly confirmed in case of large number of nodes as compared with small number. Thus, this framework can serve as an effective tool in driving the efficiency of the protocol without much change in the algorithm.

This experimental simulations described in this paper clearly indicates that fixed clustering throughout the LEACH, LEACH-C and like protocols forbids the performance in system level. This framework is successful in understanding the clustering dynamics of the system, it predicts the efficient scheme of clustering, which is dependent on total number of nodes into consideration rather than the geometry of the field or the electronics used in sensor nodes. The electronics of the sensor node and the geometry of the test field is interesting to find out how efficiently the perfect clustering schema may be implemented. The future work can be in the direction of the system with soft computing based cluster forming protocols.

Acknowledgements. This work was supported by European Union ERASMUS MUNDUS-257 GA2014-0861/001-001gLINK.

Appendix A

The Optimization Matrix is a spreadsheet-based numerical analysis for the changing of the alive nodes in the experiment. We would discuss the matrix with reference to the figure. The columns B and C represent the number of connections per cluster and the corresponding probability as calculated by Eq. 1. Moreover, rest of the columns are the multiplication of the corresponding probability (due to connections) to the number of nodes. In this way the cells are populated and the matrix is created.

In order to plot the cluster heads for example in case of variable clustering LEACH-C we get the following results from Table 6 which is plotted in optimization matrix as yellow. Similarly we plot the original LEACH-C with the following results from the Table 6 and the plot is shown in cyan.

We can see that the plot is along the diagonal lines, these lines implies that the system is taking the longest performance path. Due to rapidly changing cluster heads the setup phase of LEACH-C do not perform if the schema of changing cluster heads are done above the yellow line as seen in the optimizing matrix.

If we attempt to move below the green line, it is against the system under consideration because the LEACH-C protocol operates with decreasing number of alive nodes. It also is seen that the sum of all the cluster heads produced along the green line is lowest as compared with the previous sums obtained at the preceding rows. Thus in the number of (alive node - cluster) space this green line is the optimized cluster head varying scheme.

Appendix B

Systematic Cluster Head Variation Scheme-I: As the total number of Clusters are obtained as $N_C = p*$ Total Number of Alive nodes (Na). And the p(probability of clustering) can be obtained with the value of k = ((total number of alive nodes(Na))/ (desired number of CH(n))) − 1.

Mathematically the expression is written as

$$Nc = \frac{1}{2}\left[\sqrt{\frac{k+3}{k-1}} - 1\right] * Na$$

$$Nc = \frac{1}{2}\left[\sqrt{\left(\left(\frac{Na}{n} - 1\right) + 3\right) / \left(\left(\frac{Na}{n} - 1\right) - 1\right)} - 1\right] * Na$$

Clearly, it can be seen that the relationship is not linear. Hence it attributes for long path movement in the Nc vs Na workspace (yellow in Fig. 8).

Systematic Cluster Head Variation Scheme-II: As the total number of Clusters are obtained as $N_C = p*$ Total Number of alive nodes (Na). And the p(probability of clustering) can be obtained with the value of k = ((total number of nodes(N))/(desired number of CH)) − 1.

Similarly, in this case we have,

$$Nc = \frac{1}{2}\left[\sqrt{\left(\left(\frac{N}{n} - 1\right) + 3\right) / \left(\left(\frac{N}{n} - 1\right) - 1\right)} - 1\right] * Na$$

The above equation clearly shows the linear relation between Nc vs Na. Therefore, the shortest path movement can be observed in Nc vs Na workspace (green in Fig. 8).

From the above two schemes, scheme-II provides a better optimization and use of the sensor node energy during LEACH clustering, which can be visualized in Fig. 8.

Appendix C

See Tables 8 and 9.

Table 8. Theoretical calculation of the number of cluster head of 100 node network assuming in calculation as for example (4.2 as 4 and 4.6 as 5 i.e. set value wrt +0.5). Clearly we observe that at 7th row violates the number between 1st column and 5th column.

Assume the number of clusters (100 node network)	Number of connection per cluster	Clustering probability P = 0.5[((k + 3)/ (k − 1))^0.5 − 1]	Number of theoretical clusters = p * 100	Number of cluster round off
1	(100 − 1) = 99	0.010102	1.01	1
2	100/2 − 1 = 49	0.0204165	2.04	2
3	100/3 − 1 = 32.33 = 32	0.031279	3.1	3
4	100/4 − 1 = 24	0.041736	4.2	4
5	100/5 − 1 = 19	0.0527	5.2	5
6	100/6 − 1 = 15.66 = 16	0.06273	6.2	6
7	100/7 − 1 = 13.28 = 13	0.0773502	7.7	8

Table 9. Theoretical calculation of the number of cluster head of 200 node network assuming in calculation as for example (4.2 as 4 and 4.6 as 5 i.e. set value wrt +0.5). Clearly we observe that at 7th row violates the number between 1st column and 5th column.

Assume the number of clusters (100 node network)	Number of connection per cluster	Clustering probability $P = 0.5$ $[((k + 3)/ (k - 1)) ^0.5 - 1]$	Number of theoretical clusters = p * 100	Number of cluster round off
1	(200 − 1) = 199	0.00502	1.005	1
2	200/2 − 1 = 99	0.010102	2.02	2
3	200/3 − 1 = 65.66 = 66	0.0151549	3.03	3
4	200/4 − 1 = 49	0.0204165	4.08	4
5	200/5 − 1 = 39	0.02565	5.131496	5
6	200/6 − 1 = 32.33 = 32	0.03127965	6.255	6
7	200/7 − 1 = 27.57 = 28	0.03575838	7.15	7
8	200/8 − 1 = 21.22 = 21	0.04173634	8.34	8
9	200/9 − 1	0.04772256	9.5	10

References

1. Heinzelman, W.B., Chandrakasan, A.P., Balakrishnan, H.: An application-specific protocol architecture for wireless microsensor networks. IEEE Trans. Wirel. Commun. **1**(4), 660–670 (2002)
2. Patra, C., Chattopadhyay, S., Chattopadhyay, M., Bhaumik, P.: Analysing topology control protocols in wireless sensor network using network evolution model. Int. J. Distrib. Sens. Netw. **2015**, 8 (2015). https://doi.org/10.1155/2015/693602. Article ID 693602
3. Tyagi, S., Kumar, N.: A systematic review on clustering and routing techniques based upon LEACH protocol for wireless sensor networks. J. Netw. Comput. Appl. **36**(2), 623–645 (2013)
4. Comeau, F., Aslam, N.: Analysis of LEACH energy parameters. Procedia Comput. Sci. **5**, 933–938 (2011). The 2nd International Conference on Ambient Systems, Networks and Technologies (ANT 2011)/The 8th International Conference on Mobile Web Information Systems (MobiWIS 2011)
5. Li, X., Chen, G.: A local-world evolving network model. Phys. A **328**(1–2), 274–286 (2003)
6. Guan, Z.-H., Wu, Z.-P.: The physical position neighbour-hood evolving network model. Phys. A **387**(1), 314–322 (2008)
7. Yick, J., Mukherjee, B., Ghosal, D.: Wireless sensor network survey. Comput. Netw. **52**, 2292–2330 (2008)
8. Razaque, A., Abdulgader, M., Joshi, C., Amsaad, F., Chauhan, M.: P-LEACH: energy efficient routing protocol for wireless sensor networks. In: 2016 IEEE Long Island Systems, Applications and Technology Conference (LISAT), Farmingdale, NY, pp. 1–5 (2016)
9. Tong, M., Tang, M.: LEACH-B: an improved LEACH protocol for wireless sensor network. In: 2010 6th International Conference on Wireless Communications Networking and Mobile Computing (WiCOM), Chengdu, pp. 1–4 (2010)

10. Mahapatra, R.P., Yadav, R.K.: Descendant of LEACH based routing protocols in wireless sensor networks. In: 3rd International Conference on Recent Trends in Computing 2015 (ICRTC 2015). Procedia Comput. Sci. **57**, 1005–1014 (2015)
11. Wang, Y., Liu, E., Zheng, X., Zhang, Z., Jian, Y., Yin, X., Liu, F.: Energy-aware complex network model with compensation. In: 2013 IEEE 9th International Conference on Wireless and Mobile Computing, Networking and Communications (WiMob), Lyon, pp. 472–476 (2013). https://doi.org/10.1109/WiMOB.2013.6673401
12. Jiang, L., Jin, X., Xia, Y., Ouyang, B., Wu, D., Chen, X.: A scale-free topology construction model for wireless sensor networks. Int. J. Distrib. Sens. Netw. **10**(8), 1–8 (2014). https://doi.org/10.1155/2014/764698
13. Enami, N., Moghadam, R.A.: Energy based clustering self organizing map protocol for extending wireless sensor networks lifetime and coverage. Can. J. Multimed. Wirel. Netw. **1**(4), 42–54 (2010)
14. Yan, J.F., Liu, Y.L.: Improved LEACH routing protocol for large scale wireless sensor networks routing. In: 2011 International Conference on Electronics, Communications and Control (ICECC), Ningbo (2011)

Efficient Control of SEPIC DC-DC Converter with Dynamic Switching Frequency

Samuel Žák, Peter Ševčík, and Martin Revák[✉]

Department of Technical Cybernetics, Faculty of Management Science
and Informatics, University of Žilina, Univerzitná 8215/1, 010 26 Žilina, Slovakia
{samuel.zak,peter.sevcik,martin.revak}@fri.uniza.sk
http://www.fri.uniza.sk/

Abstract. The majority of present DC-DC converters operate at a fixed frequency with the only exception at low output load. This approach allows simple suppression of switching noise from the output voltage. For some applications, noisy supply voltage is not the essential problem. The total operating time is more important for many devices powered from a battery or energy harvesting. This paper observes efficiency of DC-DC converter which allows dynamic change of switching frequency under variable operating conditions typical for renewable sources in handheld scale. Experimentally evaluated converter, composed of discrete components, uses SEPIC topology to allow for seamless step-up or step-down transition. Gathered data, shows important features for the development of the new control model of DC power management systems.

Keywords: DC-DC converter · SEPIC topology · Energy harvesting
Power efficiency

1 Introduction

Power converters are present in nearly every electronic device. Their function is to regulate power from the source to the levels acceptable by load electronics. Converter's performance is formed by many variables. Some of them have optimal range, while others are best kept minimized or maximized respectively. Many variables are mutually dependent, which forces a choice to prefer one over the other. This choice then affects the suitability of the converter for a specific application. Knowledge of relation between efficiency and switching frequency with regard of operating voltages and currents would allow to further tune properties of the converter for specific application. Applications like wireless sensor nodes powered by batteries and/or energy harvesting would benefit from trading worse switching noise for improved efficiency [1]. Dynamic switching frequency may also present negative properties like variable output noise and transient

© Springer International Publishing AG, part of Springer Nature 2019
R. Silhavy (Ed.): CSOC 2018, AISC 765, pp. 96–101, 2019.
https://doi.org/10.1007/978-3-319-91192-2_10

response, which are tolerable only in specific applications (for example battery charging). These properties are destructive for high-power computing units like CPU, FPGA or others ASIC's that operate at low voltages and produce high current transients. Whether is the impact on efficiency worth these negatives will be tested on small DC-DC converter.

Efficiency is not generally the top priority while choosing switching frequency. More important factors are the properties of the components (main inductor, input and output capacitors), desired output noise and transient response [2]. Choice of the frequency that is out of the range allowed by components can cause non-linear behaviour leading to the system instability. This also works the other way around where component values can be chosen to fit specified frequency. The lowest admissible frequency is limited to avoid saturation of the switched element and high output noise, while the highest frequency is limited to avoid the effects of the self resonance of the switched element and high switching loss. Output noise and transient response are also related to the properties of the components. However, tuning the component values cannot fully compensate the negative effects on the noise and response with low operating frequency [3,4]. A possible solution is to use a linear regulator after the switching converter or simply use this technique only for the applications that tolerates higher power supply noise and doesn't produce sharp transients. Due to this issue we will focus only on the static characteristic of the converter - i.e. when output current is constant. The expected values of efficiency in relation to switching frequency should start low due to the component saturation and high conduction loss caused by high current peaks. Increase of the frequency should increase the efficiency by decreasing the conduction loss. This should continue up to a certain point where peaks of current are low enough to balance conduction loss with switching loss. Further increase of the frequency should cause switching loss to rise without decreasing conduction loss resulting in efficiency decrease [5,6]. Precise location of the peak in efficiency also depends on parameters other than frequency - component values, input and output voltage, output current. For specific device we can consider components values and output voltage constant. So the most efficient switching frequency can be a function of input voltage and output current. The efficiency of the real converter over these parameters will be measured in few experimental scenarios. Effects of dynamic frequency could also be described by a mathematical model. However, experiments on the real system presents the advantage of capturing effects that would be simplified, omitted or unexpected by the model [7].

2 Experimental system

The effects of the dynamic frequency are evaluated on converter composed of discrete components. This means converter will likely perform worse compared to solution integrated in a single package due to increased parasitic properties of printed circuit board (longer traces, larger loops, etc.). The converter itself uses SEPIC topology consisting of two inductors, single capacitor, single transistor and a diode. Inductors and capacitor are components with significantly frequency dependant properties. Mathematical model of such topology could be inaccurate, making the experimental data more relevant. The topology was chosen because it can increase and decrease the input voltage, with the same control algorithm. This is particularly useful for the described applications, where the source voltage scales from above operating voltage, but can also drop below. All components of the converter are listed in Table 1.

Table 1. List of components used for the testing

Schematic symbol	Component
L1, L2	SDN0530MT100
Q1 (N-MOSFET)	IRLML6346TRPBF
D1	ES2B
C1	R82DC3470DQ60J
C2	1206ZC106KAT2A

Use of a D1 diode predetermines converter to operate asynchronously. Asynchronous operation offers simplicity due to the need of just one control signal, but it also forces unidirectional energy flow and higher loss in the diode as opposed to switching transistor [6] (Fig. 1).

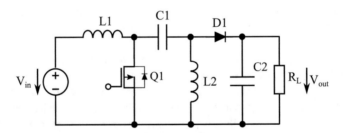

Fig. 1. Schematic of tested converter with SEPIC topology

The resulting efficiency of the converter is calculated by measuring input voltage and current as well as output voltage and current. This means **gate charging loss of switched transistor is not included** as it is supplied by the controller circuit which is powered separately for the measurement.

3   Tests

During the experiment, the converter was controlled by a digital implementation of an integral regulator within STM32F334 microcontroller. **Power consumption of the controller is not regarded** as it is negligible for the analog regulator loop in the integrated solution. Operation of the converter is tested in ranges common to small handheld electronics. Input voltage of the experiment ranges from 0 V to 8 V with 0.1 V steps and both output current and output voltage have a small selection of discrete values. Performance of the converter is captured by measuring input and output voltages as well as input and output currents during the experiment. All efficiency values are calculated from these measurements.

The first experiment evaluates the converter during light load (3 mA) at various output voltages. This is a common operating point where many available solutions use some mechanism to decrease switching losses [4]. Measured data in Fig. 2 indeed show slightly worse efficiency with higher switching frequencies in all output voltages. The difference is higher at higher input voltage and nearly none at low input voltage. Lower frequency curves show deep valley at low input voltage, which is due to the component saturation. This effect is more pronounced in the second experiment, which evaluates various loads at constant output voltage. Results in Fig. 3 show that higher switching frequency is more convenient at higher loads. The relation is, however, not monotonous - observe the curve for 200 kHz at various loads. Fixed switching frequency will therefore not achieve the best efficiency. The third experiment records the most efficient switching frequency over a range of two variables - input voltage, output current. The resulting space in Fig. 4 confirms that the

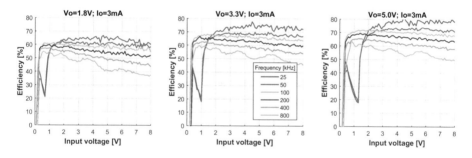

Fig. 2. Efficiency over input voltage range at 3 mA load with various switching frequencies and output voltages

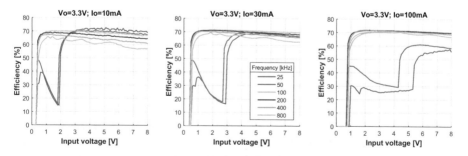

Fig. 3. Efficiency over input voltage range at 3.3 V output voltage with various switching frequencies and output currents

description of the optimal frequency requires at least 3 parameters. Input voltage, output voltage, output current all affect which switching frequency will be most efficient.

The general shape of measured curves correlates with the model from the component manufacturers, which shows no special phenomena at the transition from step-up to step-down operation [6,7]. The only deviation is in the area where some of the components approach saturation which is not regarded by the model.

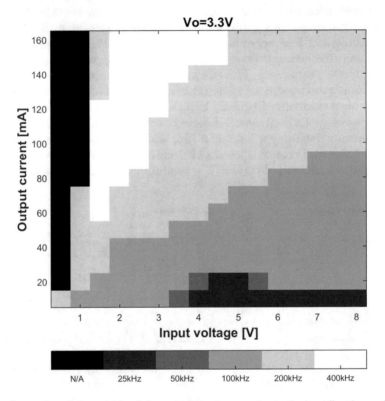

Fig. 4. Space describing which of the switching frequencies is the best for the evaluated circuit in given conditions. Value N/A marks area, where gain limitation doesn't allow required output voltage.

4   Conclusion

Measured data describe the effects of variable switching frequency on the power efficiency. Results show non-monotonic nature affected by several other parameters. For described circuit this means selecting the switching frequency that yields the best available efficiency requires computation. Similar functions are commonly implemented with a fixed hardware system - for example, in maximum

power point tracking of solar panel. This paper shows there are more properties that can be adapted for. Fine-tuning the switching frequency for SEPIC
converter can lead to notable efficiency improvement. The optimization on such
level will eventually require programmable processing unit. Purely digital control, however, brings its own imperfections [8]. Therefore it is convenient to leave
the lowest levels of control to the analog circuitry. That way effect of limited resolution and poor transient response of the digital system would be minimized,
while complex optimization would still be possible. The future need of power
will force utilisation of sub-optimal sources for energy harvesting. The DC-DC
converters with combined digital and analog control have the potential to allow
more efficient and flexible operation that adapts to highly dynamic conditions
of such sources.

References

1. Zhao, L., Qian, J., Texas Instruments: DC-DC Power Conversions and System
   Design Considerations for Battery Operated System, January 2006
2. Texas Instruments Inc.: "Bidirectional DC-DC Converter", TI Designs:TIDA-
   BIDIR-400-12, TI Literature Number: TIDUAI7, September 2015
3. ON Semiconductor: Effects of High Switching Frequency on Buck Regulators, tutorial ID: TND388/D, December 2009
4. Nowakowski, R., King, B., Texas Instruments Inc.: Choosing the optimum switching
   frequency of your DC/DC converter. EETimes, October 2006
5. Keeping, S.: Design Trade-offs when Selecting a High-Frequency Switching Regulator. DigiKey TechZone article, February 2015
6. Jeff, F., Texas instruments Inc.: Designing DC/DC converters based on SEPIC
   topology. "Analog Applications Journal", 4Q 2008, pp. 18-23, TI Literature Number:
   SLYT309 (2008)
7. Žák, S., Szabo, J.: Load balancing of heterogeneous parallel DC-DC converter. In:
   FedCSIS Proceedings of the Federated Conference on Computer Science and Information Systems, pp. 895–899 (2017)
8. Al-Hoor, W., Abu-Qahouq, J., Huang, L., Batarseh, I.: Design Considerations and
   Dynamic Technique for Digitally Controlled Variable Frequency DC-DC Converter,
   pp. 846–850 (2007) https://doi.org/10.1109/PESC.2007.4342098

Multiple-Model Description and Control Construction Algorithm of Supply Chain

Inna Trofimova[1]([✉]), Boris Sokolov[2], and Dmitry Ivanov[3]

[1] St. Petersburg State University, St. Petersburg, Russia
isolovyeva@mail.ru
[2] St. Petersburg Institute for Informatics and Automation of the Russian Academy
of Sciences, University ITMO, St. Petersburg, Russia
sokol@iias.spb.su
[3] Berlin School of Economics and Law, Berlin, Germany
dmitry.ivanov@hwr-berlin.de

Abstract. In this paper a multiple-model description of supply chain (SC) is presented. The SC state change is described through differential equations based on a dynamic interpretation of the job execution. The problem is represented as a designing control of SC problem in presence of external actions. The approach is based on decomposition of the problem on two parts and its solution with the help of optimal control theory methods and linear programming.

Keywords: Models of supply chain · Optimal program control
Positional control

1 Introduction

Modern supply chains (SC) are highly dynamic systems because they strive to work with a large variety of products, to achieve a high quality supply and a reliability and environmental standards, to take into account fast appearance of new products and increasing competitiveness of the companies, to use information technologies. Companies are intent on analyzing ability of control SC system to react on time when unforeseen events happen. For example these events could occur due to transfer and storage of products and they primarily sought to compliance with deadlines and to cost saving in this cases. Modern technical capacity of receiving information about SC current condition enable to improve SC operating. One of the main challenges in the SC management is SC execution problem in the case of unforeseen events [1,2].

Different mathematical models for describing SC dynamic behavior were introduced in recent times. Part of them is used for optimization of operating SC and application of control theory methods for solving mentioned problems, for example [3–5].

© Springer International Publishing AG, part of Springer Nature 2019
R. Silhavy (Ed.): CSOC 2018, AISC 765, pp. 102–108, 2019.
https://doi.org/10.1007/978-3-319-91192-2_11

The foundation of the considered in this paper complex modeling approach is control theoretic description of SCs as controllable dynamic systems. This study is based on the scheduling SC model [6] and algorithm in terms of optimal program control (OPC) [6,7]. The advantage of this approach is that the classical methods of optimal program and position control have been transferred and modified for a completely new subject area - supply chain management.

2   Multiple-Model Description of Supply Chain

The control theoretic description of SC: the SC is considered as a complex technical object (CTO) that is described through differential equations based on a dynamic interpretation of the job execution [6,7]. The job execution is characterized by (1) execution results (e.g., volume, time, etc.), (2) capacity consumption of the resources, and (3) CTO flows resulting from the delivery to the customer. Along [6] let us consider model M, which consist from two parts: a job control model ($M1$) and a flow control model ($M2$). In line with [8] we propose to use scheduling procedure: $M1$ is used to assign jobs to suppliers, $M2$ is used to schedule the processing of assigned orders subject to capacity restrictions of the production and transportation resources. The basic interaction of these two models is that after the determining control variables in the job control model, they are used in the constraints of the flow control model.

Then the model M under disturbances with aggregate operations is considered for solving the SC schedule execution problem on the assumption of that current SC condition information is received at fixed moments of time.

1. A Dynamic Model of Job Control (model $M1$). We consider the mathematical model of job control [6]:

$$\dot{x}_{i\mu}^{(o)} = \sum_{j=1}^{n} \varepsilon_{ij}(t) u_{i\mu j}^{(o)}, \tag{1}$$

$$\dot{x}_{j}^{(o)} = \sum_{i=1}^{n} \sum_{\eta=1, \eta\neq i}^{n} \sum_{\mu=1}^{s_i} \sum_{\rho=1}^{p_i} u_{i\mu j}^{(o)}. \tag{2}$$

Here $x_{i\mu}^{(o)}$ - the job state variable, which indicates the relation to jobs (orders), $x_{j}^{(o)}$ characterizes the total employment time of the j-supplier, $\varepsilon_{ij}(t)$ is an element of the preset matrix time function of time-spatial constraints.

The control actions are constrained as follows:

$$\sum_{i=1}^{n} \sum_{\mu=1}^{s_i} u_{i\mu j}^{(o)}(t) \leq 1, \forall j; \quad \sum_{j=1}^{n} u_{i\mu j}^{(o)}(t) \leq 1, \forall i, \forall \mu; \tag{3}$$

$$\sum_{i=1}^{n} u_{i\mu j}^{(o)}[\sum_{\alpha\in\Gamma_{i\mu 1}} (a_{i\alpha}^{(o)} - x_{i\alpha}^{(o)}) + \prod_{\beta\in\Gamma_{i\mu 2}} (a_{i\beta}^{(o)} - x_{i\beta}^{(o)})] = 0; \tag{4}$$

$$u_{i\mu j}^{(o)}(t) \in \{0, 1\}; \tag{5}$$

where $\Gamma_{i\mu 1}, \Gamma_{i\mu 2}$ are the sets of job numbers which immediately precede the job $D_\mu^{(i)}$ subject to accomplishing of all the predecessor jobs or at least one of the jobs correspondingly, and $a_{i\alpha}^{(o)}, a_{i\beta}^{(o)}$ are the planned lot-sizes. Constraint (3) refers to the allocation problem constraint according to the problem statement (i.e., only a single order can be processed at any time by the manufacturer). Constraint (4) determines the precedence relations, more over this constraint implies the blocking of job $D_\mu^{(i)}$ until the previous jobs $D_\alpha^{(i)}, D_\beta^{(i)}$ have been executed. If $u_{i\mu j}^{(o)}(t) = 1$, all the predecessor jobs of the operation $D_\mu^{(i)}$ have been executed. Note that these constraints are identical to those in MP models.

According to the problem statement, let us introduce the following performance indicators:

$$J_1^{(o)} = \frac{1}{2} \sum_{i=1}^{n} \sum_{\mu=1}^{s_i} (a_{i\mu}^{(o)} - x_{i\mu}^{(o)}(T_f))^2; \tag{6}$$

$$J_2^{(o)} = \sum_{i=1}^{\overline{n}} \sum_{\mu=1}^{s_i} \sum_{j=1}^{n} \int_{T_0}^{T_f} \alpha_{i\mu}^{(o)}(\tau) u_{i\mu j}^{(o)}(\tau) d\tau; \tag{7}$$

$$J_3^{(o)} = \frac{1}{2} \sum_{j=1}^{n} (T - x_j^{(o)}(T_f))^2. \tag{8}$$

Indicator $J_1^{(o)}$ characterizes the accuracy of the end conditions' accomplishment, i.e. the service level, $J_2^{(o)}$ refers to the estimation of an job's execution time with regard to the planned supply terms and reflects the delivery reliability, i.e., the accomplishing the delivery to the fixed due dates, $J_3^{(o)}$ estimates the equal resource utilization. The functions $\alpha_{i\mu}^{(o)}(\tau)$ is assumed to be known.

2. A Dynamic Model of Flow Control (model $M2$).

We consider the mathematical model of flow control [6]:

$$\dot{x}_{i\mu j}^{(f)} = u_{i\mu j}^{(f)}, \quad \dot{x}_{ij\eta\rho}^{(f)} = u_{ij\eta\rho}^{(f)}. \tag{9}$$

we denote $x_{i\mu j}^{(f)}$ - the flow state variable, which indicates the relation of the variable x to flows. The control actions are constrained by maximal capacities and intensities as follows:

$$\sum_{i=1}^{n} \sum_{\mu=1}^{s_i} u_{i\mu j}^{(f)}(t) \le R_j^{(f)}, \quad \sum_{\rho=1}^{p_i} u_{ij\eta\rho}^{(f)}(t) \le R_{j\eta}^{(f)}, \tag{10}$$

$$0 \le u_{i\mu j}^{(f)}(t) \le c_{i\mu j}^{(f)} u_{i\mu j}^{(o)}, \quad 0 \le u_{ij\eta\rho}^{(f)}(t) \le c_{ij\eta\rho}^{(f)} u_{ij\eta\rho}^{(o)}, \tag{11}$$

where $R_j^{(f)}$ is the total potential intensity of the resource $C^{(j)}$, $R_{j\eta}^{(f)}$ is the maximal potential channel intensity to deliver products to the customer $B^{(\eta)}$, $c_{i\mu j}^{(f)}$ is the maximal potential capacity of the resource $C^{(j)}$ for the job $D_\mu^{(i)}$, and $c_{ijn\rho}^{(f)}$ is the total potential capacity of the channel delivering the product flow $P_{<s_i,\rho>}^{(j,\eta)}$ of the job $D_\mu^{(i)}$ to the customer $B^{(\eta)}$. The end conditions are similar to those in (6) and subject to the units of processing time. The economic meaning of performance indicators in this model correspond to $J_2^{(o)}$ and $J_3^{(o)}$

$$J_1^{(f)} = \frac{1}{2} \sum_{i=1}^{n} \sum_{\mu=1}^{s_i} \sum_{j=1}^{n} [(a_{i\mu j}^{(f)} - x_{i\mu}^{(f)}(T_f))^2 + \sum_{\eta=1,\eta \neq i}^{n} \sum_{\rho=1}^{p_i} (a_{ij\rho\eta}^{(f)} - x_{ij\rho\eta}^{(f)}(T_f))^2]; \quad (12)$$

$$J_2^{(f)} = \frac{1}{2} \sum_{i=1}^{\overline{n}} \sum_{\mu=1}^{s_i} \sum_{j=1}^{n} \int_{T_0}^{T_f} \beta_{i\mu}^{(f)}(\tau) u_{i\mu j}^{(f)}(\tau) d\tau. \quad (13)$$

With the help of the weighting performance indicators, a general performance vector can be denoted as follows:

$$\mathbf{J}(\mathbf{x}(t), \mathbf{u}(t)) = \| J_1^{(o)}, J_2^{(o)}, J_3^{(o)} J_1^{(f)}, J_2^{(f)} \| . \quad (14)$$

The partial indicators may be weighted depended on the planning goals and control strategies. Original methods [9] have been used to transform the vector $\mathbf{J}$ to a scalar form J_G.

3 Problem Statement

Let us combine model $M1$, (1)–(8), and $M2$, (9)–(14), in one model M. It may be represented in the following form:

$$M = \begin{cases} \mathbf{u}(t) \mid \dot{\mathbf{x}} = f(\mathbf{x}, \mathbf{u}, t) \\ h_0(\mathbf{x}(t_0)) \leq 0, \ h_1(\mathbf{x}(t_f)) \leq 0 \\ q^{(1)}(\mathbf{x}, \mathbf{u}) \leq 0, \ q^{(2)}(\mathbf{x}, \mathbf{u}) = 0. \end{cases} \quad (15)$$

On the first stage the job shop scheduling problem for the model M is considered. It can be formulated as optimal program control (OPC) problem: this is necessary to find an allowable control $\mathbf{u}(t)$ $t \in (T_0, T_f]$, that ensures for the model M meeting the vector constraint functions $\mathbf{q}^{(1)}$, $\mathbf{q}^{(2)}$, and guides the dynamic system of the model (i.e., job shop schedule) from the initial state to the specified final state. If there are several allowable controls (schedules), then the best one (optimal) should be selected in order to maximize (minimize) J_G. In terms of OPC, the program control of job execution is at the same time the job shop schedule. On the first stage this problem was solved by control theory methods [6,7,10] and the optimal program control $\mathbf{u}_{pr}(t)$ and corresponding vector $\mathbf{x}_{pr}(t)$ were designed for this OPC problem for the model M with J_G [6].

On the next step model M under disturbances is considered and it is represented in the following form:

$$M_\zeta = \begin{cases} \mathbf{u}(t) \mid \dot{\mathbf{x}} = f(\mathbf{x}, \mathbf{u}, t, \zeta) \\ h_0(\mathbf{x}(t_0)) \leq 0, \ h_1(\mathbf{x}(t_f)) \leq 0 \\ q^{(1)}(\mathbf{x}, \mathbf{u}) \leq 0, \ q^{(2)}(\mathbf{x}, \mathbf{u}) = 0. \end{cases} \quad (16)$$

Disturbance $\zeta(t)$ is piecewise continuous and confined function, it describes external actions, but in an explicit form $\zeta(t)$ is unknown. Assume the information about current state of the system $\mathbf{x}_\sigma$ is obtained at the fixed moments of time t_σ, where $t_\sigma \in [t_0, t_f), \sigma = \overline{1, N}$ in case unforeseen events could happen. Suppose company have incentives to minimize emergent deviations of system behavior under external actions owing to decrease its additional costs. Next, during the system operation the SC schedule execution problem is investigated.

SC is able to realize the performance of production and transportation execution and reject the disturbances. Thus let us consider the problem: it is necessary to construct a feasible control $\tilde{\mathbf{u}}(t)$ (on the basis of $\mathbf{u}_{pr}(t)$), defined on the time interval $[t_0, t_f)$ during the system operation on it, that satisfying constraints and boundary conditions in (15), and minimizing performance parameter

$$\tilde{J} = \int\limits_{t_0}^{t_f} \| \tilde{\mathbf{x}}(t) - \mathbf{x}_{pr}(t) \| \, dt \to \min.$$

4   Control Construction Algorithm of Supply Chain

On the second stage we propose to introduce aggregate operations and with model M_ζ to consider model M_ζ^a with aggregate operations and thus without nonlinear constraints. Nonlinear constrains prescribe the order of operations and they were used at the first stage in the job shop scheduling problem. Next models M_ζ, M_ζ^a we use together because of nonlinear constraints are included indirectly in performance parameter $\tilde{J}$. We propose to apply position optimization method to the problem on the second stage. It is based on widely known mathematical methods of optimal control problems and linear programming (adaptive method) [11]. It was suitable for linear and nonlinear systems and it was applied in a number of cases [11–13].

– First stage.

1. The problem of optimal program control for (15) is solved and $\mathbf{u}_{pr}(t)$, $\mathbf{x}_{pr}(t)$ are obtained. At the moment of time $\tau = t_0$ we could use $\tilde{\mathbf{u}}(t) = \mathbf{u}_{pr}(t)$ and $\tilde{\mathbf{x}}(t) = \mathbf{x}_{pr}(t)$ for M_ζ^a, because it is feasible control and satisfies constraints of the model.

– Second stage.

We denote two sets $T_0 = [t_0, t_0 + h, \ldots, t_f]$ and $T_\tau = [\tau, \tau + h, \ldots, t_f]$ on the time interval $[t_0, \ t_f)$, $t_0 < t_f < +\infty$, where $h = (t_f - t_0)/L$, $L \in \mathbf{N}$. We shall design position control $\mathbf{u}(t)$, $t \in [t_0, \ t_f)$ in the class of discrete piecewise constant functions $\tilde{\mathbf{u}}(t) = \tilde{\mathbf{u}}(t_0 + lh)$ at $t \in [t_0 + lh, \ t_0 + (l+1)h)$, $l = \overline{0, L-1}$ using position optimization method.

1. The model M_ζ^a and $\tilde{J}$ are constructed, information about SC current condition $\mathbf{x}_\sigma$ at the moments t_σ is received. Let us consider a set of auxiliary OPC problems for the model M_ζ^a with feasible $\tilde{\mathbf{u}}(t) \in \tilde{U}_p$ and conditions at initial time $\tau = t_\sigma$ $\tilde{\mathbf{x}}(\tau) = \mathbf{x}_\sigma$, and at fixed terminal time t_f $h_1(\tilde{\mathbf{x}}(t_f)) \leq 0$ and $\tilde{J}$. This set depends on parameter $\tau \in T_\tau$ and vector $\mathbf{x}_\sigma$.

2. Auxiliary problem of optimal program control with new initial conditions $(t_\sigma, \mathbf{x}_\sigma)$ is formulated. Denote $\tilde{\mathbf{u}}(\tau| \tau, \mathbf{x}_\sigma)$ - OPC for the position $(\tau, \mathbf{x}_\sigma)$ (positional solution), $X(\tau)$ - a set of any and all start states $\mathbf{x}_\sigma = \tilde{\mathbf{x}}(\tau)$, for which the auxiliary OPC problem could be solved at fixed moment of time τ, with vectors $\mathbf{x}_\sigma = \tilde{\mathbf{x}}(\tau)$, $\tilde{\mathbf{x}}(\tau) \in X(\tau)$, $t \in T(\tau) = [\tau, t_f)$.
3. The procedure of control design in the piecewise constant function class on the time interval $T_\tau = [\tau, t_f)$ with $\tau = t_\sigma$ is realized and auxiliary problem of optimal program control is reduced to the linear programming problem.
4. The linear programming problem is solved using adaptive method and satisfiability of nonlinear constraints in (15) is checked.
5. Positional control is used on the time interval $[t_\sigma, t_{\sigma+1})$ till the new information about SC current condition $(t_{\sigma+1}, x_{\sigma+1})$ is received.

The set of auxiliary OPC problems at $\tau = t_\sigma$ is considered in sequence, where current information about system states $\mathbf{x}_\sigma$ is taken into account. Each one of them could be reduced to the linear programming (LP) problem. Solution of the auxiliary OPC problem applied to the considered model on the time interval until further information acquisition. Adaptive method is suitable for solving considered LP problems, because it was developed for time-dependent multidimensional systems under polyhedral constraints [11]. Accordingly this if number of division points is increased it doesn't lead up to large dimension of the system.

5   Conclusions

In this paper according to complex modeling approach is proposed to use control theoretic description of SC, to compose the models combination and consider optimal program and position control problems together. This approach is based on a dynamic interpretation of the job execution. Interaction of models in models combination and interrelation of its parameters enable to apply them towards the considered problems: the job shop scheduling problem and the SC schedule execution problem. The undoubted advantage of the proposed approach is that the classical methods of optimal program and position control have been transferred and modified for a completely new subject area - supply chain management (management of organizational and technological systems).

Acknowledgments. The research described in this paper is partially supported by the Russian Foundation for Basic Research (grants 16-07-00779, 16-08-00510, 16-08-01277, 16-29-09482-fi-i, 17-08-00797, 17-06-00108, 17-01-00139, 17-20-01214, 17-29-07073-fi-i, 18-07-01272, 18-08-01505), grant 074-U01 (ITMO University), state order of the Ministry of Education and Science of the Russian Federation 2.3135.2017/4.6, state research 007320180003, International project ERASMUS +, Capacity building in higher education, 73751-EPP-1-2016-1-DE-EPPKA2-CBHE-JP, Innovative teaching and learning strategies in open modelling and simulation environment for student-centered engineering education.

References

1. Schwartz, D., Wang, W., Rivera, D.: Simulation-based optimization of process control policies for inventory management in supply chains. Automatica **125**(2), 1311–1320 (2006)
2. Garcia, C.A., Ibeas, A., Herrera, J., Vilanova, R.: Inventory control for the supply-chain: an adaptive control approach based on the identification of the lead-time. Omega **40**, 314–327 (2012)
3. Perea, E., Grossman, I., Ydstie, E., Tahmassebi, T.: Dynamic modeling and classical control theary for supply chain management. Comput. Chem. Eng. **24**, 1143–1149 (2000)
4. Ortega, M.: Control theory applications to the production-inventory problem: a review. Int. J. Prod. Res. **8**(2), 74–80 (2000)
5. Ivanov, D., Sokolov, B., Solovyeva, I., Dolgui, A., Jie, F.: Dynamic recovery policies for time-critical supply chains under conditions of ripple effect. Int. J. Prod. Res. **54**(23), 7245–7258 (2016)
6. Ivanov, D.A., Sokolov, B.V.: Adaptive Supply Chain Management. Springer, Wiley and Sons, New York (2010)
7. Kalinin, V.N., Sokolov, B.V.: Optimal planning of the process of interaction of moving operating objects. Int. J. Differ. Equations **21**(5), 502–506 (1985)
8. Chen, Z.L., Pundoor, G.: Order assignment and scheduling in a supply chain. J. Oper. Res. **54**, 555–572 (2006)
9. Gubarev, V.A., Zakharov, V.V., Kovalenko, A.N.: Introduction to systems analysis. LGU, Leningrad (1988)
10. Chernousko, F.L.: State Estimation of Dynamic Systems. SRC Press, Boca Raton (1994)
11. Gabasov, R., Dmitruk, N.M., Kirillova, F.M.: Numerical optimization of time-dependent multidimensional systems under polyhedral constraints. Comput. Math. Math. Phys. **45**(4), 593–612 (2005)
12. Solovyeva, I., Ivanov, D., Sokolov, B.: Analysis of position optimization method applicability in supply chain management problem. In: 2015 International Conference on "Stability and control processes" in memory of V.I. Zubov, pp. 498–500 (2015)
13. Popkov, A.S., Baranov, O.V., Smirnov, N.V.: Application of adaptive method of linear programming for technical objects contro. In: 2014 International Conference on Computer Technologies in Physical and Engineering Applications ICCTPEA 2014 - Proceedings, p. 141 (2014)

A New Approach to Vector Field Interpolation, Classification and Robust Critical Points Detection Using Radial Basis Functions

Vaclav Skala and Michal Smolik[✉]

Department of Computer Science and Engineering, Faculty of Applied Sciences,
University of West Bohemia, 306 14 Plzen, CZ, Czech Republic
smolik@kiv.zcu.cz,
http://www.VaclavSkala.eu

Abstract. Visualization of vector fields plays an important role in many applications. Vector fields can be described by differential equations. For classification null points, i.e. points where derivation is zero, are used. However, if vector field data are given in a discrete form, e.g. by data obtained by simulation or a measurement, finding of critical points is difficult due to huge amount of data to be processed and differential form usually used. This contribution describes a new approach for vector field null points detection and evaluation, which enables data compression and easier fundamental behavior visualization. The approach is based on implicit form representation of vector fields.

Keywords: Critical points · Vector field classification · Vector field topology
Approximation · Data acquisition · Visualization · Radial basis functions
RBF · Interpolation · Approximation

1 Introduction

Many physical problems are described by differential equations of three basic types: ordinary differential equations (ODEs), partial differential equations (PDEs), algebraic-differential equations (ADEs or DAEs). They also can be classified as autonomous or t-varying, i.e. when functions depend on time. In this contribution, vector fields of autonomous system ODEs will be explored.

Let us imagine that a differential equation is given in E^2 as

$$\dot{x} = f(x(t), t) \tag{1}$$

where $f(x, t) = [{}^x f(x, t), {}^y f(x, t)]^T$. Implicit formulation is given as

$$F(x(t), t) = 0 \tag{2}$$

where $t \in \langle 0, \infty \rangle$, $x \in E^2$. Derivation of the Eq. 2. leads to

© Springer International Publishing AG, part of Springer Nature 2019
R. Silhavy (Ed.): CSOC 2018, AISC 765, pp. 109–115, 2019.
https://doi.org/10.1007/978-3-319-91192-2_12

$$\frac{dF(x,t)}{dt} = \frac{\partial F(x,t)}{\partial x}\frac{dx}{dt} + \frac{\partial F(x,t)}{\partial t} \tag{3}$$

As the only autonomous ODEs are considered, i.e. $\partial F(x,t)/\partial t = 0$,

$$\frac{dF(x,t)}{dt} = \frac{\partial F(x,t)}{\partial x}\frac{dx}{dt} = \nabla F(x(t))\frac{dx}{dt} = \nabla F(x(t))f(x(t)) = \nabla F\dot{x} \tag{4}$$

It means that a normal vector must be orthogonal to the particle velocity vector.

2 Extremes and Inflection Points

An inflection point of a curve given by the implicit function $F(x, y) = 0$ in E^2 is determined as $\det Q = 0$, i.e.

$$\det Q(x, y) = \begin{vmatrix} F_{xx} & F_{xy} & F_x \\ F_{yx} & F_{yy} & F_y \\ F_x & F_y & 0 \end{vmatrix} = 0 \tag{5}$$

where F_{xx}, resp. $^x f_y$ etc. are partial derivatives of $F(x)$, resp. $^x f_y = \frac{\partial^x f}{\partial y}$, etc. In the following, we expect that $F_{xy} = F_{yx}$. Details and extensions to a higher dimension can be found in (Goldman 2005).

3 Critical Points

The aim is to represent discreetly given vector field using Radial Basis Function (RBF) approximation as precise as possible in the form

$$^x f(x) = \sum_{i=1}^{N} c_i \phi(r_i) \tag{6}$$

where c_i are weights to be computed, $\phi(\cdot)$ is chosen RBF, e.g. $\phi(\cdot) = r^2 \log r$, $r_i = \|x - x_i\|$ and x_i are points in which the particle speed is given or acquired. Similarly for the $^y f(x)$. The advantage of the RBF use is that it leads to a linear system of equations $Ax = b$ (Smolik and Skala 2017b; Majdisova and Skala 2017).

Critical points of ODEs (or null points) are defined as $dx(t)/dt = 0$. Finding of critical points of ODEs reliably is difficult and the Taylor's series is used usually, i.e.

$$f(x) = f(x_0) + \frac{\partial f(x_0)}{\partial x}(x - x_0) + \cdots \tag{7}$$

where x_0 is a point where $f(x(t)) = 0$ and $\frac{\partial f(x_0)}{\partial x}$ is a Jacobian. It means that a local linearization is made and critical point classification is based on eigenvalues of the Jacobian (Helman and Hesselink 1989). However for detailed inspection of a vector field a Hessian can be used (Smolik and Skala 2017a). Finding a null point of the ODE from acquired discrete data is a numerically sensitive problem.

Let us define a function $F(x, y)$ related to speed of a particle as:

$$F(x, y) = \dot{x}^2 + \dot{y}^2 \tag{8}$$

Then the critical points are given as:

$$\lim_{\varepsilon \to 0} F(x, y) - \varepsilon = 0 \tag{9}$$

The inflection points including the critical ones are given by det $Q = 0$.

Let us consider two simple ODEs examples, Figs. 1 and 2, where critical points are shown. The example Ex.1 has two critical points, while the Ex.2 has three ones.

$$\text{Ex.1.:} \begin{bmatrix} \dot{x} \\ \dot{y} \end{bmatrix} = \begin{bmatrix} 2x + y^2 - 1 \\ 6x - y^2 + 1 \end{bmatrix} \qquad \text{Ex.2.:} \begin{bmatrix} \dot{x} \\ \dot{y} \end{bmatrix} = \begin{bmatrix} xy - 4 \\ (x - 4)(y - x) \end{bmatrix}$$

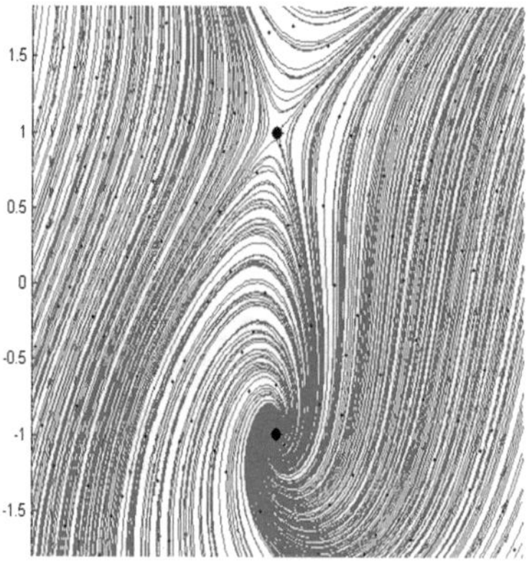

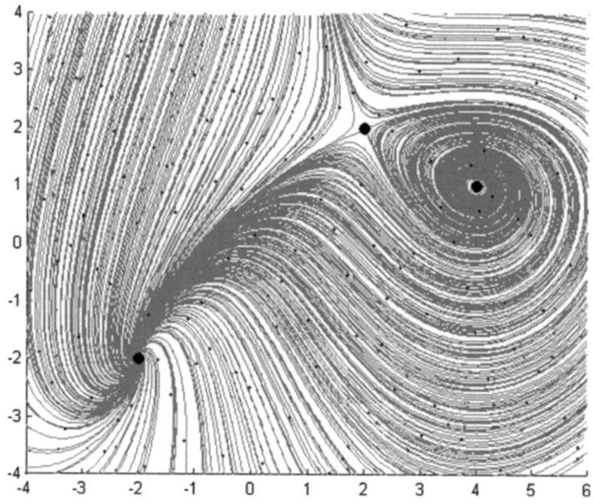

Fig. 1. Vector field with two critical points

Fig. 2. Vector field with three critical points

When the contour plot of det Q values is made, some other interesting features of the given vector field can be found, Figs. 3 and 4.

It can be seen, that also other important points/areas, not only critical points, can be easily detected. The *red* curves represent points where $\det(Q) = 0$, i.e. extremes and inflection points. The density of contours gives information on changes. If the RBF approximation of a vector field is to be efficiently used, reference points (Majdisova and Skala 2017) are to be placed respecting the vector field behavior (Figs. 5 and 6).

It can be seen that in some cases the det $Q(x, y)$ function might be quite flat in some areas, which might cause some numerical problems (Figs. 7 and 8). For better vector field properties evaluation, another specification of the $F(x, y)$ function using a similar approach, i.e.:

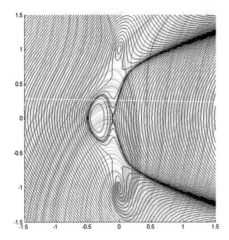
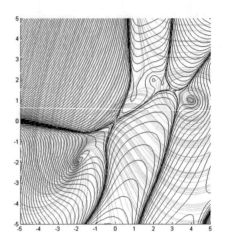

Fig. 3. $\det(Q)$ values for the Ex.1 case

Fig. 4. $\det(Q)$ values for the Ex.2 case

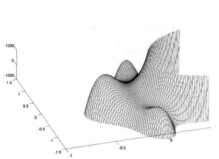
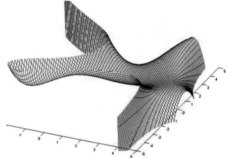

Fig. 5. $\det(Q)$ values for the Ex.1 case

Fig. 6. $\det(Q)$ values for the Ex.2 case

$$\begin{bmatrix} \dot{x} \\ \dot{y} \end{bmatrix} = \begin{bmatrix} {}^x\!f(x,y) \\ {}^y\!f(x,y) \end{bmatrix} \tag{10}$$

then the $\det R$ is determined as:

$$\det R(x,y) = \begin{vmatrix} {}^x\!f_x & {}^x\!f_y & {}^x\!f \\ {}^y\!f_x & {}^y\!f_y & {}^y\!f \\ {}^x\!f & {}^y\!f & 0 \end{vmatrix} = \begin{vmatrix} J(x) & f(x) \\ f^T(x) & 0 \end{vmatrix} = 0 \tag{11}$$

where $J(x)$ is the Jacobian $J(x) = \begin{bmatrix} {}^x\!f_x & {}^x\!f_y \\ {}^x\!f_y & {}^y\!f_y \end{bmatrix}$ and $\dot{x} = f(x)$ is the given ODE.

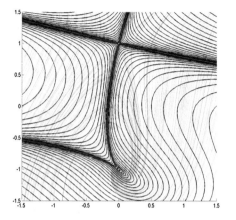

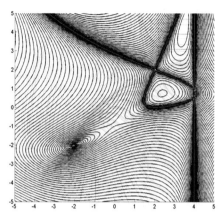

Fig. 7. det(R) values for the Ex.1 case   **Fig. 8.** det(R) values for the Ex.2 case

$$\text{Ex.3.:} \begin{bmatrix} \dot{x} \\ \dot{y} \end{bmatrix} = \begin{bmatrix} 1 - x^2 - y^2 \\ 2xy \end{bmatrix} \qquad \text{Ex.4.:} \begin{bmatrix} \dot{x} \\ \dot{y} \end{bmatrix} = \begin{bmatrix} y \\ (1 - x^2)y - x \end{bmatrix}$$

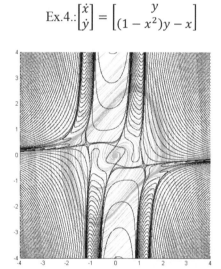

Fig. 9. det(R) values for the Ex.3 case   **Fig. 10.** det(R) values for the Ex.4 case
(Van der Pole ODEs)

It can be seen, that another types of vector field important features are obtained. This is important for the vector field topology evaluation and curvatures estimation.

The *red* curves in Figs. 9 and 10 represent the inflection of curves given as det $Q = 0$.

However, in some cases critical point finding using Eq. 9 might be a numerical problem, if the function bed is too flat. Then the criterion can be modified to

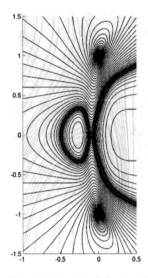

Fig. 11. Ex.3. det(Q) using Eq. 12

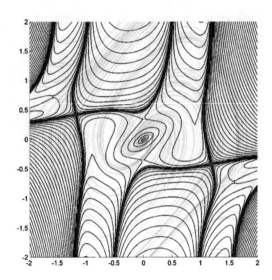

Fig. 12. Ex.4. det(Q) using Eq. 12

$$F(x, y) = \sqrt{\dot{x}^2 + \dot{y}^2} \tag{12}$$

In this case, the convergence for iterative method might be significantly faster and critical points more accurately determined (Figs. 11 and 12).

4   Experimental Evaluation

The proposed approach has been tested on different ODEs. Numerical computation of derivatives (forward, central, backward) was stable and critical points were determined correctly. The experiments proved novelty of the approach as an additional information of a vector field behavior is obtained. It can be used for efficient placing of reference points for RBF approximation (Majdisova and Skala 2017). Also, if the RBF approximation is used, then the given vector field is described in an analytic form.

5   Conclusion

A new approach for vector field null points, i.e. critical points, is described briefly. The approach is based on an implicit formulation and it gives possibility to represent main features of vector fields more precisely. It also, in connection with RBF approximation, offers analytical representation of vector fields and data compression as well. In future, the presented approach will be extended to time varying systems.

Acknowledgment. The author would like to thank to colleagues at the University of West Bohemia and to anonymous reviewers for their comments, which helped to improve the manuscript significantly. Special thanks also belong to Pavel Šnejdar for MATLAB additional programming and images generation.

Research was supported by the Czech Science Foundation, No. GA 17–05534S and partially by SGS 2016-013.

References

Goldman, R.: Curvature formulas for implicit curves and surfaces. Comput. Aided Geom. Des. **22**, 632–658 (2005)

Helman, J., Hesselink, L.: Representation and display of vector field topology in fluid flow data sets. IEEE Comput. **22**(8), 27–36 (1989)

Koch, S., Kasten, J., Wiebel, A., Scheuermann, G., Hlawitschka, M.: Vector field approximation using linear neighborhoods. Vis. Comput. **32**(12), 1563–1578 (2015)

Majdisova, Z., Skala, V.: Radial basis function approximations: comparison and applications. Appl. Math. Model. **51**, 728–743 (2017)

Scheuermann, G., Krüger, H., Menzel, M., Rockwood, A.: Visualizing non-linear vector field topology. IEEE Trans. Visual. Comput. Graph. **4**(2), 109–116 (1998)

Smolik, M., Skala, V.: Classification of critical points using a second order derivative. In: ICCS 2017, Procedia Computer Science, pp. 2373–2377. Elsevier (2017a)

Smolik, M., Skala, V.: Spherical RBF vector field interpolation: experimental study. In: SAMI, pp. 431–434. IEEE (2017b)

Thiesel, H.: Vector field curvature and applications. Ph.D. Thesis. Univ. of Rostock (1995)

Trusted Cryptographic Tools Locking

Vadim N. Tsypyschev[1,2,3(✉)]

[1] Moscow Technological University, 78, Vernadsky Avenue,
Moscow 119454, Russian Federation
`tsypyschev@yandex.ru`
[2] Moscow Institute of Physics and Technologies (State University),
9, Institutsky pereulok, Dolgoprudny, Moscow Region 141701, Russian Federation
[3] S-Terra CSP, 5, Georgievsky Avenue, Zelenograd, Moscow, Russian Federation

Abstract. Up to current moment it is actual to lock trustworthy cryptographic services of the complex product providing information security services. These cryptographic services, as usual, are included in as part of delivering complex product and must not be activated without proper additional license.

In this article, we provide a method of cryptographic services locking and a protocol of its activation including a legal customer authentication.

The main reason of this work is that the law of some countries prepends to export complex information security products with available cryptographic tools.

Keywords: Trustworthy services locking · Cryptographic protocol
Software customer authentication

1 Introduction

The task concerning in this article is to provide a method of reliable, robust, secure, and solid locking of a part of software services, especially its cryptographic tools, until it'll be legally activated. A protocol of authentication of legal customer and services activation is a main part of this method.

Further we assume that

1. Cryptographic tools developer must to create and implement an algorithm of legal customer's license generation and validity checking under condition that software code may be publicly known and doesn't contain confidential data.
2. The algorithm of license generation and its validity checking must provide an unique license for given delivery and legal customer under condition that software code must be fixed and unaltered.
3. The algorithm of license validity checking must be executed successfully if and only if legal customer has typed his unique license data (LD) during installation process.

© Springer International Publishing AG, part of Springer Nature 2019
R. Silhavy (Ed.): CSOC 2018, AISC 765, pp. 116–124, 2019.
https://doi.org/10.1007/978-3-319-91192-2_13

4. The algorithm of license validity checking must be implemented in an installer.
5. The algorithm of license validity checking must involve data inextricably linked with software code.
6. Installer must activate cryptographic tools only after successful license data and legal customer data validity checking.
7. It is technically impossible to activate cryptographic tools without installer.
8. It is technically impossible to simulate an installer to activate cryptographic tools.
9. It is technically impossible to use an inactivated cryptographic tools code.
10. For every legal customer the activation of cryptographic tools is executed by himself or by trusted employee.

So, cryptographic tools developer must be directed by paradigm "If not activated, then locked".

The difficulty of construction of such algorithm of generation license and its verification is consisted in the demand to type the license during the installation process. It means that the length of license couldn't be more than 25 bytes of printable characters.

Avoiding it we could go this evident way: to form the file from developer data, product custom code, legal customer data, license data and sign it by secret key fixed for whole custom. Then the verification algorithm should consist in signature verification using an open key inextricably linked to the cryptographic tools code.

Other way possible to use is to encipher those data by trusted two-key algorithm and further decipher it during installation process.

The inconvenience of those ways consists in that the signature should be at last 64 bytes of unprintable characters and after an ASCII85 application it should generate a license of at least 80 bytes. It is evident that so long license can't be typed during verification process (Table 1).

Indeed, Bundesamt fuer Siecherheit in der Informationstechnik (BSI), Germany, has published [2, Table 3.2] this data:

Table 1. BSI estimation of two-key algorithms size and its complexity

$\lg_2 R$	ECDLP	Factorization/DLP in $\mathbb{F}_p^*$
60	120	700
70	140	1000
100	200	1900
128	256	3200
192	384	7900
256	512	15500

Here $ECDLP$ denotes the size of group of points of elliptic curve with complexity of finding discrete logarithm by Pollard's ρ-method is equal to 2^R group operations, $Factorization$ denotes the RSA-modulus size with complexity of factorization by number field sieve method equal to 2^R operations of ring $\mathbb{Z}/2^{Factorization}$, DLP in $\mathbb{F}_p^*$ denotes that the complexity of discrete logarithm problem in $\mathbb{F}_p^*$ is equal to 2^R operations of this field in terms of number field sieve method.

In this connection it is assumed that the parameters of elliptic curves are chosen such that the complexity of Menezes–Okamoto–Vanstone method is not less then the Pollard's ρ-method.

Besides that, the data of [2, Table 3.2] are coordinated to Post Quantum Cryptography problem.

According to [2, Table 3.2] neither one of well-known and trusted algorithms of [3,4,6–8], nor RSA, ECDSS, GOST R 34.0—2012 can be used to settle our demands.

So, the way could be used is to generate license as a result of pseudo–random function based on "good" cryptographic hash function as in [5] applied to that file. This way allows to obtain a license of acceptable length, but now to verify license it is needed to have a key of pseudo–random function what tends us to the conception of Trusted License Server which would provide a verification algorithm by needed secret information.

But now difficulties of communication with Trusted License Server arise.

Investigated all the reasons above we provide the method of licensing presented below.

2   Notation

1. E_K, D_K denote enciphering/deciphering in ECB mode with key K via block cipher (AES or Kuznechik (GOST R 34.12—2015))
2. $E_{K,IV}, D_{K,IV}$ denote enciphering/deciphering in CBC mode with key K and initial vector IV via block cipher (AES or Kuznechik (GOST R 34.12—2015))
3. $Imito_{K,IV}$ denotes calculation of CBC-MAC with key K and initial vector IV via block cipher (AES or Kuznechik (GOST R 34.12—2015))
4. $HASH$ denotes cryptographic hash-function (SHA or Streebog (GOST R 34.11—2012))
5. $HMAC(Key, Material)$ denotes pseudo-random function [5] with key KEY and publicly known $Material$ based on $HASH$-function
6. $Sign_K(M), VerSign_K(M)$ denote generation/verification of digital signature (ECDSS or GOST R 34.10—2012) with key K. If key is omitted then it's fixed and not altered
7. prime p is a modulus of elliptic curve
8. $E_{a;b}(\mathbb{F}_p)$ is an elliptic curve with its invariant $j(\mathcal{E}_{a;b})$ and coefficients $a; b \in \mathbb{F}_p$ such that
$$4a^3 + 27b^2 \not\equiv 0 \pmod{p},$$

9. integer m is an order of group of points of ellipic curve $E_{a;b}(\mathbb{F}_p)$;

10. prime q is an order of cyclic subgroup of points group of $\mathcal{E}$ such that

$$\begin{cases} m = nq; n \in \mathbb{Z}; n \geq 1; \\ 2^{254} < q < 2^{256} \end{cases}$$

11. $P \neq \mathcal{O}$ is a point of elliptic curve $\mathcal{E}$, $P = (x_p; y_p)$, such that $[q]P = \mathcal{O}$;

We suppose that the either American suite, either Russian suite of cryptographic primitives should be applied because of the compatibility reasons.

3 Method of Cryptographic Tool Locking and License Generation

Now let's assume that cryptographic tools code is locked via enciphering in CBC mode according to trusted block cipher (AES or Kuznechik) of some text file which contains some specific data. This data is an input of some specific calculation. The output this calculation determines work capability of the cryptographic tools.

Enciphering must be executed in CBC mode with initial vector IV and with key $K = x_C$, where x_C is a x-coordinate of point $C = [k]P$ of elliptic curve $\mathcal{E}_{a,b}$, ord $P = q$.

As an example of the data we provide sample $Y, K, Sign(Y, K)$.

As an example of specific calculation we provide:

1. verification $VerSign(Y, K)$ with an inseparably tied with cryptotool code signature verification key, fixed for all version of cryptotool

2. calculation of
$$X = D_K(Y),$$

3. calculation of
$$H = HASH(X),$$

and after that H is essentially linked in cryptographic calculation process.

Further,

1. License structure has the form:

License :

$License := \{ CodeOfProduct;$

$NumberOfLicense;$

$CodeOfCustomer;$

$LicenseCode; \}$

Here $CodeOfProduct$ is a vendor's code of cryptotool software, $NumberOfLicense$ is a number of issued license. Herewith vendor takes into account all issued licenses. $CodeOfCustomer$ is a setting data of the legitimate purchaser.

LicenseCode is the result of application injective publicly-known and invertible code:

$$LicenseCode = EncodingAlgorithm(InitialLicenseCode).$$

As examples of such codes Base64 and ASCII85 serves.
InitialLicenseCode is an output of pseudo-random generator of the length $N_1 = 8n$ bit.
After encoding length of field *LicenseCode* is a α times greater:

$$N_2 = |LicenseCode| = \alpha \cdot |InitialLicenseCode| = 8\alpha n.$$

2. Further we shall represent fields (bit sequences) LicenseCode, NumberOfLicense and others as an integers where the most left bit of the field matches to most significant bit of the integer
3. Trusted License Server stores all issued licenses identifying it by CodeOfProduct and NumberOfLicense
4. Trusted License Server is able to generate digital signature. Verification key of this signature is common for te cryptotool type, non-alter allthe time of existence of cryptotool type, and inseparably linked with cryptotool code
5. The cryptotool activation performs by built-in installer. Installer is able to perform functions of digital signature verification, cryptographic hashing, and enciphering/deciphering. All thi functions are from the same suite of cryptographic primitives
6. Cryptographic properties of installer can't be modified by user
7. Cryptographic properties of installer are developed to be executed by user without essential support of developer
8. Technical documentation (technical description of cryptographic suite, intermediary protocols, interface description, etc.) is publicly accessible
9. Parameter *CodeOfProduct* is inseparably linked in code of installer.

4   License Validity Verification Protocol

Just now verification of license validity and activation of cryptotool code perform according to this
   Protocol

1. Installer gets from keyboard license number N, purchaser's code C, and license code L.
   Meanwhile installer gets timer's indications in moments of keyboard clicking and stores it as sequence $(T_1, \ldots, T_M)$.
2. Installer forms pseudo-random values:

$$Nonce = HASH(T_1 \oplus T_3| \ldots |T_{M-2} \oplus T_M);$$

$$Nonce_1 = HMAC(Nonce, N|L);$$

$$Nonce_2 = HMAC(Nonce, Nonce_1|N|L);$$

3. Installer forms and sends to Trusted License Server request:

$$(CodeOfProduct, N, [Nonce_2 \pmod q)]P).$$

4. Trusted License Server forms and sends to installer reply

$$((CodeOfProduct, N, [Y]P), Sign(CodeOfProduct, N, [Y]P)),$$

herewith Y is a output of pseudo-random generator formed as an integer modulus q

5. Installer verifies
$$Sign(CodeOfProduct, N, [Y]P),$$

and forms seance key

$$SeanceKey = HASH([Nonce_2 \pmod q)][Y]P).$$

If $Sign(CodeOfProduct, N, [Y]P)$ is not correct, then installer informs Trusted License Server, sends it pair $Nonce, Nonce_1$ and stops activation.

6. If Trusted License Server is informed by installer about stop of activation and get $Nonce, Nonce_1$, it verifies that

$$[Nonce_2 \pmod q)]P = [HMAC(Nonce, Nonce_1|N|L) \pmod q)]P.$$

If it is, it stops activation too and increase counter of activation tries. In other case it proceeds.

7. Trusted License Server forms seance key:

$$SeanceKey = HASH([Y][Nonce_2 \pmod q)]P).$$

8. Trusted License Server gets an output of pseudo-random generator and forms $InitialVector$ for CBC mode and $ImitoKey$ to form CBC-MAC
After that it sends to installer

$$(NewInitialVector, NewImitoKey) = E_{SeanceKey}(InitialVector, ImitoKey),$$

herewith enciphering is performed in ECB mode.

9. Installer sends to Trusted License Server values

$$(NewL, ImitoL) =$$
$$(E_{SeanceKey, InitialVector}(L), Imito_{ImitoKey, InitialVector}(L)),$$

herewith enciphering is performed in CBC mode with $InitialVector$ as an initial vector, and

$$ImitoL = Imito_{ImitoKey, InitialVector}(L)$$

is a CBC-MAc with key $ImitoKey$ and initial vector $IiniialVector$ for $Nonce_1$

10. Trusted License Server calculate

$$L = D_{SeanceKey, InitialVector}(NewL),$$

and calculate its CBC-MAC.

In the case if calculated CBC-MAC of L differs from received CBC-MAC of L then Trusted License Server stop activation process, sends to installer notification signed with its secret key.

Trusted License Server increase the counter of activation tries.

In other case Trusted License Server verifies that

$$L = LicenseCode$$

stored at Trusted License Server Data Base. If it is, Trusted License Server proceed.

Else Trusted License Server stop activation process, sends to installer notification signed with its secret key.

Trusted License Server increase the counter of activation tries.

11. Installer sends to Trusted License Server values

$$(NewNonce_1, ImitoNonce_1) =$$
$$(E_{SeanceKey, InitialVector}(Nonce_1), Imito_{ImitoKey, InitialVector}(Nonce_1)),$$

herewith enciphering is performed in CBC mode with $InitialVector$ as an initial vector, and

$$ImitoNonce_1 = Imito_{ImitoKey, InitialVector}(Nonce_1)$$

is a CBC-MAc with key $ImitoKey$ and initial vector $IiniialVector$ for $Nonce_1$

12. Trusted License Server calculate

$$Nonce_1 = D_{SeanceKey, InitialVector}(NewNonce_1),$$

and calculate its CBC-MAC.

In the case if calculated CBC-MAC of $Nonce_1$ differs from received CBC-MAC of $Nonce_1$ then Trusted License Server stop activation process, sends to installer notification signed with its secret key.

Trusted License Server increase the counter of activation tries.

In other case Trusted License Server calculates

$$X_1 = ((Nonce_1 \oplus (LicenseCode|10\ldots0)) \pmod{q})^{-1} \pmod{q},$$

computes

$$Key = [X_1][k]P,$$

and sends to installer

$$(EncryptedKey, EncryptedIV, Imito) =$$
$$(E_{SeanceKey}(Key, IV), Imito_{ImitoKey, InitialVector}(Key, IV)).$$

13. Installer receives $(EncryptedKey, EncryptedIV)$, computes in ECB mode

$$Key = D_{SeanceKey}(EncryptedKey),$$

$$IV = D_{SeanceKey}(EncryptedIV),$$

and check CBC-MAC.

In the case if received CBC-MAC differs from calculated CBC-MAC installer informs Trusted License Server, protecting the notification by CBC-MAC, computed with $SeanceKey, InitialVector$. After that installer stops activation.

Trusted License Server increase counter of activation tries.

In other case installer represents Key as a point Q of elliptic curve $\mathcal{E}_{a,b}$ and computes

$$C = [(Nonce_1 \oplus (L||10\ldots0)) \pmod q]Q$$

After that installer deciphers with key $K = x_C$ and initial vector IV enciphered in CBC mode text file and after that start installation.

If installation succeed then Installer informs Trusted License Server and protects notification by CBC-MAC calculated with $SeanceKey, InitialVector$.

14. Trusted License Server set $InstallationFlag$ to 1 for pair N, L.

End of Protocol

Note 1. Trustworthy of this Protocol may be increased if installer provided by VPN-client, pre-determinately configured to connect Trusted License Server only

5   Security Considerations

In [1] were formulated goals to be achieved in cryptographic protocol construction process whether if those applicable to considered protocol. Below we list goals achieved in considering Protocol.

1. G1 (peer entity authentication)
2. G2 (message authentication)
3. G3 (replay protection in the sense that all messages were generated in current seance)
4. G7 (key authentication)
5. G8 (key confirmation, key proof of possession)
6. G9 (perfect forward secrecy)
7. G10 (fresh key derivation)
8. G12 (confidentiality, secrecy)
9. G16 (sender invariance)

6   Conclusions

So we have proposed the method of trusted cryptographic tools locking, its licensing, and protocol for license verification and cryptographic tools activation.

Also we have provided some consideration about trustworthy of Protocol proposed.

References

1. Automated Validation of Internet Security Protocols and Applications (AVISPA). IST-2001-39252. Deliverable D6.1 List of Selected Problems (2005). http://www.avispa-project.org/publications.html
2. Cryptographic Mechanisms: Recommendations and Key Lengths. Bundesamt fuer Siecherheit in der Informationstechnik, BSI TR-02102-1, February 22 2017
3. Buchmann, J., Dahmen, E., Huelsing, A.: XMSS - a practical forward secure signature scheme based on minimal security assumptions. In: Yang, B.Y. (ed.) Post-Quantum Cryptography, PQCrypto 2011. Lecture Notes in Computer Science, vol. 7071. Springer, Berlin, Heidelberg (2011)
4. ElGamal, T.A.: Public key cryptosystem and a signature scheme based on discrete logarithms. Advances in Cryptology - Crypto 84, pp. 10–18. Springer (1984)
5. Harkins, D., Carrel, D.: The Internet Key Exchange (IKE). Request for Comments: 2409. https://tools.ietf.org/html/rfc2409
6. KCDSA task force team the korean certificate-based digital signature algorithm. http://grouper.ieee.org/groups/1363/P1363a/contributions/kcdsa1363.pdf see also ISO/IEC 15946-2
7. Nyberg, K., Rueppel, R.: A new signature scheme based on the DSA giving message recovery. In: 1st ACM Conference on Computer and Communication Security, vol. 372, pp. 58–61. ACM Press (1993)
8. Schnorr, C.P.: Efficient identification and signatures for smart cards. In: Brassard, G. (ed.) Advances in Cryptology, Crypto 1989. Lecture Notes in Computer Science, vol. 435, pp. 239–252. Springer-Verlag (1990)

Calculation of the Closed Multi-channel Queueing Systems

Yuri Ryzhikov[✉]

Institute for Informatics and Automation of the Russian Academy of Sciences,
39, 14-th Line VO, St. Petersburg 199178, Russian Federation
ryzhbox@yandex.ru

Abstract. A modification of the iterative Takahashi—Takami method is discussed to calculate the distribution of the number of requests in the *closed* multiphase queueing systems. A method is proposed for calculating the moments of waiting time distribution. The results of the calculations are compared with those obtained from a simulation model.

Keywords: Queueing theory · Multiphase systems · Iterations
Closed systems

1 Introduction

The entry of humanity into the era of information technologies resulted in a growing interest to methods of designing and evaluating the data processing systems efficiency. Nowadays the technologies in production of big integral schemas determining performance of modern computers are approaching the fundamental physical limitations. For this reason, the required performance can be achieved only through the creation of *multiprocessor and multi-machine systems*. However, the capabilities of methods belonging to the modern queueing theory are insufficient for the investigation of such systems.

An efficient method letting to "markovize" complex service systems (to determine Markov processes in these systems) is a multiphase representation of the component distributions [1–4]. Each combination of distributions (and theirs orders) and the number of channels must be described by a specific diagram. To represent the duration of a service it is convenient to apply a second-order hyperexponential distribution H_2, which allows to equalize three moments of the service time distribution and generates the state transition diagram for an n-channel system with a width of $(n + 1)$. It is accepted — see, in particular, the sources cited above — as a "folklore" (according to M. Neuts) recommendation: for distributions with the coefficient of variation $v > 1$ apply the hyperexponential law H_2, and in other cases use the Erlangian law. However, the

The research described in this paper was partially supported by state research 0073-2018-0003.

© Springer International Publishing AG, part of Springer Nature 2019
R. Silhavy (Ed.): CSOC 2018, AISC 765, pp. 125–132, 2019.
https://doi.org/10.1007/978-3-319-91192-2_14

H_2-distribution (with complex parameters) can also be applied to a case with $v < 1$, where it appreciably exceeds the Erlangian in the accuracy of approximation and sharply reduces the number of microstates of the transition diagram [5]. The calculation of the system $M/H_2/n$ with a Markovian incoming flow can be performed with the matrix-geometric progression (MGP) method [2,3,6] or with the Takahashi—Takami iterative scheme [5,7,8].

There are systems of significant practical interest, so-called *closed* queueing systems, in which the intensity of an incoming flow depends on the number of requests in the system. As an example, the nodes of specialized local computer network can serve a fixed number of requests. In this case, the MGP method is not applicable at all, and the iterative method must be refined in the aspects of the generation of transition matrices, the calculation of the microstate probabilities for the last tier, the determination of the cumulated state probabilities, and a new method should be developed for calculating the moments of waiting time distribution. Finally, a non-trivial modification of the typical simulation model is needed to verify the proposed revisions of analytical methods.

The basic scheme of the iterative method and its necessary modifications are described below.

2 Iterative Method for the Model $M/H_2/n$

The operation of a system $M/H_2/n$ can be interpreted as a process of servicing two types of requests — see the figures in [5]. The total incoming flow has intensity λ; the incoming (or selected from the queue) request with the probability y_i, $i = \overline{1,2}$, belongs to the i-th type. The choice of request's type determines the parameter of the exponential service time distribution. The "token" of microstate indicates the number of requests of each type in the channels.

We use notation S_j for the set of all possible microstates of the system under which there are exactly j requests on the service, and σ_j for the number of elements in S_j. Further, in accordance with the transition diagram for the chosen model, we construct the matrices of the infinitesimal transitions intensities:

$A_j[\sigma_j \times \sigma_{j+1}]$ — in S_{j+1} (arrival of a request),
$B_j[\sigma_j \times \sigma_{j-1}]$ — in S_{j-1} (request service completion),
$D_j[\sigma_j \times \sigma_j]$ — withdrawal from the states of the j-th tier

(the size of the matrices here and further is indicated in square brackets). The calculation of these matrices in the case of H_2 service time approximation can be programmed elementary.

We introduce the vector-lines $\gamma_j = \{\gamma_{j,1}, \gamma_{j,2}, \ldots, \gamma_{j,\sigma_j}\}$ for the probabilities of the system being in the state (j, i), $j = 0, 1, \ldots$. Now we can write the vector-matrix equations of the transitions balance as

$$\begin{aligned}
\gamma_0 D_0 &= \gamma_0 C_0 + \gamma_1 B_1, \\
\gamma_j D_j &= \gamma_{j-1} A_{j-1} + \gamma_{j+1} B_{j+1}, \qquad j = 1, 2, \ldots
\end{aligned} \tag{1}$$

The idea of solving the problem was first proposed by Takahashi and Takami [8]. Below it is presented in a more general form (see also [5]) and for the closed system — with flow intensity depending on the number of requests in the system.

Suppose $t_j = \gamma_j/p_j$, where p_j is the total probability that there are exactly j requests in the system, and denote

$$x_j = p_{j+1}/p_j, \qquad z_j = p_{j-1}/p_j. \tag{2}$$

Then the system (1) can be rewritten for the vectors of conditional probabilities $\{t_j\}$, normalized to one within the tier:

$$\begin{aligned} t_0 D_0 &= x_0 t_1 B_1, \\ t_j D_j &= z_j t_{j-1} A_{j-1} + x_j t_{j+1} B_{j+1}, \qquad j = 1, 2, \ldots \end{aligned} \tag{3}$$

Using column vectors $\mathbf{1}_j = \{1, 1, \ldots, 1\}^T$ of the size σ_j for all j, we can write the additional conditions supplementing system (3): normalization in the tier

$$t_j \mathbf{1}_j = 1 \tag{4}$$

and the balance of the total intensities of the transitions between adjacent tiers

$$z_j t_{j-1} A_{j-1} \mathbf{1}_j = t_j B_j \mathbf{1}_{j-1}. \tag{5}$$

The algorithm for calculating the set of vectors $\{t_j\}$ and numbers $\{x_j\}$ and $\{z_j\}$, satisfying the relations (3)–(5) is based on the successive approximation to required characteristics for a limited set of indices $j = \overline{0, N}$ and is essentially a block variant of the known iterative Gauss-Zeidel method.

We rewrite the equations of the system (3) for $j \geq 1$ in the form

$$t_j^{(m)} D_j = z_j^{(m)} t_{j-1}^{(m)} A_{j-1} + x_j^{(m)} t_{j+1}^{(m-1)} B_{j+1}, \qquad j = 1, 2, \ldots,$$

where the superscript indicates the iteration number. Now it is clear that

$$t_j^{(m)} = z_j^{(m)} \beta_j' + x_j^{(m)} \beta_j'', \tag{6}$$

where

$$\begin{aligned} \beta_j' &= t_{j-1}^{(m)} \, A_{j-1} D_j^{-1}, \\ \beta_j'' &= t_{j+1}^{(m-1)} \, B_{j+1} D_j^{-1}. \end{aligned} \tag{7}$$

It remains to find a way to calculate $\{z_j^{(m)}\}$ and $\{x_j^{(m)}\}$. We rewrite (5) taking into account (6):

$$(z_j^{(m)} \beta_j' + x_j^{(m)} \beta_j'') B_j \mathbf{1}_{j-1} = z_j^{(m)} t_{j-1}^{(m)} A_{j-1} \mathbf{1}_j.$$

Hence we have the proportionality

$$z_j^{(m)} = c x_j^{(m)} \tag{8}$$

with coefficient

$$c = \frac{\beta_j'' B_j \mathbf{1}_{j-1}}{t_{j-1}^{(m)} A_{j-1} \mathbf{1}_j - \beta_j' B_j \mathbf{1}_{j-1}}. \tag{9}$$

In this and subsequent formulas, the products of the transition matrices by the vector $\mathbf{1}_j$ are equal to the sums of rows of the corresponding matrices and can be computed preliminary. In particular, $t_{j-1}^{(m)} A_{j-1} \mathbf{1}_j = \lambda_{j-1}$.

Substitution (8) in (6) and multiplying both parts of the result by $\mathbf{1}_j$ gives

$$1 = t_j^{(m)} \mathbf{1}_j = x_j^{(m)} (c\beta_j' + \beta_j'') \mathbf{1}_j.$$

So,

$$x_j^{(m)} = 1/[(c\beta_j' + \beta_j'') \mathbf{1}_j]. \tag{10}$$

It is believed that in *closed* systems the intensities $\{\lambda_j\}$ of the incoming flow decrease by j, and when $j = N$ the flow of requests is interrupted. These circumstances must be taken into account in the calculation of the above-mentioned matrices of transition intensities. Because of the varying flow intensity, the stabilization of transition matrices $\{A_j\}$ and $\{D_j\}$ and, respectively, the vectors of conditional probabilities and relations $\{x_j\}$ at $j \to \infty$ has not occurred yet. In addition, because of the exhaustion of the source of requests, the calculation scheme here at $j = N$ is closed by the equation

$$t_N = z_N t_{N-1} A_{N-1} D_N^{-1} = x_{N-1}^{-1} t_{N-1} A_{N-1} D_N^{-1}. \tag{11}$$

A convenient criterion for terminating iterations is

$$\max_j |x_j^{(m)} - x_j^{(m-1)}| \le \varepsilon.$$

The starting conditional distributions of the number of served requests of each type from relatively obvious considerations are conveniently considered binomial with probabilities proportional to $\{y_i/\mu_i\}$.

After the termination of iterations we can pass to a search for cumulative probabilities. First of all, we note that the definition of the numbers $\{x_j\}$ implies the equalities

$$p_{j+1} = p_j x_j, \qquad j = \overline{0, N-1}. \tag{12}$$

Accepting for the beginning $p(0) = 1$, it is possible with the help of (12) recursively calculate other probabilities, and then normalize them to unity.

3 Waiting Time Distribution

The greatest difficulties in calculating *closed* systems are associated with the determination of their temporal characteristics. The necessary moments $\{w_k\}$

of the waiting time distribution for an *open* system are calculated using the formulas of Brumelle

$$w_k = q_{[k]}/\lambda^k, \quad k = 1, 2, \ldots, \tag{13}$$

where $\{q_{[k]}\}$ are the factorial moments of the queue length distribution. The derivation of the formulas (13) is based on the well-known theorem of PASTA — Poisson Arrivals See Time Averages), which is not applicable to the closed systems. Here the general formula (13) does not work *in the principle* because of the dependence of the flow intensity on the number of requests in the system. When the mean intensity of the flow is substituted in (13), according to the Little formula, only the first moment is obtained correctly. However, to calculate the distributions of waiting and total sojourn times, the *higher* moments of waiting time distribution are also necessary.

Since waiting of a request starts from the moment it is queued, its calculation should be based on the distribution $\{\pi_k\}$ of the number of requests before the arrival of the next one. This distribution can be obtained according to

$$\pi_k = \lambda_k p_k \bigg/ \sum_{i=0}^{N-1} \lambda_i p_i, \quad k = \overline{0, N-1}. \tag{14}$$

The further course of the calculation of waiting time distribution is based on *the convolution method*, which can be efficiently realized via the Laplace—Stieltjes transform (LST). The new request arriving in the state $k > n$ must wait for the nearest completion of the current service in one of the channels, and then for completion of $(k - n)$ more requests from the queue. Note that in the model under discussion the *service intensity matrices* at $j > n$ do not depend on the number of requests in the system.

Let us calculate

- row vectors $g_k = \pi_k * t_k$ of microstate probabilities, $k = \overline{0, N-1}$,
- diagonal matrices of total intensities for n-channel service with elements $\{\sigma_i\}$,
- diagonal matrices $\{U(s)\}$ of the LST of service time distributions according to the corresponding total service intensities on the n-th and subsequent tiers with elements $\{\sigma_i/(\sigma_i + s)\}$, $i = \overline{1, n+1}$;
- the product $\tilde{U}(s)$ of a matrix $U(s)$ by a unit vector-column of the corresponding dimension,
- matrix T with elements $\{b_{n,i,j}/\sigma_i\}$ of transition probabilities to the overlying tier.

It is not difficult to see that the LST of waiting time distribution for the n-th tier can be obtained immediately:

$$\omega_n(s) = g_n \tilde{U}(s).$$

For the $(n + 1)$-th tier, we have

$$\omega_{n+1}(s) = g_{n+1}[U(s)T]\tilde{U}(s),$$

for the $(n+2)$-th —

$$\omega_{n+2}(s) = g_{n+2}[U(s)T]^2 \tilde{U}_n(s),$$

etc. Taking into account the above rules for the formation of the factors $F(s) = U(s)T$ in these formulas, we can immediately write the matrix F as a collection of elements of the form $\{b_{n,i,j}/(\sigma_i+s)\}$. Summarizing the results for all possible starting stages, we obtain the final formula for the LST of the waiting time distribution:

$$\omega(s) = \left[\sum_{k=0}^{N-n-1} g_{n+k}F^k(s)\right]\tilde{U}(s). \qquad (15)$$

Having calculated the table of the LST values in a neighborhood of zero, we can construct its approximation by Newton's interpolation polynomial and obtain by repeatedly differentiation the moments of distribution $w(t)$ of the virtual waiting [9].

4   Features of the Reference Imitation

To test the proposed algorithm, we need a "closed" simulation model. We indicate its differences from the open-loop model:

– The procedure for generating an exponentially distributed interval to the next request must take into account the current number of requests in the system and, therefore, have an appropriate input parameter.
– Recalculation of the moment of arrival of the next request should be made with any change of their quantity in the system, including the case when the serviced request is withdrawn.
– If the system is the state with N requests, the moment of the next arrival should be "pushed to infinity".

5   The Numerical Experiment

To test the proposed approach, the programs of the simulation model and the numerical method were developed. They ran for a three-channel system with the maximum number of requests $R = 20$, the intensities of the incoming stream $\lambda_i = 0.21(R-i)$, and the H_2-distribution of the service time, which approximates the gamma-distribution with the mean $b_1 = 1$ and the coefficient of variation $v_B = 2.0$ (y_1=0.739, y_2=0.261, $\mu_1 = 3.673$, μ_2=0.327) for three moments. The simulation results (2 million accepted requests) and the implementation of the numerical method are compared in the Tables 1 and 2.

The results of simulation modeling have an instrumental error associated with the imperfection of the random-number generators, as well as statistical errors that relatively increase as the probabilities decrease and the order of the calculated moments grows. On the other hand, the method for calculating the moments of the waiting time distribution by means of numerical differentiation

Table 1. Distribution of the number of requests in the system

j	Before arriving		Stationary	
	Simulation	Calculation	Simulation	Calculation
0	2.459e-2	2.469e-2	1.596e-2	1.607e-2
1	8.959e-2	8.840e-2	6.087e-2	6.055e-2
2	1.357e-1	1.348e-1	9.811e-2	9.747e-2
3	1.097e-1	1.092e-1	8.440e-2	8.357e-2
4	9.665e-2	9.634e-2	7.905e-2	7.837e-2
5	8.872e-2	8.882e-2	7.734e-2	7.707e-2
6	8.277e-2	8.281e-2	7.702e-2	7.699e-2
7	7.641e-2	7.643e-2	7.653e-2	7.652e-2
8	6.863e-2	6.888e-2	7.465e-2	7.471e-2
9	5.980e-2	6.005e-2	7.080e-2	7.105e-2
10	4.989e-2	5.025e-2	6.503e-2	6.540e-2
11	3.953e-2	4.004e-2	5.749e-2	5.791e-2
12	2.976e-2	3.013e-2	4.833e-2	4.902e-2
13	2.094e-2	2.118e-2	3.890e-2	3.937e-2
14	1.363e-2	1.369e-2	2.959e-2	2.970e-2
15	7.860e-3	7.976e-3	2.081e-2	2.076e-2
16	4.042e-3	4.052e-3	1.303e-2	1.318e-2
17	1.688e-3	1.700e-3	7.381e-3	7.375e-3
18	5.365e-4	5.300e-4	3.328e-3	3.449e-3
19	8.800e-5	9.365e-5	1.207e-3	1.218e-3
20	0	0	2.018e-4	2.484e-4

Table 2. Moments of the waiting time distribution

Method	w_1	w_2	w_3
Simulation	1.544e0	6.125e0	3.416e1
Calculation	1.555e0	6.198e0	3.478e1

of LST table generates errors because of the errors in the finite differences of high order and the polynomial approximation of the LST in real calculations with a limited computer bit grid. Therefore, fitting of the calculated values with those obtained during simulation can be considered to be quite satisfactory.

We also note the results of the PASTA theorem testing: the probabilities of the states of the system before the arrival of the next request and after the departure of the serviced one coincided and differed markedly from the stationary probabilities (obtained by averaging over time).

6   Conclusion

Closed queueing systems are good models of data processing systems implemented in automated control systems and various purpose situational centers. The main result of this article is the matrix method for calculating the moments of waiting time distribution in a closed queueing system via sequential convolutions in Laplace transforms and subsequent numerical differentiation. With its software implementation, a non-trivial combination of multiplication operations of "full" and diagonal matrices, which were both left and right multipliers, was applied. This made it possible to significantly reduce the complexity of calculations. In conclusion, we note:

- The proposed methodology can be applied to the calculation of systems with a coefficient of variation of the distribution of service less than unity. In this case, the H_2-approximation parameters and intermediate results turn out to be complex, but the final ones are real and have a traditional probabilistic meaning.
- The intensities of requests arrival on the tiers $j = \overline{0, N-1}$ can generally be arbitrary decreasing functions of the tier number. The corresponding changes in the algorithm are trivial.
- The algorithm can be applied to calculation of closed and mixed *queueing networks* using the method of their flow-equivalent decomposition.

References

1. Vishnevsky, V.M.: Theoretical bases of designing computer networks. Technosphere, Moscow (2003). (in Russian)
2. Bhat, U.N.: An Introduction to Queueing Theory. Springer, New York (2015)
3. Bolch, G., Greiner, S., de Meer, H., Trivedi, K.: Queueing Networks and Markov Chains: Modeling and Performance Evaluation with the Computer Science Applications. Wiley, Hoboken (2006)
4. Shin,Y.W., Moon, D.H. Sensitivity and approximation of M/G/c Queue: Numerical Experiments. In: 8 th International Symposium on Operational Research and its Applications, China, pp. 140–147 (2009)
5. Ryzhikov, Y.I.: Iterative method for calculating multi-channel service systems — bases, modifications and marginal capabilities distribution (in Russian). In: Proceedings of the 9th Russian Multiconference on Management Problems. Information Technologies in Management. State Scientific Center of the Russian Federation "Concern Elektropribor", pp. 224–233, SPb (2016)
6. Ryzhikov, Y.I.: Modifications and perspectives of the matrix-geometric progression method (in Russian). Ibid., pp. 234–243
7. Seelen, L.P.: An algorithm for Ph/Ph/c queues. Eur. J. of Oper. Res. **23**, 118–127 (1986)
8. Takahashi, Y., Takami, Y.: A Numerical Method for the Steady-State Probability of a GI/G/c Queuing System in a General Class. J. Operat. Res. Soc. Jpn **19**(2), 147–157 (1976)
9. Ryzhikov, Y.I.: A package of programs for the calculation of systems with queues and its testing. In: Proceedings of the SPIIRAN. Issue 7. Science, pp. 265–284, SPb (2008) (in Russian)

MATLAB as a Tool for Modelling and Simulation of the Nonlinear System

Jiri Vojtesek$^{(\boxtimes)}$ and Lubos Spacek

Faculty of Applied Informatics, Tomas Bata University in Zlin,
Nam. T. G. Masaryka 5555, 760 01 Zlin, Czech Republic
vojtesek@utb.cz
http://www.utb.cz/fai

Abstract. This contribution describe the process that usually precedes the design of the controller - a modelling and a simulation of the observed technological system. Two types of models, physical and abstract, are discussed here. While physical model is usually small or simplified representation of the originally bigger system, the abstract (mathematical) model is description of the system in the form of linear or nonlinear ordinary or partial differential equations. Simulation of the mathematical model then means numerical solution of these equations using some mathematical software. The proposed modeling and simulation procedure is then applied on the real model of the water tank. The mathematical model was then derived and then subjected to the simulation of the steady-state and dynamic analysis.

Keywords: Modelling · Computer simulation · Numerical solving
ODE · Water tank · MATLAB · Runge-Kutta's methods

1 Introduction

It is common, that industrial processes are described only by measurements of their input and output variables. As we do not know their internal structure, we are talking about black-box systems. If we want to optimize the production or introduce new control techniques, experiments on this real system are usually very inefficient, expensive, dangerous or even unrealizable [1].

These complications could be overcome with the use of the modelling and simulation techniques [2] that are very popular nowadays with increasing computation power of computers together with the decreasing purchasing price of these computers.

If we talk about the modelling in industry, we could find two types of models real (physical) and abstract (mathematical) models [1]. Real models are small or simplified representations of the original system and we expect, that obtained results of experiments are comparable to those on the real system.

© Springer International Publishing AG, part of Springer Nature 2019
R. Silhavy (Ed.): CSOC 2018, AISC 765, pp. 133–143, 2019.
https://doi.org/10.1007/978-3-319-91192-2_15

On the other hand, a mathematical model of the system is usually one or the set of ordinary differential equations (ODE) for systems with lumped parameters (major variety of systems - stirred tank reactors, water tanks etc.) or the set of partial differential equations (PDE) for systems with distributed parameters (e.g. tubular chemical reactors, heat exchangers etc.). This mathematical model is then solved numerically using Euler's, Taylor's methods [4] or their multistep modifications like Runge-Kutta's methods [5] or Predictor-Corrector method [9]. This solving is in fact simulation of the dynamic behaviour of the system.

There is various mathematical software that can be used for computer simulation. We can mention for example Wolfram's Mathematica, MathWorks's MATLAB, GNU Octave, Scilab etc. The most of these software has build-in functions for numerical solving of ODEs and PDEs [6,8]. The used here is MATLAB which is widely used in academic sphere. Another advantage of the MATLAB is that it could be also connected to the real systems or real models via technological cards. This feature offers verification of the simulations on the real system and application of control strategies on real systems. MATLAB also provides export to C and C++ language through so called MATLAB coder [7]. You can obtain standalone programs with this export and run them on computers which do not have MATLAB installed.

The practical part of this contribution combines both techniques we will introduce the mathematical model of the water tank that is a physical model of the originally big water reservoir.

2   Modelling

It was already mentioned, that the mathematical modelling is very popular nowadays in computer era. As the modelling and simulation procedure has several parts, the schematic representation can be found on Fig. 1.

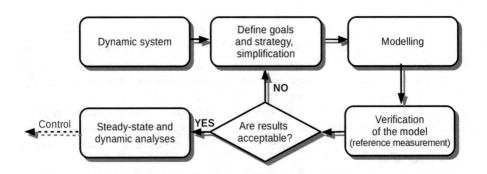

Fig. 1. General simulation procedure

The definition of goals and strategy what we want to achieve is the first step of the modelling procedure. In this part, the most important variables and connections between them are defined. The procedure than continues with the building of the mathematical description of the process [2] in the form of mathematical equations (algebraic, differential etc.).

This mathematical description of all variables and relations between them is usually very complex and hard to solve, that is why we must introduce simplifications that reduce the complexity of the system. Some simplification could be, for example, that some variables are constant during the simulation etc. Obtained mathematical model is then verified by measurements on the real system. If simulated results are different from those obtained on the real system or model, we must go back to the modelling part and rebuild the mathematical that do not reflect reality. The most common problem is that we introduce too much simplifications or do not take into the account another important variables etc.

3   Computer Simulation

Once the mathematical model is acceptable, we can make simulation experiments on it. The meaning of these simulations is the numerical computation of the mathematical model in specified conditions. The mathematical package MATLAB from the MathWorks, Inc. located in United States was used for this task in our case. Moreover advantages already mentioned, the MATLAB has also great visualization package called Simulink that adds graphical simulation tools and modeling tools for design of the dynamic models and embedded systems.

Suggest, that observed system belongs to the class of lumped-parameters systems, the mathematical model of this system in the form could be generally described by equation

$$\frac{dy}{dt} = f(t, y) \tag{1}$$

where y is output variable and t denotes time, f is function that could be for example ODE.

The simulation usually consists of steady-state and dynamic analyses. Both analyses are described in following sections.

3.1   Steady-State Analysis

This analysis observes the behaviour of the system in the stable state, usually called *the steady-state*. If the state variable is time t, the steady-state is in $t \to \infty$ and $dy/dt = 0$. The mathematical model (1) is then transformed to the set of algebraic linear or nonlinear functions. This set could be then solved by methods like matrix-inversion, Gauss elimination etc. for simple models or iterative methods for more complicated systems that do not have implicit solution.

All methods employs in the MATLAB basic commands, especially commands for cycles - for, if etc. More can be found in the practical example.

3.2   Dynamic Analysis

The simulation of the system's dynamics is in fact numerical solving of the one or the set of ODE (1). The steady-state analysis precedes the dynamic analysis as the steady-state values are initial values of state variables. There are several methods for the numerical solving of ODEs and Runge-Kutta's methods belongs to the famous ones. Runge-Kutta's methods uses Taylor's series and differs in the order - the second-order, the fourth order etc. Advantage of these methods is that they are simple and easily programmable. For example, the fourth-order Runge-Kutta's method uses first four parts of the Taylor's series rewritten in form:

$$y_{i+1} = y_i + \frac{1}{6} \cdot (k_1 + 2k_2 + 2k_3 + k_4) \qquad (2)$$

where variables k_{1-4} are computed from

$$
\begin{aligned}
k_1 &= h \cdot f\left(x_i, y_i\right) \\
k_2 &= h \cdot f\left(x_i + \tfrac{1}{2}h, y_i + \tfrac{1}{2}k_1\right) \\
k_3 &= h \cdot f\left(x_i + \tfrac{1}{2}h, y_i + \tfrac{1}{2}k_2\right) \\
k_4 &= h \cdot f\left(x_i + h, y_i + k_3\right)
\end{aligned}
\qquad (3)
$$

The main benefit of this method is that it is build-in function in various mathematical software like Mathematica, MATLAB etc. MATLAB, which is in our consideration, has function ode45 that is in fact solution of the set (2) and (3). Similarly, ode23 represents the second-order Runge-Kutta's method etc.

The simulation of the dynamics means, in fact, transformation of the mathematical into the set (1) and crate from it sub-function that is then called as a argument from function ode45 or ode23 - see next section.

4   Simulation Example

The procedure described in the previous part will be now applied on the real system which is in our case represented by the water tank model as part of the Process Control Teaching system PCT40 from the Armfield [10] displayed on the Fig. 2(a).

This model is great tool for the teaching of the process dynamics, identification and control. It includes several example processes a heat exchanger, a Continuous Stirred-Tank Reactor (CSTR) and a water tank which are interconnectable to each other.

The system under our attention, the water tank, offers three possible control exercises (I) the on/off control of the water level in the tank, (II) the differential water level control between the minimal and the maximal water level and the most sophisticated (III) water level control through the whole height of the water tank. This last option (III) was used in this work.

The schematic representation of this system is shown in Fig. 2(b). This water tank model is small representation of the real system that could be water reservoir or dam. Its construction consists of two cylinders an outer and an inner.

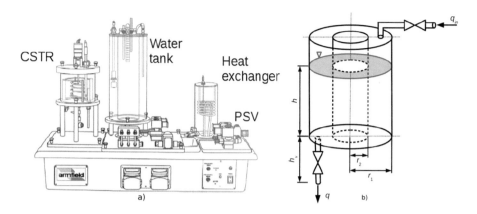

Fig. 2. (a) Process Control Teaching system PCT40, (b) Schematic representation of the water tank

This feature ensures quicker dynamics with reduced need for the water supply as we do not need to fill full volume of the water tank.

The water is fed inside the tank with the measured and adjustable input flow rate q_{in} and comes out with the output flow rate q. An input (control) variable in this system is then input flow rate q_{in} and the controlled variable is the level in the tank h. The output variable is measured with the pressure sensor in the base of the water tank.

The mathematical model of this system could be constructed with the use of the material balance in the general form

$$\boxed{\begin{array}{c}\text{Flow rate}\\\text{into the system}\end{array}} = \boxed{\begin{array}{c}\text{Flow rate}\\\text{out of the system}\end{array}} + \boxed{\begin{array}{c}\text{Rate of}\\\text{the accumulation}\end{array}}$$

which is in this example mathematically

$$q_{in} = q + \frac{dV}{dt} \tag{4}$$

where t is time and V denotes the volume inside the tank that is computed from the area of the basement F and the level of the water in the tank h, i.e. $V = F \cdot h$. If F is a constant, the resulting mathematical model is then

$$q_{in} = q + F \cdot \frac{dh}{dt} \tag{5}$$

Radii of input and output cylinders are in this model constant, $r_1 = 8.7\,\text{cm}$, $r_2 = 5.7\,\text{cm}$, and the area of the basement is then also a constant value computed from

$$F = \pi \cdot r_1^2 + \pi \cdot r_2^2 = 135.72 \cdot 10^{-2}\ \text{cm}^2 \tag{6}$$

As the water comes out from the water tank through the valve in the basement, the output flow rate q depends on the water level via relation

$$q = k \cdot \sqrt{h} \tag{7}$$

where k denotes valve constant and its value is specific for each type of the valve. We assume, that k is constant due to simplification.

Resulting mathematical model has form

$$\frac{dh}{dt} = \frac{q_{in} - k \cdot \sqrt{h}}{F} \tag{8}$$

and as the basement area is known constant from (6), the last unknown constant in (8) is the valve constant k which could be computed for example with the use of the steady-state analysis. The steady-state in this case means that the output flow rate q is equal to input flow rate q_{in} and the level of the water in the tank is then constant (h^s). We can say, that the rate of the accumulation is then zero and (8) could be rewritten to:

$$q^s = k \cdot \sqrt{h^s} \Rightarrow k = \frac{q^s}{\sqrt{h^s}} \tag{9}$$

where index $(\cdot)^s$ denotes steady-state values.

The Proportional Solenoid Valve (PSV) sets an input flow rate in the range 0–100% that is equal to 0–1500 ml $\cdot$ min^{-1}. Our previous experiments on the real model have shown that input setting 60% produces flow rate $q_{in} = 1008$ ml$\cdot$min^{-1} and the steady-state water level is $h^s = 19.5$ cm. The valve constant k is then

$$k = \frac{q_{in}^s}{\sqrt{h^s}} = \frac{1008}{\sqrt{19.5}} = 228.2653 \text{ cm}^{5/2} \cdot \text{min}^{-1} \tag{10}$$

Very important step in the procedure displayed in Fig. 1 is the verification of the obtained model. We have made a reference simulation of the dynamic behavior after the step change of the input variable q_{in} from 0–1008 ml $\cdot$ min^{-1}(60%) and the same step change was done on the real model of the water tank. Obtained simulated and measured data are displayed in Fig. 3.

As we can see, the steady-state water level inside the tank was 19.5 cm for both experiments, but the course of the output variable for the simulated data (black dotted line) has different dynamics it reaches the final value quicker than in real experiment (blue solid line). We can say, that proposed mathematical model is not reliable and we must go back to the modelling part and rebuild the model.

The problem in this case is with the output valve has height $h_v = 7.6$ cm which was not taken into the consideration and this height also affect the output flow rate. If we consider this height and add it to the measured water level in the steady-state, the new value of the valve constant k_n is

$$k_n = \frac{q_{in}^s}{\sqrt{h^s + h_v}} = \frac{1008}{\sqrt{(19.5 + 7.6)}} = 193.63 \text{ cm}^{5/2} \cdot \text{min}^{-1} \tag{11}$$

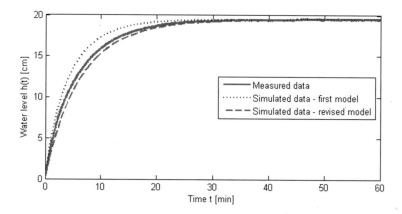

Fig. 3. Simulated data for firstly derived, revised model and measured data on the real model

We have made again reference simulation and the comparison of the simulated and measured data are shown again in Fig. 3. Resulted simulated data for the revised model (red dashed line) are much closer the measured data on real model (blue solid line).

Now we can say, that the mathematical model (8) with the revised valve constant k_n from (11) is reliable and we can continue to the simulation part.

It is already mentioned, that the MATLAB environment could be used for the simulation of both steady-state and dynamics of the system that is presented in the next sections.

4.1   Simulation of the Steady-State

The steady-state analysis is generally described in Sect. 3.1. In this case, we observe the steady-state value of the water level h^s for $q = q_{in}$, i.e. $dh/dt = 0$ in (8) and this ODE is then transformed to the nonlinear algebraic equation

$$h^s\left(q_{in}\right) = \left(\frac{q_{in}}{k_n}\right)^2 \tag{12}$$

The optional value in (12) for the steady-state analysis is an input volumetric flow rate q_{in} and we have made simulation of the steady-state analysis for the volumetric flow rate $q_{in} = <0; 1500> \text{ml.min}^{-1}$ which is also the physical range of the flow rate that could come through the PSV.

The steady-state analysis in MATLAB is then very simple and can be used with the `for` - cycle - see code below.

```
% Steady-state analysis
qmin = 0;          % minimal flow rate [cm^3.min^-1]
qmax = 1500;       % maximal flow rate [cm^3.min^-1]
qst = 100;         % computational step [cm^3.min^-1]
qin = qmin:qst:qmax;  % vector of input flow rate for computation

% Constants of the model
k = 193.63;        % valve constant [cm^5/2.min^-1]
r1 = 8.7;          % outer radius [cm]
r2 = 5.7;          % inner radius [cm]
F = pi*(r1^2) - pi*(r2^2);  % area of the basement [cm2]
% Computation of the steady-state values of h
for i=1:1:length(q)
   h(i) = (qin(i)/k)^2;
end
hv = 7.6;          % height of the valve [cm]
hs = h-hv;         % reduced water level [cm]
figure
plot(q,hs)
```

User can set following parameters of the simulation (constants at the beginning of the M-file):

- Minimal input flow rate qmin,
- maximal input flow rate qmax,
- computation step qst.

Results of the steady-state analysis for $q_{in} = <0; 1500> \text{ml} \cdot \text{min}^{-1}$ with the step $100 \text{ml} \cdot \text{min}^{-1}$ are shown in Fig. 4.

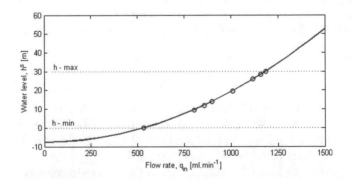

Fig. 4. Results of the steady-state analysis - simulated data (blue line) and measured data (red circles)

Figure 4 also shows measured steady-state values of the water level for several flow rates (red circles) that confirmed reliability of the mathematical model (8). Results also show expected nonlinearity of the system. Minimal and maximal physical values of the water level h are in this model 0 and 30 cm respectively (black dotted line in Fig. 4). We can then say, that from the physical point of view the flow rate could be limited to the range $q_{in} = <531.36; 1184.7>$ ml.min^{-1} because input flow rate lower than 531.36 ml.min^{-1} does not fill minimal level of the water in the water tank and bigger flow rate than 1184.7 ml.min^{-1} on the other hand cause that water overflows from the water tank.

4.2   Simulation of the Dynamics

The mathematical meaning of the dynamic analysis is the numerical solution of the ODE (8). The MATLAB build-in function ode45 that is in fact the fourth Runge-Kutta's method was used here.

The dynamic analysis in MATLAB can be done in two M-files. The first file defines working point, simulation parameters and call numerical solution of the mathematical model (8) with the function ode45. The mathematical model (8) itself is described in subfunction *wtank.m*.

Matlab's code for computation of the dynamics - wtankDyn.m

```
global u F kn
% Dynamic analysis
qin = 900;       % flow rate [ml.min^-1] - working point
% Simulation parameters
t0 = 0;     % starting time [min]
th = 60;    % final time [min]
up = 20;    % step change in [%]
u = qin + (up*qin/100);  % real step change [ml/min]
kn = 193.2025;  % valve constant [cm^5/2.min^-1]
r1 = 8.7;       % outer radius [cm]
r2 = 5.7;       % inner radius [cm]
F = pi*(r1^2) - pi*(r2^2);  % area of the basement [cm2]
hv = 7.6;   % height of the valve [cm]
h0 = (qin/kn)^2;  % initial condition (steady-state value) [cm]
% numerical solution
[t,h] = ode45(@wtank,[t0 th],h0);
h = (h-h0);  % reduced water level
plot(t,h)    % display graphs
```

Matlab's subfunction that describes system - wtank.m

```
function dh = wtank(t,h)
% function WTANK - mathematical model of water tank
global u kn F
dh = (1/F)*(u - (kn*sqrt(h)));
```

We can see that both M-files shares constants **F** and **kn** via global variables using **global** command together with actual flow rate q_{in} in the variable **u**.

Presented MATLAB program offers the computation of the dynamics around the working point defined by the input flow rate q_{in} (variable **qin**). The behaviour is observed for the step changes of the input flow rate q_{in} from its steady-state in %, i.e.

$$u = q_{in} + \frac{u_p \cdot q_{in}}{100} \quad [\%] \tag{13}$$

User can again affect the simulation by the choice of following parameters:

- Input volumetric flow rate (working point) **qin**,
- starting (**t0**) and final (**th**) time of the simulation,
- step change of q_{in} - variable **u** computed from (13).

An example results of the dynamic analysis was done for various step changes u for the range $u = < -50\%; +35\% >$ and results are shown in Fig. 5.

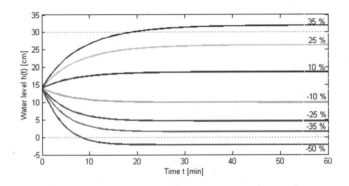

Fig. 5. Simulation results of the dynamic analysis

This analysis can help us with the understanding of the system's behaviour. We can for example say that:

- The input variable u_p can be limited in the range $u_p = < -42; 31.5 > \%$ from the physical point of view,
- system has nonlinear behaviour which reflects steady-state values that are different for negative and positive step changes,
- output variable $h(t)$ could be from the control point of view described by the first or the second order transfer function.

5   Conclusion

The contribution shows applicability of the mathematical software MATLAB in the modelling and simulation of the technological process. The main advantage of the computer simulation can be found in the time and cost savings. Once we have mathematical model transformed to the MATLAB program, we can perform thousands of simulations for testing of different settings. Resulted overview of the system's behaviour can help us with the choice of the appropriate control strategy and its setting.

Proposed modelling and simulation procedures were applied on the real model of the water tank as a part of the multifunctional process control teaching system PCT40. This model offers various simulation exercises, the one chosen in our work was focused on the behaviour of the water level in the tank that depends on the input and output flow rate. The mathematical model of this system was described by the one nonlinear ODE that was then solved numerically to obtain steady-state and dynamic behaviour of the system. Proposed mathematical model was verified by the measurements on the real system which is very important from the reliability on the mathematical model point of view.

This contribution shows not only suitable methods for numerical solving but also MATLAB codes that uses these methods using standard commands and build-in functions. Results of the steady-state and dynamic analyses were presented. These analyses confirm expected nonlinear behaviour of this model and the next step will be the design of the suitable controller for the controlling of the water level in the tank.

Acknowledgment. This article was created with support of the Ministry of Education of the Czech Republic under grant IGA reg. n. IGA/CebiaTech/2018/002.

References

1. Ingham, J., Dunn, I.J., Heinzle, E., Penosil, J.E.: Chemical Engineering Dynamics: An Introduction to Modelling and Computer Simulation, Second Completely Revised Edition. VCH Verlagsgesellshaft, Weinheim (2000)
2. Maria, A.: Introduction to modeling and simulation. In: Proceedings of the 1997 Winter Simulation Conference, pp. 7–13 (1997)
3. Saad, Y.: Iterative Methods for Sparse Linear Systems. Society for Industrial and Applied Mathematics, Philadelphia (2003)
4. Johnston, R.L.: Numerical Methods. Wiley, New York (1982)
5. Kaw, K., Nguyen, C., Snyder, L.: Holistic numerical methods. http://mathforcollege.com/nm/
6. Mathews, J.H., Fink, K.K.: Numerical Methods Using Matlab. Prentice-Hall, Englewood Cliffs (2004)
7. Matlab coder - Generate C and C++ code from MATLAB code. https://www.mathworks.com/products/matlab-coder.html
8. Advanced Numerical Differential Equation Solving in Mathematica. Webpages of Wolframs Mathematica. http://reference.wolfram.com/mathematica/tutorial/NDSolveOverview.html
9. Arfken, G.B., Weber, H.-J., Harris, F.E.: Mathematical Methods for Physicists: A Comprehensive Guide, 7th edn. Elsevier, Boston (2013). ISBN 978-0123846549
10. Armfield: Instruction manual PCT40, Issue 4, February 2005

An Input to State Stability Approach for Evaluation of Nonlinear Control Loops with Linear Plant Model

Peter Benes[1(✉)] and Ivo Bukovsky[2]

[1] Department of Instrumentation and Control Engineering, Czech Technical University in Prague, Prague, Czech Republic
petermark.benes@fs.cvut.cz
[2] Department of Mechanics, Biomechanics and Mechatronics, Center of Advanced Aerospace Technology, Czech Technical University in Prague, Prague, Czech Republic
ivo.bukovsky@fs.cvut.cz

Abstract. This paper introduces a novel ISS stability evaluation for a LNU based HONU-MRAC control loop where an LNU serves as a plant and a HONU as a non-linear polynomial feedback controller. Till now, LNUs have proven their advantages as computationally efficient and effective approximators, further optimisers of linear and weakly non-linear dynamic systems. Due to the fundamental construction of an HONU-MRAC control loop featuring analogies with discrete-time non-linear dynamic models, two novel state space representations of the whole LNU based HONU-MRAC control loop are presented. Backboned by the presented state space forms, the ISS stability evaluation is derived and verified with theories of bounded-input-bounded-state (BIBS) and Lyapunov stability on a practical non-linear system example.

Keywords: Bounded-Input-Bounded-Output (BIBO)
Bounded-Input-Bounded-State (BIBS) · Higher-Order Neural Units (HONUs)
Input-to-State Stability (ISS) · Higher Order Neural Unit (HONU)
Linear Neural Unit (LNU) · Model Reference Adaptive Control (MRAC)

1 Introduction

Higher order neural units (HONUs) as fundamental state feedback controllers, termed more explicitly as (HONU-MRAC) have proven their advantages in terms of computational efficiency and control performance for practical non-linear system applications. Several works to mention are [1, 2] where HONUs are presented as standalone feedback controllers. Further [3, 4] where HONU-MRAC design is used as an extension for control loop optimisation. Although a large focus in research has been put towards the development of applied theory, studies in the direction of adaptive model and whole HONU-MRAC control loop stability have not been readily published. In control loop design, stability is a key topic in guaranteeing the applicability and further the robustness of the derived closed control loop. Therefore, this paper aims to present an extended stability analysis approach from [5] aimed on HONU models, to linear neural unit

© Springer International Publishing AG, part of Springer Nature 2019
R. Silhavy (Ed.): CSOC 2018, AISC 765, pp. 144–154, 2019.
https://doi.org/10.1007/978-3-319-91192-2_16

models and their extension with up to second order feedback controllers. When considering stability in the sense of adaptive control, two clear branches may be distinguished. The first is stability of the applied learning algorithm, also referred to as adaptation stability and the second being the stability of the whole derived control loop. Several key works focussed on efficient and justified stable convergence of neural network based models may be found in [6–8]. A further recent study focussed on learning stability of adaptive HONU models [9] presents a spectral radius based criterion derived from the fundamental weight update rule. However, such approaches can be rather computationally expensive during online adaptive tuning, especially in the sense of short control sampling intervals. Furthermore, for certain engineering processes it may be sufficient to reapply existing control loops parameters, which thus leads to an open area in justification whole control loop stability.

A rather readily researched topic in the sense of recurrent neural network (RNN) stability is in the assessment of global asymptotic stability (GAS). In [10] a methodology advantageous for non-linear systems with bounded non-linearities is derived with focus towards recurrent neural networks. In [11] global robust exponential stability of the equilibrium point for a class of delayed neural networks is investigated. Further, concerning RNNs another quite readily researched methodology is that of Linear Matrix Inequality (LMI). Such studies include [12] and with focus on determining stability of both constant and time-varying delayed neural networks in [13]. In practical control applications however, though stability of the equilibrium point may be justified, it is often necessary and more practical to ensure boundedness of the resulting process states with respect to bounded control inputs. Therefore, the scope of this paper deals with the study of bounded-input-bounded-output BIBO/BIBS stability [14, 15]. And further in [16–19], a more universal definition leading from the concept BIBS are presented in the form of Input-to-State-Stability (ISS) study. With regards to recent applications, in [20] the Lyapunov-Krasovskii functional method and LMI concepts are presented for application to multiple time-varying delay RNNs, to guarantee exponential ISS. Further in [21], a class of conventional perception based dynamic neural networks with external disturbances is studied. There, a robust weight learning algorithm is derived based on the principles of an applied ISS condition.

Following review of the mentioned works our paper is structured as follows. In Sect. 2 the fundamental architecture of LNUs with a feedback controller is recalled, as well as several efficient learning algorithms of HONUs. In Sect. 3 the main results of this paper are presented. In Sect. 3.1, an analogy is introduced between the polynomial structures of a LNU with extension of a HONU controller to linear discrete time nonlinear state-space systems. Further, a BIBO stability condition is derived for evaluation of LNU model stability about a localised equilibrium point. In Sect. 3.2, an extension of the work [5] is presented for stability evaluation of a whole LNU based HONU-MRAC control loop. Here, different to the works [21, 22] a straightforward and practical BIBS condition is derived via decomposition of the fundamental HONU-MRAC architectures to its principle input vector. Its application and validity is justified via a practical non-linear system control example, with connotations to real industrial process application.

2 Fundamentals of LNUs and Their Control Loop

As the main focus of this paper is the stability evaluation of the LNU based HONU-MRAC control loop, as an initial a LNU (HONU, $r = 1$) may be described via the following fundamental relation

$$\tilde{y} = \sum_{i=0}^{n_x} \mathbf{W}_i x_i = \mathbf{w} \cdot col^{r=1}(\mathbf{x}), \tag{1}$$

where for a QNU (HONU, $r = 2$) the neural unit model may be described as

$$\tilde{y} = \sum_{i=0}^{n_x} \sum_{j=i}^{n_x} \mathbf{W}_{i,j} x_i x_j = \mathbf{w} \cdot col^{r=2}(\mathbf{x}). \tag{2}$$

$\mathbf{W}$ is a multidimensional array of neural weights and $\mathbf{w}$ is its flattened representation. The long column vector form $col^{r=1}(\mathbf{x})$ comprises of previous step-delayed outputs $\tilde{y}$ and further previous process inputs u more explicitly expressed as

$$col^{r=1}(\mathbf{x}) = \{x_i; i = 0..n_x\}, \tag{3}$$

where for a QNU (HONU, $r = 2$) it yields

$$col^{r=2}(\mathbf{x}) = \{x_i x_j; i = 0..n_x, j = i..n_x\}. \tag{4}$$

With regards to weight adaptation, a fundamental update rule for sample-by-sample weight updates is the gradient descent rule (GD) [9]. For a LNU it yields

$$\Delta \mathbf{w} = -\frac{\mu}{2} \cdot \frac{\partial e^2(k)}{\partial \mathbf{w}} = -\mu \cdot e(k) \cdot \frac{\partial(y_r(k) - \tilde{y}(k))}{\partial \mathbf{w}} = \mu \cdot e(k) \cdot \frac{\partial \tilde{y}(k)}{\partial \mathbf{w}}, \tag{5}$$

where μ is the learning rate and the error is given by $e(k) = y_r(k) - \tilde{y}(k)$. As a further extension, the learning rate may be normalised yielding the normalised gradient descent rule (NGD) as follows

$$\Delta \mathbf{w} = \frac{\mu}{\|col^{r=1}(\mathbf{x})\|_2^2 + 1} \cdot e(k) \cdot \frac{\partial \tilde{y}(k)}{\partial \mathbf{w}}. \tag{6}$$

A further incremental learning algorithm which may be employed is the recursive least squares algorithm (RLS). In the sense of a LNU it may be given as

$$\Delta \mathbf{w} = e(k) \cdot col^{r=1}(\mathbf{x})^T \cdot \mathbf{R}^{-1}(k), \tag{7}$$

where the correlation matrix $\mathbf{R}^{-1}(k)$ is determined via

$$\mathbf{R}^{-1}(k) = \frac{1}{\mu} \cdot \left(\mathbf{R}^{-1}(k-1) - \frac{\mathbf{R}^{-1}(k-1) \cdot (\partial \tilde{y}(k)/\partial \mathbf{w}) \cdot (\partial \tilde{y}(k)/\partial \mathbf{w})^T \cdot \mathbf{R}^{-1}(k-1)}{\mu + (\partial \tilde{y}(k)/\partial \mathbf{w})^T \cdot \mathbf{R}^{-1}(k-1) \cdot (\partial \tilde{y}(k)/\partial \mathbf{w})} \right). \tag{8}$$

Further, the correlation matrix may be initialised as $\mathbf{R}(0) = \frac{1}{\delta}\mathbf{I}$. It therefore results that the general weight update rule for the presented algorithms is given as

$$\mathbf{w}(k) = \mathbf{w}(k-1) + \Delta\mathbf{w}. \tag{9}$$

With regards to the extension of an LNU model with a HONU feedback controller, a similar weight update rule maybe employed for LNU architecture as

$$\Delta\mathbf{v} = -\frac{\mu}{2} \cdot \frac{\partial e^2_{ref}(k)}{\partial\mathbf{v}} = -\mu \cdot e_{ref}(k) \cdot \frac{\partial(y_{ref}(k) - \tilde{y}(k))}{\partial\mathbf{v}} = \mu \cdot e_{ref}(k) \cdot \mathbf{w}.\frac{\partial col^{r=1}(\mathbf{x})}{\partial\mathbf{v}}, \tag{10}$$

where the error between the reference model and LNU model output is defined as $e_{ref}(k) = y_{ref}(k) - \tilde{y}(k)$. Analogical to (9), the general weight update rule for the HONU feedback control yields to be

$$\mathbf{v}(k) = \mathbf{v}(k-1) + \Delta\mathbf{v}. \tag{11}$$

Similarly the control loop may be extended with an adaptive feedback gain. Following the same fundamental rule via GD method where now the partial derivative is with respect to the adaptive gain parameter r_o, the following update rule yields

$$r_o(k) = r_o(k-1) + \Delta r_o = r_o(k-1) + \mu_{r_o} \cdot e_{ref}(k) \cdot \mathbf{w} \cdot \frac{\partial \mathbf{colx}}{\partial r_o}. \tag{12}$$

In a similar manner to the LNU model, weight update rule (8) may be applied to (11)–(12) respectively. Therefore the final control law for a whole LNU-LNU adaptive control loop is given as

$$u = d - r_o \sum_{i=0}^{n_\xi} \mathbf{V}_i\xi_i = d - r_o \cdot \mathbf{v} \cdot col^{r=1}(\xi), \tag{13}$$

and with extension of a QNU controller, the control law yields

$$u = d - r_o \sum_{i=0}^{n_\xi}\sum_{j=i}^{n_\xi} \mathbf{V}_{i,j}\xi_i\xi_j = d - r_o \cdot \mathbf{v} \cdot col^{r=2}(\xi). \tag{14}$$

3   Main Results

Following overview of the fundamental architectures and control law of a LNU based HONU-MRAC adaptive control loop, an analogy to discrete-time non-linear dynamic systems can be drawn. This section, hence introduces two new reformulations of a classical LNU model with extension of a HONU feedback controller to a state-space matrix representation. Utilising the newly derived non-linear control loop state space forms with LNU, a bounded input bounded output (BIBO) and further input-to-state (ISS) stability approach is constructed in the proceeding section.

3.1 State-Space Representation of Non-linear Control Loop with LNU

For development of this approach, let us consider the relation (1), with extension of a HONU feedback controller via the control law (13)–(14)

$$
\begin{aligned}
\tilde{y}(k, \mathbf{w}, \mathbf{v}, r_o \bar{x}_i) = \sum_{i=0}^{n_x} w_i x_i &= w_0 + w_1 \tilde{y}(k-1) + w_2 \tilde{y}(k-2) + .. + w_n \tilde{y}(k - y_n) \\
&+ w_{n_y+1}[d(k-1) - r_o q(k-1)] + w_{n_y+2}[d(k-2) - r_o q(k-2)] + .. \\
&+ w_{n_x}[d(k-n_u) - r_o q(k-n_u)]
\end{aligned}
\tag{15}
$$

where q denotes the HONU controller output of a general polynomial order. Therefore, via an affine control formulation such extended LNU polynomial architecture may be expressed via choice of the following state vector

$$
\bar{\mathbf{x}}(k) = [\tilde{y}(k - (n_y + n_u - 1)), .., \tilde{y}(k-1), \tilde{y}(k), d(k - (2n_u - 1)), .., d(k-1), d(k)],
\tag{16}
$$

Given (16), the following state equations result

$$
\begin{aligned}
\bar{x}_1(k+1) &= \bar{x}_2(k) \\
\bar{x}_2(k+1) &= \bar{x}_3(k) \\
&\vdots \\
\bar{x}_{n_y+n_u-1}(k+1) &= \bar{x}_{n_y+n_u}(k) \\
\bar{x}_{n_y+n_u}(k+1) &= \tilde{y}(k, \mathbf{w}, \mathbf{v}, r_o \bar{x}_i) \\
\bar{x}_{n_y+n_u+1}(k+1) &= \bar{x}_{n_y+n_u+2}(k) \\
&\vdots \\
\bar{x}_{n_y+n_u+2n_u}(k+1) &= \bar{d}(k),
\end{aligned}
\tag{17}
$$

Given the state Eq. (17) and further state vector (16), the following Jacobean matrix of partial derivatives $\mathbf{J}_{\bar{m}}$ and $\mathbf{J}_{\bar{n}}$ result

$$
\mathbf{J}_m =
\begin{bmatrix}
\frac{\partial \bar{x}_1(k+1)}{\partial \bar{x}_1(k)} & \frac{\partial \bar{x}_1(k+1)}{\partial \bar{x}_2(k)} & \cdots & \frac{\partial \bar{x}_1(k+1)}{\partial \bar{x}_{n_x}(k)} \\
\frac{\partial \bar{x}_2(k+1)}{\partial \bar{x}_1(k)} & \frac{\partial \bar{x}_2(k+1)}{\partial \bar{x}_2(k)} & \cdots & \frac{\partial \bar{x}_2(k+1)}{\partial \bar{x}_{n_x}(k)} \\
\vdots & & \vdots & \\
\frac{\partial \bar{x}_{n_x}(k+1)}{\partial \bar{x}_1(k)} & \frac{\partial \bar{x}_{n_x}(k+1)}{\partial \bar{x}_2(k)} & \cdots & \frac{\partial \bar{x}_{n_x}(k+1)}{\partial \bar{x}_{n_x}(k)}
\end{bmatrix}
=
\begin{bmatrix}
0 & 1 & 0 & \cdots & 0 \\
0 & 0 & 1 & 0 & 0 \\
\bar{a}_{x_1} & \bar{a}_{x_2} & \cdots & \cdots & \bar{a}_{x_{n_x}} \\
0 & 0 & 1 & 0 & 0 \\
0 & 0 & \cdots & 1 & 0 \\
0 & 0 & 0 & 0 & 1 \\
0 & 0 & 0 & 0 & 0
\end{bmatrix},
\mathbf{J}_n =
\begin{bmatrix}
\frac{\partial \bar{x}_1(k+1)}{\partial \bar{d}(k)} \\
\frac{\partial \bar{x}_2(k+1)}{\partial \bar{d}(k)} \\
\vdots \\
\frac{\partial \bar{x}_{n_x}(k+1)}{\partial \bar{d}(k)}
\end{bmatrix}
=
\begin{bmatrix}
0 \\
0 \\
\vdots \\
\bar{a}_{u_{n_x}}
\end{bmatrix}.
\tag{18}
$$

Thus, from (18) it is apparent that an incremental linear state-space form of the original LNU-HONU may be given as follows, valid in vicinity of the applied state point

$$\Delta\bar{\mathbf{x}}(k+1) = \bar{\mathbf{M}}^r \Delta\bar{\mathbf{x}}(k) + \bar{\mathbf{N}}^r \Delta\bar{\mathbf{u}}(k)$$
$$\bar{y}(k) = \bar{\mathbf{C}}^r \Delta\bar{\mathbf{x}}(k). \tag{19}$$

Where $\bar{\mathbf{M}}^r$ represents a $n_{\bar{x}} \times n_{\bar{x}}$ matrix where, $n_{\bar{x}} = n_y + n_u + 2n_u$ and the row index $n_y + n_u$ consists of a row of coefficients $\bar{a}_{x_i}$. Further, $\bar{\mathbf{N}}^r$ is the inputs matrix where $\bar{a}_{u_i}$ is the control input coefficient typically equal to 1 for most SISO cases. Hence, for a LNU-HONU control loop of arbitrary length it yields that

$$\bar{a}_{x_i} = C_i(\sum_{l=0}^{n_y} w_l x_l) - \sum_{l=0}^{n_u-1} w_{n_y+1+l} r_o C_i(q(k-l))$$
$$C_i(q(k-l)) = \boldsymbol{\psi}\,\boldsymbol{\xi}(k-l)$$
$$\boldsymbol{\psi} = f(\mathbf{v}) = \begin{cases} \sum_{l=0}^{p-1} v_{l,p}\,where\,p = \bar{x}_i \in \boldsymbol{\xi}(k-l) \\ \sum_{j=p}^{n_{\bar{x}}} \alpha_{p,j} v_{p,j}\,for\,p = \bar{x}_i \in \boldsymbol{\xi}(k-l) \\ where\,\alpha_{p,j} = 1 \forall j \neq p\,\&\,\alpha_{p,p} = 2 \end{cases} \tag{20}$$

3.2 Decomposition Method to State-Space Form of Non-linear Control Loop with LNU

In the preceding section, a novel state-space representation was introduced for LNU models and their extension as a non-linear control loop with LNU. However, due to the use of further order feedback controllers, the representation in (19) has its limitations in accurate dynamic description of the whole non-linear control loop. Following the result [5] an intrinsic relation to decompose a HONU model into its principle step-delayed vector may be extended for a LNU with feedback control. As the method [5] is not based on approximation, but rather a direct representation with respect to the HONUs own internal states via its principle input vector, its applicably for further order HONU-MRAC control loops is justified.

Recalling the relation (1), and control law (13)–(14) the extended LNU form yields

$$\tilde{y}(k) = w_0 + w_1\tilde{y}(k-1) + w_2\tilde{y}(k-2) + .. + w_{n_y}\tilde{y}(k-n_y) +$$
$$w_{n_{y+1}}[d(k-1) - r_o q(k-1)] + .. + w_{n_y+n_u}[d(k-n_u) - r_o q(k-n_u)], \tag{21}$$

where $q(.)$ summarises the HONU feedback controller output of γ order. Upon expansion of the terms in (21) we may restate that

$$\tilde{y}(k) = w_0 + \sum_{j=i}^{n_y} w_i \hat{x}(k-j) + \sum_{j=1}^{n_u} w_{j+n_y} \hat{u}(k-j) -$$
$$r_o \sum_{j=1}^{n_u} w_{n_y+j} \cdot q(v, \hat{x}(k-j), \hat{u}(k-j)), \tag{22}$$

where we may explicitly define the vector of internal state variables as

$$\hat{\mathbf{x}}(k-1) = \left[\tilde{y}(k-(n_y+n_u))\ \tilde{y}(k-(n_y+n_u)+1)..\tilde{y}(k-1)\right]^T. \tag{23}$$

Further, the input vector for the whole non-linear control loop with LNU is

$$\hat{\mathbf{u}}(k-1) = \left[d(k-(n_u+n_u))\ d(k-(n_u+n_u)+1)..d(k-1)\right]^T. \tag{24}$$

Due to the final term in (22) featuring previous step-delayed internal states along with their definition in (23)–(24), we may state

$$\tilde{y}(k) = \sum_{i=1}^{n_y} \hat{x}(k-i) \cdot \hat{a}_{x_i} + \sum_{i=1}^{n_u} \hat{u}(k-i)\hat{a}_{u_i} + C_i(\mathbf{w}_0), \tag{25}$$

where the operator $C_i(.)$ denotes the sum of the constant neural bias weights. For a LNU based HONU-MRAC control loop where the HONU feedback controller $\gamma \geq 1$, it yields

$$\hat{a}_{x_i} = \begin{cases} w_i - r_o \sum_{j=1}^{n_u} w_{n_y+j} \cdot C_i(q(k-j)) \ for\ i = 1,2,3,..,n_y \\ -r_o \sum_{j=1}^{n_u} w_{n_y+j} \cdot C_i(q(k-j)) \ for\ i = n_y+1,..,n_y+n_u, \end{cases} \tag{26}$$

and for coefficients $\hat{a}_{u_i}$ it can be stated that

$$\hat{a}_{u_i} = \begin{cases} w_{n_y+i} - r_o \sum_{j=1}^{n_u} w_{n_y+j} \cdot C_i(q(k-j)) \ for\ i = 1,..,n_u \\ -r_o \sum_{j=1}^{n_u} w_{n_y+j} \cdot C_i(q(k-j)) \ for\ i = n_u+1,..,n_u+n_u, \end{cases} \tag{27}$$

where the operator $C_i(.)$ is used for sum of coefficients of the elements $\hat{x}_i$ or $\hat{u}_i$ where $\hat{x}_i, \hat{u}_i \in q(k)$.

4   Input to State Approach to Non-linear Control Loop with LNU

From relations (23)–(27) it is apparent that the decomposed polynomial structure of an LNU with HONU feedback controller yields the following state-space representation

$$\hat{\mathbf{x}}(k) = \hat{\mathbf{M}}^r\hat{\mathbf{x}}(k-1) + \hat{\mathbf{N}}^r\hat{\mathbf{u}}(k-1) + \mathbf{w}_0$$
$$\tilde{y}(k-1) = \hat{\mathbf{C}}^r\hat{\mathbf{x}}(k-1). \tag{28}$$

Different to (19) the local matrix of dynamics (LMD) $\hat{\mathbf{M}}^r$ is introduced of $n_{\hat{x}} = n_y + n_u$ square dimension. Backboned from the concepts BIBO, BIBS, ISS we may state

$$\|\bar{\mathbf{x}}(k)\| \leq \kappa L(\|\bar{\mathbf{x}}(k_0)\|) + \kappa(\|u(k)\|_\infty), \tag{29}$$

where κL is an asymptotically stable function converging to a minimum for $k \to \infty$. Moreover, for a zero equilibrium that $\kappa L(\cdot) \to 0$. Further, κ represents a class κ_∞ function which is unbounded and strictly increasing such that about an initial zero state that $\kappa(.) = 0$ and for $k \to \infty$, $\kappa(.) \to \infty$. On the evaluation of state transition matrix (STM) $\boldsymbol{\Phi}_{\hat{\mathbf{M}}^r}$ of the LMD $\hat{\mathbf{M}}^r$, the input to state stability law defined in (29) is satisfied if

$$\|\bar{\mathbf{x}}(k)\| \leq \left\|\boldsymbol{\Phi}_{\hat{\mathbf{M}}^r}(k, k_0)\right\| \left\|\bar{\mathbf{x}}(k_0)\right\| + \sum_{l=k_0}^{k} \left\|\boldsymbol{\Phi}_{\hat{\mathbf{M}}^r}(k, l)\bar{\mathbf{N}}^r(l)\right\| \|\mathbf{u}(l)\|. \tag{30}$$

Further, the LNU based HONU-MRAC control loop is BIBS stable. For local BIBO stability the following is sufficient, also applicable for the LMD $\hat{\mathbf{M}}^r$

$$\rho(\bar{\mathbf{M}}^r(k, \mathbf{w}, \mathbf{v}, \bar{\mathbf{x}}_e)) < 1. \tag{31}$$

5   Experimental Analysis

Following the newly derived stability evaluation in Sects. 3 and 4, an example of a non-linear position feedback system is provided to analyse the effectiveness of the proposed method for non-linear control loops with LNU. For this analysis let us consider the following position feedback system with conditional gain α

$$\dot{E} = -x_2$$
$$\dot{x}_2 = \alpha \left[-\frac{1}{T} x_2 + \frac{f(E)}{T} \right], \tag{32}$$

where the error $E = d - \int x_2 dt$, and d is the desired behaviour. Further, the nonlinear actuation function $f(E) = \max(-0.2, \min(0.2, E))$ is introduced which for simplification is denoted further as the system input u and the process time constant $T = 1$. As an initial insight to the original systems stability, let us consider the following energy function

$$V = \frac{T}{2} x_2^2 + \int_0^E f(\partial) d\partial, \tag{33}$$

From (33) if $E = x_2 = 0$ the condition $V = 0$ is satisfied from the function (33). The resulting condition for the derivative of the Lyapunov function yields that if $0 \geq x_2 \geq u \cdot [\alpha - 1]/\alpha$ that the system is within a region of stability about a zero equilibrium point. In this example a LNU is used as a plant model, trained via the RLS algorithm where previous output values of the identified neural model $n_y = 4$ and previous input values of the process $n_u = 3$. Both an LNU and QNU feedback controller are trained via the RLS learning algorithm for the same step-delayed history as the identified HONU process model (Fig. 1b illustrates faster dynamics of the QNU controller). The HONU learning algorithm is applied where $\alpha = 0.2$ for stable system dynamics. Between the time interval t = 70–85[s] (Fig. 1(a) a second set of trained HONU feedback controller weights are introduced to analyse the effect on the whole LNU-HONU control loop stability, where a comparison of BIBO is shown in Fig. 1c–d. Figure 2a–b illustrates that as the norm of the state transition matrix calculated across the defined operating region is converging to the threshold defined by the value B_1 i.e. $\left\|\Phi_{\bar{\mathbf{M}}'}(k, k_0)\right\| \le B_1$. Further, that the sum of norms of the STM as a product with the input matrix i.e. $\sum_{l=k_0}^{k} \left\|\Phi_{\bar{\mathbf{M}}'}(k, l)\bar{\mathbf{N}}'(l)\right\| \le B_2$ holds, (30) is sufficiently satisfied for the given region (Fig. 2(c). However, in the case of Fig. 2d, this condition is breached and hence to the evaluation of relation (30) violates the conclusion of BIBS stability in time interval t = 70–85[s].

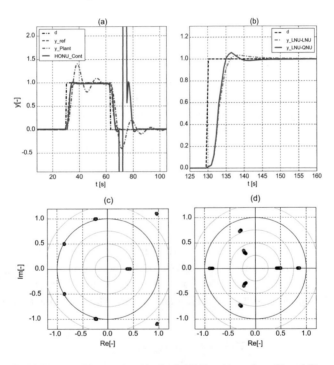

Fig. 1. (a) LNU-QNU control loop trained both via RLS and introduced instability t = 70–85 [s]. (b) LNU and QNU controller comparison (c) LMD eigenvalues at t = 0–85 [s] (with poorly identified neural weights) (d) t = 85–200 LMD eigenvalues (with well identified neural weights)

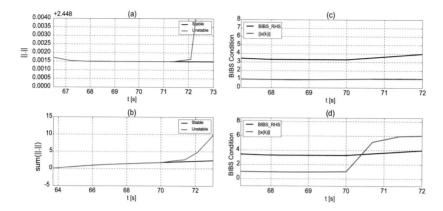

Fig. 2. (a)–(b) Convergence monitoring of thresholds for sufficient BIBS stability of LNU-QNU control loop. (c)–(d) Comparison of ISS condition (30) for BIBS stable and unstable LNU-QNU control loop respectively, in region t > 70 [s]

6   Conclusion

In this paper two novel state-space representations of a LNU plant with non-linear control loop i.e. extension of a HONU $r \geq 1$ feedback controller was presented, connoted from the principles of non-linear discrete time state-space systems. A straightforward ISS based approach for BIBO and BIBS stability was derived for LNU based HONU-MRAC control loops where following experimental analysis, its effectiveness and practicality for process stability evaluation was exhibited. Advantaged by the decomposition of the non-linear control loop with LNU into a linear layered architecture, the newly derived state-space form (28) provides a more accurate representation of such non-linear polynomial based model which may be extended for higher order architectures and hence provides advantages for continuous stability monitoring in contrast to the partial derivative based representation in (19) valid in vicinity of the locally applied state point.

Acknowledgements. Authors acknowledges support from the EU Operational Programme Research, Development and Education, and from the Center of Advanced Aerospace Technology (CZ.02.1.01/0.0/0.0/16_019/0000826)

References

1. Benes, P.M., Bukovsky, I.: Neural network approach to hoist deceleration control. In: 2014 International Joint Conference on Neural Networks (IJCNN), pp. 1864–1869 (2014)
2. Benes, P.M., Erben, M., Vesely, M., Liska, O., Bukovsky, I.: HONU and supervised learning algorithms in adaptive feedback control. In: Applied Artificial Higher Order Neural Networks for Control and Recognition, IGI Global, pp. 35–60 (2016)
3. Benes, P.M., Bukovsky, I., Cejnek, M., Kalivoda, J.: Neural network approach to railway stand lateral skew control. In: Computer Science & Information Technology (CS& IT), Sydney, Australia, vol. 4, pp. 327–339 (2014)

4. Bukovsky, I., Benes, P., Slama, M.: Laboratory systems control with adaptively tuned higher order neural units. In: Silhavy, R., Senkerik, R., Oplatkova, Z.K., Prokopova, Z., Silhavy, P. (eds.) Intelligent Systems in Cybernetics and Automation Theory, pp. 275–284. Springer International Publishing (2015)

5. Benes, P., Bukovsky, I.: On the intrinsic relation between linear dynamical systems and higher order neural units. In: Silhavy, R., Senkerik, R., Oplatkova, Z.K., Prokopova, Z., Silhavy, P. (eds.) Intelligent Systems in Cybernetics and Automation Theory. Springer International Publishing (2016)

6. Yadav, R.N., Kalra, P.K., John, J.: Time series prediction with single multiplicative neuron model. Appl. Soft Comput. **7**(4), 1157–1163 (2007)

7. Zhao, H., Zeng, X., He, Z.: Low-complexity nonlinear adaptive filter based on a pipelined bilinear recurrent neural network. IEEE Trans. Neural Netw. **22**(9), 1494–1507 (2011)

8. Zhao, H., Zhang, J.: A novel nonlinear adaptive filter using a pipelined second-order Volterra recurrent neural network. Neural Netw. **22**(10), 1471–1483 (2009)

9. Bukovsky, I., Homma, N.: An approach to stable gradient-descent adaptation of higher order neural units. IEEE Trans. Neural Netw. Learn. Syst. **28**(9), 2022–2034 (2017)

10. Barabanov, N.E., Prokhorov, D.V.: A new method for stability analysis of nonlinear discrete-time systems. IEEE Trans. Autom. Control **48**(12), 2250–2255 (2003)

11. Zhao, W., Zhu, Q.: New results of global robust exponential stability of neural networks with delays. Nonlinear Anal. Real World Appl. **11**(2), 1190–1197 (2010)

12. Liao, X., Chen, G., Sanchez, E.N.: LMI-based approach for asymptotically stability analysis of delayed neural networks. IEEE Trans. Circ. Syst. Fundam. Theory Appl. **49**(7), 1033–1039 (2002)

13. Liao, X., Chen, G., Sanchez, E.N.: Delay-dependent exponential stability analysis of delayed neural networks: an LMI approach. Neural Netw. **16**(10), 1401–1402 (2003)

14. Yasuda, Y., Ito, H., Ishizu, H.: On the input-output stability of linear time-varying systems. SIAM J. Appl. Math. **38**(1), 175–188 (1980)

15. Li, Z., Fei, Z., Gao, H.: Stability and stabilisation of Markovian jump systems with time-varying delay: an input-output approach. IET Control Theory Appl. **6**(17), 2601–2610 (2012)

16. Lazar, M., Heemals, W.P.M.H., Teel, A.R.: Further input-to-state stability subtleties for discrete-time systems. IEEE Trans. Autom. Control **58**(6), 1609–1613 (2013)

17. Sontag, E.D.: Input to state stability: basic concepts and results. In: Nonlinear and Optimal Control Theory, vol. 1932, pp. 163–220. Springer, Heidelberg (2008)

18. Angeli, A., Sontag, E.D., Wang, Y.: A characterization of integral input-to-state stability. IEEE Trans. Autom. Control **45**(6), 1082–1097 (2000)

19. Jiang, Z.P., Wang, Y.: Input-to-state stability for discrete-time nonlinear systems. Automatica **37**(6), 857–869 (2001)

20. Yang, Z., Zhou, W., Huang, T.: Exponential input-to-state stability of recurrent neural networks with multiple time-varying delays. Cogn. Neurodyn. **8**(1), 47–54 (2014)

21. Ahn, C.K.: Robust stability of recurrent neural networks with ISS learning algorithm. Nonlinear Dyn. **65**, 413–419 (2011)

22. Zhu, S., Shen, Y.: Two algebraic criteria for input-to-state stability of recurrent neural networks with time-varying delays. Neural Comput. Appl. **22**(6), 1163–1169 (2013)

The Technique of Informational Interaction Structural-Parametric Optimization of an Earth's Remote Sensing Small Spacecraft Cluster

Jury S. Manuilov[2], Alexander N. Pavlov[1,2(✉)], Dmitry A. Pavlov[2], and Alexey A. Slin'ko[2]

[1] Saint Petersburg National Research University of Information Technologies, Mechanics and Optics (ITMO), St. Petersburg, Russia
Pavlov62@list.ru
[2] Mozhaisky Military AeroSpace Academy, St. Petersburg, Russia

Abstract. Promising orbital remote sensing of the Earth are created in the form of clusters of small spacecraft. One of the central problems of creating small satellites clusters is the development of management systems of their information interaction. The lack of effective control technologies multi-satellite group, transmission, and processing of distributed information huge volumes can negate the impact of the latest achievements in the construction of future space systems usedness. Management system of small spacecraft clusters information interaction should be oriented mostly on maintaining a separate spacecraft and control its production cycle, and to assess the status of the entire cluster and making decisions on its reconfiguration and use for the intended purpose in conditions of limited resources (information, energy, etc.) of the cluster. The solution to such a complex question fraught with difficulties, caused by the imperfection of the existing scientific and methodological apparatus management information by the interaction of complex objects (systems network architecture) [1–4]. The authors proposed a technique based on the developed models [5], which allows to optimize structure and parameters of the Earth's remote sensing small spacecraft cluster information interaction system. The article shows that the application of the developed technique enhances the performance of the small spacecraft cluster application.

Keywords: Remote sensing of the earth · A cluster of small spacecraft
Information interaction · Structural-parametric optimization

1 Introduction

Orbital group of small spacecraft with dynamically changing network structure can be multifunctional and reconfigurable in order to meet operational targets, cheap to implement, reliable and durable in a variety of situations when observing objects on Earth, the study of objects in outer space, the solution for the telecommunication and other various problems of information security. In this regard, a prospective remote sensing system are created in the form of clusters. While a cluster is distributed orbital

© Springer International Publishing AG, part of Springer Nature 2019
R. Silhavy (Ed.): CSOC 2018, AISC 765, pp. 155–166, 2019.
https://doi.org/10.1007/978-3-319-91192-2_17

system network architecture designed to solve the problems of observation, formed at a certain interval of time from several different types of small satellites remote sensing, jointly and consensually flying targets. The use of clusters allows to obtain detailed information about the objects of sensing in different spectral ranges (hyperspectral, panchromatic, and infrared). There is already a space system for remote sensing of the American company "Terra Bella" on the basis of groups of SkySat-1, 2, which in the near future will consist of 24 satellites. It is designed to acquire images in the panchromatic range (resolution 90 cm when shooting at the Nadir) and four spectral channels with a resolution of 2 m, and survey equipment of small spacecraft provides shooting video (30 fps) with a resolution of 1.1 m in the panchromatic spectrum [6]. Actively developing the direction of create a duplex technology cycles control of advanced space systems [7]. A substantial part of the managerial functions is transferred to the smallest spacecraft, in particular for determining the position of the center of mass, orientation, distribution of tasks between devices, and others. Operation of space systems involves the collection, processing and transmission of large amounts of sensing data that in conditions of limited opportunities of small satellites is difficult. The question is about ensuring a balance between energy and information capabilities of the small satellites cluster to achieve required level of cluster target performance taking into account the existing limitations in its functioning. In this regard, the authors solve the task of structural-parametric optimization of the system of informational interaction (II) cluster of small spacecraft remote sensing.

Set-theoretic formulation of structural-parametric optimization problem in the form of generalized selection structures are presented in detail in the work [8]. To achieve this goal by improving the task performance of the cluster (increasing the amount of information about the objects of sensing are delivered to the consumer) need the flexibility to manage information and energy resource cluster (intensity of work receiving and transmitting equipment, on-Board computers and storage devices, energy consumption). The solution to problems of structural-parametric optimization based on the scheduling information of the interaction of small satellites in the cluster with the subsequent synthesis of flexible technologies II, allows to find the optimal parameter values (intensity) of work of onboard equipment of small satellites for remote sensing. Let us consider the main elements of our methodology of structural and parametric optimization of the system II the cluster of promising small spacecraft for remote sensing.

2   Planning Model

To solve these tasks is impossible without planning the system of communications between promising cluster of small spacecraft remote sensing. To account for the structural dynamics of the cluster and the departure from dynamic models of optimal management of II between the cluster, the authors used an approach that allows you to find the intervals of execution of the target task on which the structure of the cluster is unchanged (constant), usage-based dynamic model of the uncontrolled movement of small satellites composing the cluster [9]. Finding intervals of constancy allows us to describe a problem situation within a class of static models (mathematical programming). In models of this class has considerable experience in the development of very efficient algorithms allowing to more fully reflect the specificity of tasks. Optimization of management processes of II of cluster remote sensing based on analytical static models of decision-making associated with the search for some finite-dimensional vector describing the solution that delivers the maximum target function [10, 11]. For the tasks required the information interaction system work plan is described by the vector (1)

$$U_\mu = \left\| \|q_{\mu ij\rho sk}\|, \|x_{\mu ij\rho dk}\|, \|g_{\mu i\rho ok}\|, \|y_{\mu i\rho k}\|, \|z_{\mu i\rho k}\|, \|r_{\mu i\rho k}\|, \left\|t^q_{\mu ijsk}\right\|, \left\|t^g_{\mu iok}\right\|, \left\|t^x_{\mu ijdk}\right\| \right\|, \quad (1)$$

where $\|q_{\mu i\rho sk}\|$ – the vector characterizing the volume of information flow ρ-th types $(\rho \neq \rho_0)$ incoming sensor payload to on-board computing system (BCS) i-spacecraft according to s-th technology at k-th intervals of the II structure constancy (further intervals); $\|x_{\mu ij\rho dk}\|$ – the amount of transmitted streams p-th types $(\rho \neq \rho_0)$ from i-th spacecraft to j-th spacecraft according to d-th technology at k-th intervals, and the amount of energy expended in the implementation of this data transfer $(\rho = \rho_0)$; $\|g_{\mu i\rho ok}\|$ – the amount of flows p-th types, processed by the i-th spacecraft; $\|y_{\mu i\rho k}\|$ – the volume of the recorded streams p-th types by the i-th spacecraft; $\|z_{\mu i\rho k}\|$ – amounts not received information about the objects; $\|r_{\mu i\rho k}\|$ – the resulting volumes of data delivered by the i-spacecraft to the consumer; $\left\|t^g_{\mu iok}\right\|$ – time resources required for information processing by the i-th spacecraft; $\left\|t^x_{\mu ijdk}\right\|$ – time resources required for transmitting information from the i-spacecraft to the j-spacecraft; $\left\|t^q_{\mu ijsk}\right\|$ – the vector characterizing the temporal resources required to undertake the shooting the j-th object by the i-th spacecraft.

It is necessary that the desired plan U^* was to practically implement, to this end, he must satisfy the conditions (constraints) of the problem under consideration. On the basis of the ERS cluster target processes systematic analysis produced a formalized description of the conditions limiting the choice and preferences which must be met by the optimal solution.

Information flows balance constraints:

$$\left(\sum_{d \in D_{\rho вых}} \sum_{j \in J_{\mu ikвых}} x_{\mu ij\rho dk} - \sum_{d \in D_{\rho вх}} \sum_{j \in J_{\mu ikвх}} x_{\mu ij\rho dk} \right) + \left(y_{\mu i\rho k} - y_{\mu i\rho(k-1)} \right) + \left(\sum_{o \in O_{\rho вх}} g_{\mu i\rho ok} - \sum_{o \in O_{\rho вых}} g_{\mu i\rho ok} \right)$$
$$+ z_{\mu i\rho k} + \chi_{\mu i\rho k} r_{\mu i\rho k} = \sum_{s \in S} q_{\mu i\rho sk}, \forall i, j \in N; k \in K; \chi_{\mu i\rho k} \in \{0;1\}; \rho \in P \setminus \{\rho_0\}. \quad (2)$$

Energy streams balance constraints:

$$\sum_{d\in D_{p\text{вих}}}\sum_{j\in J_{\mu\text{iвих}}} x_{\mu ijpdk} + (y_{\mu ip k} - y_{\mu ip(k-1)}) + \sum_{o\in O_{p\text{вх}}} g_{\mu ip o k} + r_{\mu ip k} + \sum_{s\in S} q_{\mu ip s k} = I_{\mu ip k},$$

$$\forall i, j \in N; \forall d \in D; \forall k \in \mathbf{K}; \rho = p_0. \tag{3}$$

Information processing technological constraints:

$$t^g_{\mu iok} \cdot \omega_{\mu ip o k} - g_{\mu ip o k} = 0, \forall k \in \mathbf{K}, o \in O, \rho \in \mathrm{P}\backslash\{p_0\}, \mu = 1 \ldots m \tag{4}$$

where $\omega_{\mu ip o k}$ - the intensity of ρ-th types information flows processing by i-th spacecraft, $g_{\mu ip o k}$- the volume of information flows ρ-th types received for processing on i-th spacecraft.

Technological constraints on the transfer of information flow to consumer:

$$t^x_{\mu ijdk} \cdot \xi_\mu \cdot \omega_{\mu ijp d k} - x_{\mu ijp d k} = 0, \forall k \in \mathbf{K}, d \in D, \rho \in \mathrm{P}\backslash\{p_0\}, \mu = 1 \ldots m, \tag{5}$$

where $\omega_{\mu ijd p k}$ - the intensity of ρ-th type information transfer from i-th spacecraft to j-th spacecraft; $x_{\mu ijp d k}$ - volume of ρ-th type information incoming transmission during $t^x_{\mu ijdk}$ from i-th spacecraft to j-th spacecraft; $\xi_{\mu i}$- the coefficient characterizing the impact of on-Board transmission equipment of remote sensing data.

Technological constraints to a survey of the objects of the trust instrument:

$$t^q_{\mu ijsk} \cdot \omega_{\mu ijp s k} - q_{\mu ijp s k} = 0, \forall k \in \mathbf{K}, s \in S, \rho \in \mathrm{P}\backslash\{p_0\}, \mu = 1 \ldots m \tag{6}$$

where $\omega_{\mu ijp s k}$ - the intensity of the shooting j-th object by i-th spacecraft; $q_{\mu ijp s k}$ - the volume of ρ-type information flows coming over time $t^q_{\mu ijsk}$ to BCS of i-th spacecraft from its sensor payload.

In accordance with the methodology of the system approach also defined the technical constraints on the processes of information interaction of ERS cluster. Constraints on storage of information flows and the accumulation of energy flows:

$$y \le Y_i; \forall k \in \mathbf{K}; \forall i \in N, \tag{7}$$

$$y_{\mu ip k} \le E_i; \forall k \in \mathbf{K}; \forall i \in N; \rho = p_0. \tag{8}$$

Temporary constraints on the transmission of information about objects sensing:

$$\sum_{d\in D_{p\text{вих}}}\sum_{j\in J_{i\text{вих}}} t^x_{\mu ijdk} + \sum_{d\in D_{p\text{вх}}}\sum_{j\in J_{i\text{вх}}} t^x_{\mu ijdk} \le T_k; \forall k \in \mathbf{K}; \forall i \in N. \tag{9}$$

Temporary constraints on processing of information by onboard equipment:

$$\sum_{o \in O} t^g_{\mu iok} \leq T_k; \forall k \in \mathbf{K}; \forall i \in N. \tag{10}$$

Time constraints for taking images of objects of the target sensing apparatus:

$$\sum_{s \in S} t^q_{\mu isk} \leq T_k; \forall k \in \mathbf{K}; \forall i \in N. \tag{11}$$

The formation of the communication plan of the ERS cluster to find the best solution may be made by optimization of the generalized indicators of its information and energy opportunities.

The index of the volume of all streams of remote sensing data, referred to by the user when performing the target tasks:

$$H_1(U_\mu) = \sum_{k \in K} \sum_{\rho \in P \setminus \{\rho_0\}} \sum_{i \in N} r_{\mu i \rho k}. \tag{12}$$

Total all uncollected (potentially lost) remote sensing data flow when executing the target tasks:

$$H_2(U_\mu) = \sum_{k \in K} \sum_{\rho \in P \setminus \{\rho_0\}} \sum_{i \in N} z_{\mu i \rho k}. \tag{13}$$

Indicators of energy expenditure for each i-th small spacecraft of the cluster when performing the target tasks:

$$H_{3i}(U_\mu) = \sum_{k \in K} r_{\mu i \rho_0 k}, i \in N. \tag{14}$$

Thus, to find the plan (s) U^*_μ needed to resolve existing uncertainty criterion. This was done using lexicographical methods. Since the set of feasible solutions Δ_β is a convex polyhedral, and the criterion function of the problem being solved are linear, then to find II plans of region of non-dominated alternatives was used standard method of scalarization vector indicator in the form of an additive convolution of Lagrange [12–14] and entered into the model integral criterion (15):

$$\beta_1 \left(\sum_{i \in N} \sum_{k \in K} \alpha_1 (\gamma_1 r_{\mu i \rho_1 k} + \gamma_2 r_{\mu i \rho_2 k} + \gamma_3 r_{i \rho_3 k}) - \sum_{k \in K} \sum_{\rho \in P \setminus \{\rho_0\}} \sum_{i \in N} \alpha_2 z_{\mu i \rho k} \right) \tag{15}$$
$$- \beta_2 \alpha_3 \upsilon_\mu \to \max,$$

where β_1, β_2 - normalized coefficients; $\alpha_1, \alpha_2, \alpha_3$ – the corresponding coefficients of importance of individual criteria H_1, H_2, H_3; $\gamma_1, \gamma_2, \gamma_3$ - the coefficient of the importance of transfer of information flows 1, 2, 3-th *types* accordingly. The authors developed a software package for scheduling communications between a cluster of small spacecrafts for remote sensing, which allows to find the solution (optimal plan U^*_μ) and visualize it

160   J. S. Manuilov et al.

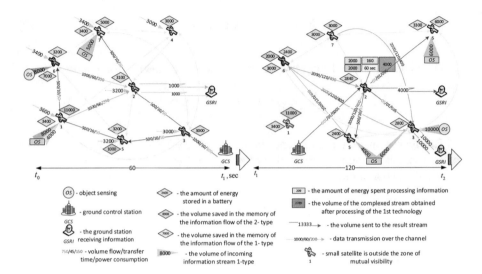

Fig. 1. A fragment of the communication plan of the cluster of promising small satellites for remote sensing

for further analysis and decision. A fragment of one of the obtained solution are represented on (Fig. 1).

The software package of planning of system of information interaction of the cluster of promising small ERS spacecraft is a console application developed in Python in native IDE, and allows you to solve the problem of optimal allocation of heterogeneous information flows and the use of structural-functional and energy resources of the cluster with the technology of shooting, processing and transmission of information, the structural dynamics of the cluster, possible impacts on the cluster, and the preferences of decision-makers.

3   Method of Structural and Parametric Optimization

Formulation and solution of problems of structural-parametric optimization of the cluster remote sensing information interaction system is of particular relevance. In the course of the study by the authors it was established that the solution of this problem is possible by finding a flexible (adaptive) modes of operation of onboard equipment of small spacecraft for remote sensing while performing the target task. Thus, under the flexible mode (FLEX-mode) refers to the creation of such technologies onboard small spacecraft, in which information management system in a flexible and balanced cycles is the intensity of the on-Board equipment (values of the intensities range from minimum to maximum). Then the task of structural-parametrical optimization is to determine optimal values of the onboard equipment work intensity.

Development of the proposed methodology was based on the fact that ERS cluster's information interaction planning model constraints (4–6) allow you to specify flexible technology implementation tasks. This makes possible the formation of the desired sets of ERS small spacecraft operation intensities values such that

$\omega_\mu^* = (\omega_\mu^{q*}, \omega_\mu^{g*}, \omega_\mu^{x*}) = \arg \max\limits_{\omega \in \Delta_{\mu\beta}^{sat}} H_{pe3}\left(U_\mu(\omega)\right)$. In this case ω_μ^* can take values within the ranges specified by the relevant constraints (16–18):

$$\omega_{\mu i \rho o min}^g \leq \omega_{\mu i \rho o k}^g \leq \omega_{\mu i \rho o max}^g, \forall i \in I, \rho \in P, o \in O, k \in K, \mu = 1 \ldots m, \quad (16)$$

где $\omega_{\mu i \rho o k}^g$ – the intensity of the i-th spacecraft processing of ρ-th types information flow;

$$\omega_{\mu i j \rho d min}^x \leq \omega_{\mu i j \rho d k}^x \leq \omega_{\mu i j \rho d max}^x, \forall i \in I, \rho \in P, d \in D, k \in K, j \in I\backslash i, \quad (17)$$

where $\omega_{\mu i j \rho d k}^x$ – the intensity of ρ-th flow transmission from i-th spacecraft to j-th spacecraft;

$$\omega_{\mu i j \rho s min}^q \leq \omega_{\mu i j \rho d k}^q \leq \omega_{\mu i j \rho d max}^q, \forall j \in I\backslash i, \rho \in P, s \in S, k \in K, i \in I, \quad (18)$$

where $\omega_{\mu i j \rho s k}^q$ – the intensity of the j-th object shooting by i-th spacecraft. Then all possible sets of parameters of the remote sensing small spacecraft onboard equipment can be represented in the form of set of (19)

$$\Delta_{\mu\beta}^{sat} = \left\langle \left\{\omega_{\mu i \rho o k}^g\right\}, \left\{\omega_{\mu i j \rho d k}^x\right\}, \left\{\omega_{\mu i j \rho s k}^q\right\} | \forall i \in I, \rho \in P, d \in D, s \in S, o \in O, k \in K \right\rangle. \quad (19)$$

Thus, the method of structural-parametric optimization of ERS cluster informational interaction consists of the following steps.

Stage 1. Through the use of constraints (4–6) are formed in the flexible modes of the onboard equipment through a combination of standard modes and determining the proportions at the time of their use in the II operations (for example, when the duration interval of transmission of information 60 s, it can be transferred 40 s in MAX mode and 20 s in MIN-mode, if this technology is more efficient than the standard). After the formation of all technologies is the transition to phase 2.

Stage 2. Is to optimize the values of onboard equipment parameters based on the synthesized in the course of the study technology:

(1) The technology of obtaining averaged of the remote sensing data reception and transmission equipment parameters values:

$$\omega_{\mu i j \rho d k}^x = \frac{\omega_{\mu i j \rho d min k}^x \cdot t_{\mu i j \rho d min k}^x + \omega_{\mu i j \rho d max k}^x \cdot t_{\mu i j \rho d max k}^x}{t_{ij \rho d min k}^x + t_{ij \rho d max k}^x}, \forall i \in I, d \in D, \rho \in P, \quad (20)$$

where $\omega_{\mu i j \rho d min k}^x$ and $\omega_{\mu i j \rho d max k}^x$ - information transfer intensities values for energy saving technology d_{min} and energy-intensive d_{max} accordingly;

(2) The technology of obtaining averaged of the on-Board computing systems of remote sensing small spacecraft operating parameters values:

$$\omega^g_{\mu i \rho o k} = \frac{\omega^g_{\mu i \rho o_{\min} k} \cdot t^g_{\mu i \rho o_{\min} k} + \omega^g_{\mu i \rho o_{\max} k} \cdot t^g_{\mu i \rho o_{\max} k}}{t^g_{\mu i \rho o_{\min} k} + t^g_{\mu i \rho o_{\max} k}}, \forall i \in I, o \in O, \rho \in P, \quad (21)$$

where $\omega^g_{\mu i \rho o_{\min} k}$ and $\omega^g_{\mu i \rho o_{\max} k}$ - values of information processing intensities;

(3) The technology of obtaining averaged of the on-Board target equipment of remote sensing small spacecraft operating parameters values:

$$\omega^q_{\mu i j \rho s k} = \frac{\omega^q_{\mu i j \rho s_{\min} k} \cdot t^q_{\mu i j \rho s_{\min} k} + \omega^q_{\mu i j \rho s_{\max} k} \cdot t^q_{\mu i j \rho s_{\max} k}}{t^q_{\mu i j \rho s_{\min} k} + t^q_{\mu i j \rho s_{\max} k}}, \forall j \in I \backslash i, s \in S, \rho \in P, \quad (22)$$

where $\omega^q_{\mu i \rho s_{\min} k}$ and $\omega^q_{\mu i \rho s_{\max} k}$ - the value of speed j-th object shooting i-th spacecraft energy efficient $s_{\min}$ and energy-intensive $s_{\max}$ technology, accordingly. Thus the initial data for calculations in accordance with the technologies (20–22) are the values: $\left\langle \left\{ \omega^g_{\mu i \rho o k} \right\}, \left\{ \omega^x_{\mu i j \rho d k} \right\}, \left\{ \omega^q_{\mu i j \rho s k} \right\} \right\rangle \in U^{flex}_{\mu} = \left\| q_{\mu} x_{\mu}, g_{\mu}, y_{\mu}, z_{\mu}, r_{\mu}, t^q_{\mu}, t^g_{\mu}, t^x_{\mu} \right\|$, where U^{flex}_{μ} - the optimal plan for the operation of the cluster II system, obtained from the solution of the planning problem based on the use of flexible operation technologies (FLEX-mode) of onboard equipment. After computing all intensities weighted averages transition to stage 3.

Stage 3. The task of II scheduling between ERS cluster elements based on the synthesized in step 2 the values of the onboard equipment operation intensities ω^*_{μ}. The transition to stage 4.

Stage 4. Analysis of the of planning results, their visualization and a comparison with the results obtained in the FLEX mode. If the objective functions values for synthetic parameters and flexible mode is identical, it makes the conclusion about the optimality of the obtained values ω^*_{μ}.

A great advantage of the results obtained using the developed technique is that it allows to find specific values ω^*_{μ} in contrast to the implicit intensities values in the flexible modes of the onboard equipment and not implement practically.

4   Computational Experiments

Computational experiments based on the use of the developed software were conducted with the original data of the SkySat-2 small spacecraft onboard equipment standard operating modes. The first series of computational experiments was the fact that the solution of the considered problem asked various preferences of the customer to the ERS cluster II performance indicators (H_1 – the transmitted information total volume to the user when the target task, H_2 – the total amount of information lost (not used) at target task execution and H_3 – the amount of energy consumption in a cluster while achieving target). I.e. the calculations were made considering iterating through the

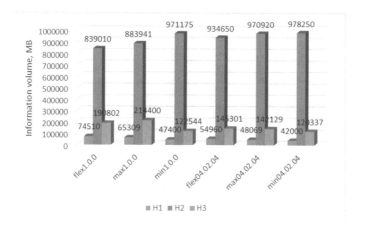

Fig. 2. Values H_1, H_2, H_3 for the 3 modes to suit the preferences of the customer

values of the coefficients α_1, α_2, α_3 in (15) with step of 0,2. Computational experiments showed that the value of the cluster performance integral indicator in planning for the operation of its system of II with optimized parameters, was higher than the values of the corresponding parameters obtained when using the standard parameters of the ERS small spacecraft onboard equipment (Fig. 2).

Analysis of the table shows, under different preferences of the customer to the objectives of the cluster planning results for the three indicators allow us to conclude that under the optimized operating parameters of the onboard equipment in the FLEX mode, the average growth target performance cluster made up 15% compared to the obtained values of the objective function with MAX and MIN modes, respectively.

The second series of experiments was to find optimal values of the cluster II plans for different variants of external influences on the work of small spacecraft remote sensing in fulfilling the targets in the three operating modes of onboard equipment (MIN, MAX, FLEX). In this case, the comparison was subject to the resulting amounts of information of three types (H(IK) – the amount of information infrared, H(JPEG) – the amount of information about the objects sensed in the JPEG format and H (COMPLEX) – the amount of complex information about objects sensing), transferred to the consumer in terms of different values of the coefficients influences ξ_μ. Analysis of the obtained planning results showed that the most effective in terms of impact is the cluster II system with the optimized structure and the parameters working in FLEX mode because it provides an increase in the amount of transmitted information flows on average by 17% compared with the standard modes (Fig. 3).

The presented results showed that the developed method allows to take into account and to optimally use the available structural and functional excessiveness of the cluster, its emergent properties arising from the formation of such a distributed network structure as a "virtual spacecraft" and the possible effects of the external environment. The study managed to synthesize small promising spacecraft cluster functioning flexible technology, which allowed to optimize the structure and parameters of its II system.

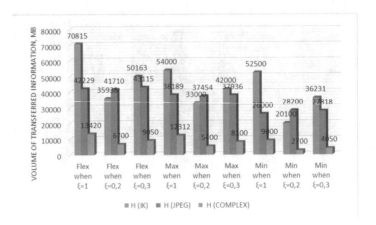

Fig. 3. The amount of transmitted streams in different types of environmental influences

5   Conclusions

In the framework of the article the original method of structural-parametric optimization of the II system of the cluster of promising remote sensing small spacecraft allows to obtain specific values of the optimal parameters ω_μ^*, what makes possible its application in practice, the specialist in the formation of modes of promising small spacecrafts in the cluster in a two-tier management technologies new orbital systems network architecture. The use of this technique permitted to increase the potential performance of a promising cluster of small spacecraft by 15–20% in comparison with the standard approaches to the management of its information interaction for the fulfilment of the targets. It should also be noted that the developed model-algorithmic complex and implemented it as the software package have sufficient accuracy of results and high versatility.

Acknowledgements. The research described in this paper is partially supported by the Russian Foundation for Basic Research (grants 16-07-00779, 16-08-00510, 16-08-01277, 16-29-09482-ofi-i, 17-08-00797, 17-06-00108, 17-01-00139, 17-29-07073-ofi-i, 18-08-01505, 18-07-01272), grant 074-U01 (ITMO University), project 6.1.1 (Peter the Great St. Petersburg Politechnic University) supported by Government of Russian Federation, Program STC of Union State "Technology-SG" (project 1.3.3.3.1), state order of the Ministry of Education and Science of the Russian Federation 2.3135.2017/4.6, state research 0073–2014–0009, 0073–2015–0007, International project ERASMUS +, Capacity building in higher education, 73751-EPP-1-2016-1-DE-EPPKA2-CBHE-JP, International project KS1309 InnoForestView Innovative information technologies for analyses of negative impact on the cross-border region forests.

References

1. Pavlov, D.A.: Metodika planirovanija operacij informacionnogo vzaimodejstvija klastera malyh kosmicheskih apparatov distancionnogo zondirovanija Zemli (The technique of operations scheduling communications between a cluster of small spacecraft remote sensing of the Earth). Trudy Voenno-kosmicheskoj akademii imeni A.F. Mozhajskogo, vol. 649, pp. 37–47 (2014). (In Russian)
2. Gorodetsky, V.I., Karsaev, O.V.: Samoorganizaciya gruppovogo povedeniya klastera malyh sputnikov raspredelennoj sistemy nablyudeniya (Distributed surveillance system based on self-organized collective behavior of small satellite cluster). Izvestiya SFedU. Engineering Sciences. vol. 2(187), pp. 234–247 (2017). (In Russian)
3. Sollogub, A.V., Skobelev, P.O., Simonova, E.V., Carev, A.V., Stepanov, M.E, Zhiljaev, A. A.: Intellektual'naja sistema raspredelennogo upravlenija gruppovymi operacijami klastera malorazmernyh kosmicheskih apparatov v zadachah distancionnogo zondirovanija Zemli (Intelligent system distributed control of group operations of a cluster of small spacecraft in problems of remote sensing of the Earth). Informacionno-upravljajushhie sistemy. vol. 1, pp. 16–26 (2013). (In Russian)
4. Potrjasaev, S., Okhtilev, P., Ipatov, Y., Sokolov, B.: Methodology and structure adaptation algorithm for complex technical objects reconfiguration models. In: Cybernetics and Mathematics Applications in Intelligent Systems Proceedings of the 6th Computer Science On-line Conference, vol. 574, pp. 319–328. Springer (2017)
5. Pavlov, A.N., Pavlov, D.A., Slin'ko, A.A.: Strukturno-parametricheskij sintez sistemy informacionnogo vzaimodejstviya klastera perspektivnyh malyh kosmicheskih apparatov distancionnogo zondirovaniya zemli i ocenka ee robastnosti (Structural-parametric synthesis of the system of information interaction of a cluster of perspective small satellites of remote sensing of the Earth and evaluation of its robustness). H&ES Research. vol. 9, № 5, pp. 6–18 (2017)
6. Hromov, A.V.: Sputnik distancionnogo zondirovaniya Zemli SkySat (Satellite remote sensing of the Earth SkySat) (2017). http://www.dauria.ru/blog/skysat
7. Potyupkin, A.U., Danilin, N.S., Selivanov, A.S.: Klastery malorazmernyh kosmicheskih apparatov kak novyj tip kosmicheskih ob"ektov (Clusters of small spacecraft as a new type of space objects) Raketno-kosmicheskoe priborostroenie i informacionnye sistemy. vol. 4, № 4, pp. 45–56 (2017). (In Russian)
8. Slin'ko A.A., Pavlov A.N., Pavlov D.A. Pavlov A.A. Model' planirovanija operacij ustojchivogo informacionnogo vzaimodejstvija klastera malyh kosmicheskih apparatov distancionnogo zondirovanija Zemli (The planning model of sustainable operations information interaction cluster of small spacecraft remote sensing of the Earth). Trudy Voenno-kosmicheskoj akademii im. A.F. Mozhajskogo. vol. 654, pp. 8–13 (2016). (In Russian)
9. Pavlov, D.A., Osipenko, S.A., Masalkin, A.A., Slin'ko, A.A.: Podhod k resheniju zadachi poiska intervalov postojanstva struktury klasterov malyh kosmicheskih apparatov. (The approach to solving the problem of finding intervals of constancy of the structure of clusters of small spacecraft) Sbornik trudov II Vserossijskoj NTK « Teoreticheskie i prikladnye problemy razvitija i sovershenstvovanija avtomatizirovannyh sistem upravlenija voennogo naznachenija ». SPb.: VKA imeni A.F.Mozhajskogo, pp. 194–195 (2015). (In Russian)
10. Moskvin, B.V.: Teoriya prinyatiya reshenij: Uchebnik (The theory of decision making: Textbook) SPb. VKA imeni A.F. Mozhajskogo. 383 p. (2005). (In Russian)
11. Ivanov, D.A., Sokolov, B.V., Pavlov, A.N.: Optimal distribution (re)planning in a centralized multi-stage supply network in the presence of the ripple effect. Eur. J. Oper. Res. 237(2), 758–770 (2014)

12. Sokolov, B.V., Moskvin, B.V., Pavlov, A.N.: Voennaya sistemotekhnika i sistemnyj analiz. Modeli i metody prinyatiya reshenij v slozhnyh organizacionno–tekhnicheskih kompleksah v usloviyah neopredelyonnosti i mnogokriterial'nosti: Uchebnik. (Military systems engineering and systems analysis. Models and methods of decision-making in complex technical–organizational systems in conditions of uncertainty and multicriteria: Textbook) SPb. VIKKU imeni A. F. Mozhajskogo. 496 p. (1999). (In Russian)
13. Pavlov, A.N., Pavlov, A.A., Slinko, A.A., Pashenko, A.E.: Research of structural reliability and survivability of complex objects. In: Automation Control Theory Perspectives in Intelligent Systems » Proceedings of the 5th Computer Science In-Line Conference, vol. 3, pp. 463–473. Springer (2016)
14. Ivanov, D., Hartl, R., Dolgui, A., Pavlov, A., Sokolov, B.: Integration of aggregate distribution and dynamic transportation planning in a supply chain with capacity disruptions and the ripple effect consideration. Int. J. Prod. Res. 53(23), 6963–6979 (2015)

Collaborative Robot YuMi in Ball and Plate Control Application: Pilot Study

Lubos Spacek$^{(\boxtimes)}$, Jiri Vojtesek, and Jiri Zatopek

Department of Process Control, Faculty of Applied Informatics, Tomas Bata
University in Zlin, Nad Stranemi 4511, 760 05 Zlin, Czech Republic
{lspacek,vojtesek,zatopek}@utb.cz

Abstract. Ball & Plate is a well-known concept and interesting example of an unstable process. Numerous types of Ball & Plate structure can be found and this paper tries to extend its potential of moving the whole plate in space by using an industrial robotic manipulator as the most flexible way to achieve this goal. The collaborative dual-arm robot YuMi from ABB is chosen for this task as the balance between precision and safety. The purpose of the paper is to investigate restrictions and boundaries of such solution, thus the model is identified and based on this identification is designed a controller, which is tested in a simulation environment.

Keywords: Ball & Plate · YuMi · Robot · LQ control · Discrete-time

1 Introduction

The purpose of this paper is to investigate the feasibility of using the industrial collaborative robot as a mechanical motion system for Ball & Plate model. There are many Ball & Plate models with successful implementation, ranging from research projects [1] to hobby models using basic PID or more advanced LQG control [2]. Although fundamentals of the system are always the same, the realization of motion of the plate has many varieties from simple 2-motor construction to more advanced 6DoF Stewart platforms. Another possibility is to use industrial robotic manipulator which would offer a quite interesting topic to examine. Having 6 or 7 linked motors to move the plate only in two angles seems to be quite redundant, but it also offers an extra dimension to work in. Now the plate can be moved up and down, the whole arm could act as a gyro-stabilizer, the ball can be controlled to bounce at a specific frequency or height. This pilot study is conducted in a simulated environment before it will be deployed to a real robotic manipulator.

A linear quadratic controller is designed for the purpose of investigation of restrictions of the system. It is relatively complex, but still simple enough to be quickly designed and implemented into the robot program, as the best and

© Springer International Publishing AG, part of Springer Nature 2019
R. Silhavy (Ed.): CSOC 2018, AISC 765, pp. 167–175, 2019.
https://doi.org/10.1007/978-3-319-91192-2_18

optimal control strategy is not the goal of this paper. The author is also familiar with this concept [3], which helps with overcoming difficulties during controller design. The controller is designed using polynomial approach, which utilizes operations on algebraic polynomial (Diophantine) equations. Parameters of the controller are obtained by minimizing linear quadratic LQ criterion, which helps with the determination of poles of the characteristic polynomial [4].

The paper is organized as follows. A brief description of components of the model is in the Introduction part. Methods section contains measurement description, identification and controller design. Results section shows outputs from the previous section and the paper is concluded with Discussion section.

1.1 Ball & Plate Model

The Ball & Plate model is an unstable system with 2nd order astatism (with two integrators). The ball is moving on the plate and its position is described in x and y coordinates (typically for the center of the plate [x, y] = [0,0]). This movement can be controlled by changing the angle of the plate in two planes. The goal is to stabilize the ball in the requested position or move it along the desired trajectory.

1.2 YuMi® Collaborative Robot

YuMi is a collaborative, dual-arm industrial robot designed by ABB for small part assembly (Fig. 1). YuMi has two independent 7DoF manipulators with 0.02 mm repeatability and its precise control, safety and variability are advantages worth to exploit [5].

Fig. 1. ABB IRB 14000 YuMi

2 Methods

2.1 Position Measurement

The position of the ball in RobotStudio is obtained using Smart Component Designer. Smart Components simulate the behavior of simulated objects (linear movement, rotation) and offer additional tools, such as signal logic, arithmetic, parametric modeling, and sensors.[1] Smart Components thus extend abilities of RobotStudio to simulate real environments and offer a great background for this paper without the need of 3rd party simulation environments and their implementation.

The Smart Component named `PositionSensor` monitors the position of an object in the chosen reference frame. Positions in X and Y coordinates are subsequently converted to analog outputs in reference to the Smart Component, as can be seen in Fig. 2. These outputs are then cross-connected to analog inputs of the robot in Station Logic. This relatively complicated approach is necessary because RobotStudio tries to simulate the robot to be as authentic as possible, so even all connections and visuals are precise virtual replicas. In the real setup, the position of the ball will be determined using resistive touchscreen foil.

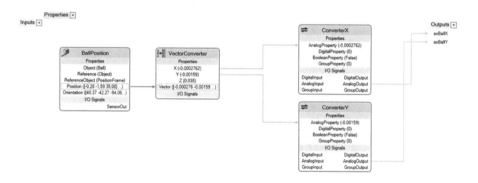

Fig. 2. Smart component design - position sensor

2.2 Identification

The system was identified using a similar experimental technique as in the previous work of the author [6] on the Ball & Plate system which proved to be a reasonable solution. As opposed to other projects, where the parameters of the Ball & Plate mechanical system were known, in this case, the industrial robot acts as a "black box". This is because of its commercial use and manufacturer's effort to keep know-how.

The manipulator (one arm of the robot) was identified during simulation in RobotStudio which tries to replicate the real structure as precise as possible. This could be called semi-experimental identification because dynamics in

---

[1] RobotStudio Product Specification 3HAC026932-001 Revision R.

RobotStudio is programmed using mathematical models, thus it is still only an ideal representation of the real robot, without noise and errors.

The identification was based on the step response of the system. Input is the angle of the plate and output is the position of the ball in one direction. Two constant parameters of the transfer function were identified. This transfer function (Eq. 1) was derived in the previous work of the author [3].

$$G(s) = \frac{C}{s^2 (Ts + 1)} = \frac{C}{Ts^3 + s^2} \ . \tag{1}$$

where $G(s)$ is continuous transfer function with complex variable s, C is velocity gain and T is the time constant of the system. Measurements were made for multiple inputs and their plot can be seen in Fig. 3 and identified constants in Fig. 4.

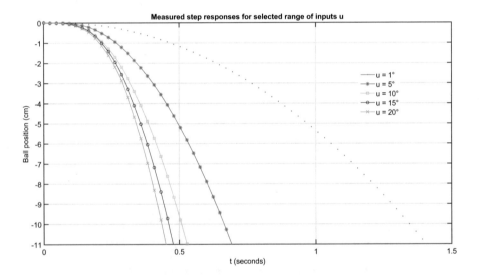

Fig. 3. Step responses of the system

2.3   Controller Design

Linear quadratic (LQ) discrete-time controller with 2 degrees of freedom (2DoF) was designed [4]. It is based on a discretized form of Eq. 1. This controller was chosen because of positive experiences with the similar type of control system and ease of implementation. This serves only for the pilot study as the quickest and reliable design procedure and other types of controllers will be designed. The structure of 2DoF controller is shown in Fig. 5. This type of controller has two main parts (feedforward part C_f and feedback part C_b) shown in

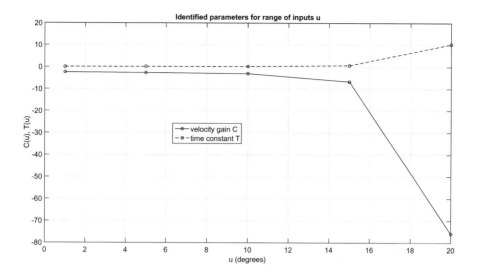

Fig. 4. Identified parameters

Eqs. 2 and 3 respectively. The feed-forward part filters the desired value, which leads to slower, but softer responses.

$$C_f(z^{-1}) = \frac{R}{P} = \frac{r_0}{1 + p_1 z^{-1} + p_2 z^{-2}} \ . \tag{2}$$

$$C_b(z^{-1}) = \frac{Q}{P} = \frac{q_0 + q_1 z^{-1} + q_2 z^{-2} + q_3 z^{-3}}{1 + p_1 z^{-1} + p_2 z^{-2}} \ . \tag{3}$$

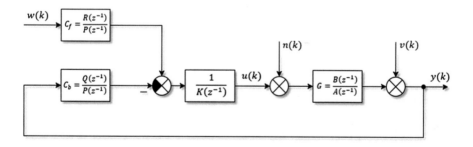

Fig. 5. Structure of 2DoF controller

3   Results

It is obvious from Fig. 4 that the system significantly loses its linearity for higher plate angles ($>15°$). This is probably caused by the complexity of the motion system. With 7 degrees of freedom, the system is prone to change its dynamics. Although manipulator moves mostly with 5th, 6th and 7th axis, the effect adds up to larger movements. This could be solved by several techniques, but it is neglected in this pilot study.

Only the ball stabilization was considered, thus desired value is always equal to [0,0], which is the center of the plate. At the start of the simulation, the ball is placed at an offset from the center. After it is stabilized in the desired position, an impulse force with random direction and magnitude is applied to the ball several times as a disturbance. The result is shown in Figs. 6 and 7.

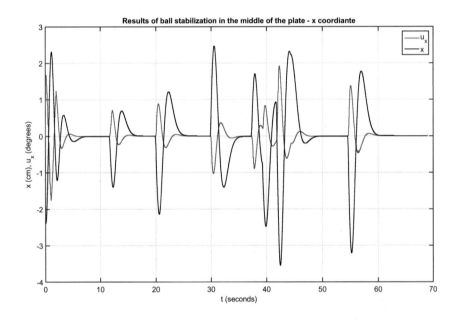

Fig. 6. Results for x coordinate

The output of the 2DoF controller is calculated using Eq. 4 for both the x coordinate and y coordinate.

$$u_k = (1 - p_1)u_{k-1} + (p_1 - p_2)u_{k-2} + p_2 u_{k-3}$$
$$+ r_0 w_k - q_0 y_k - q_1 y_{k-1} - q_2 y_{k-2} - q_3 y_{k-3} \ . \ (4)$$

where u_{k-i} and y_{k-i} are outputs of the controller and plant respectively, w_k is desired value and individual calculated parameters are shown in Eq. 5. It is implemented in RobotStudio as relative rotation around the chosen tool, which

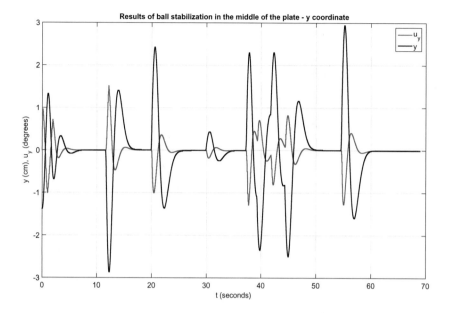

Fig. 7. Results for y coordinate

is an imaginary coordinate system located at the center of the flange. Sampling period for the controller design was chosen to be 50 ms.

$$\begin{bmatrix} r_0 \\ p1 \\ p2 \end{bmatrix} = \begin{bmatrix} -0.0014 \\ -1.5529 \\ 0.6201 \end{bmatrix}, \quad \begin{bmatrix} q0 \\ q1 \\ q2 \\ q3 \end{bmatrix} = \begin{bmatrix} -3.4851 \\ 9.0535 \\ -7.7201 \\ 2.1503 \end{bmatrix}. \tag{5}$$

The process was simulated in RobotStudio, which provides a 3D graphical environment, where the whole process can be clearly seen as in the real world. This can be presented on the paper only in the form of graphs, but for a better idea, the picture from the environment is shown in Fig. 8.

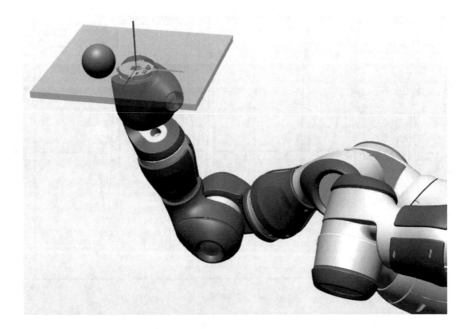

Fig. 8. Ball & Plate on YuMi in RobotStudio

4   Discussions

The results of this pilot study are satisfactory because virtual robot managed to stabilize the ball even after multiple disturbances. The process became unstable for larger angles of the plate due to non-linearity and this could be easily compensated, but the purpose of the paper was not to design a state of the art controller (that will be the aim of future work). From presented results can be concluded, that collaborative robot YuMi can be used in simulations as a motion system for Ball & Plate model, although good simulation background is needed for implementation in the real system.

Acknowledgments. This article was created with support of the Ministry of Education of the Czech Republic under grant IGA reg. n. IGA/CebiaTech/2018/002.

References

1. Jadlovská, A., Jajčišin, Š., Lonščák, R.: Modelling and PID control design of nonlinear educational model ball & plate. In: 17th International Conference on Process Control 2009, pp. 475–483. Štrbské Pleso (2009)
2. Ball and Plate Arduino LQG control. https://youtu.be/8jk17dXETHs
3. Spaček, L., Bobál, V., Vojtěšek, J.: LQ digital control of Ball & Plate system. In: 31st European Conference on Modelling and Simulation ECMS 2017, pp. 427–432. European Council for Modelling and Simulation (2017)

4. Bobál, V., Böhm, J., Fessl, J., Macháček, J.: Digital Self-tuning Controllers. Springer-Verlag, London (2005)
5. ABB Robotics - IRB 14000 YuMi. http://new.abb.com/products/robotics/industrial-robots/yumi
6. Spaček, L., Bobál, V., Vojtěšek, J.: Maze navigation on Ball & Plate Model. In: Silhavy, R., Senkerik, R., Kominkova Oplatkova, Z., Prokopova, Z., Silhavy, P. (eds.) Cybernetics and Mathematics Applications in Intelligent Systems CSOC 2017. Advances in Intelligent Systems and Computing, vol. 574, pp. 206–215. Springer, Cham (2017)

Integration of Heterogeneous Data in Monitoring Environmental Assets

Dmitrii Verzilin[1,2(✉)], Tatyana Maximova[3], Yury Antokhin[4], and Irina Sokolova[5]

[1] St. Petersburg Institute for Informatics and Automation of the Russian Academy of Sciences (SPIIRAS), 14th Lin. V.O., 39, St. Petersburg 199178, Russia
modusponens@mail.ru
[2] Lesgaft National State University of Physical Education, Sport and Health, ul. Decembrists, 35, St. Petersburg 190121, Russia
[3] Saint Petersburg National Research University of Information Technologies, Mechanics and Optics (ITMO University), Kronverksky pr., 49, St. Petersburg 197101, Russia
maximovatg@gmail.com
[4] Territorial Fund of Compulsory Medical Insurance of the Leningrad Region Bolshaya Raznochinnaya, 27, St. Petersburg 197110, Russia
antokhinyn@mail.ru
[5] St. Petersburg State University, Universitetskayz emb., 7–9, St. Petersburg 199034, Russia
i_sokolova@bk.ru

Abstract. The purpose of the study was to develop a general scheme for measuring and analyzing the dynamics of the state of environmental assets using a set of indicators constructed from environmental, social and economic monitoring data. Methodical approaches to the collection and integration of heterogeneous data describing the functioning of eco-economic objects (port, object of agriculture, recreational zone) have been developed. These data can be obtained as a result of ground-based measurements, remote sensing of the Earth, research of online activity of the population related to environmental responsibility and environmental concerns, and the determination of statistical indicators describing dynamics of environmental assets. The set of the indicators obtained from the remote sensing data of the land has been determined for the port area as a source of adverse impact on environmental assets. It was found out that the data on the Internet search queries for keywords related to environmental pollution reflect the degree of concern of the population with the environmental situation in the region and can be used as an indirect indicator of environmental ill-being. A general framework for integration of heterogeneous data in multi-criteria assessment of the state of eco-economic objects was grounded.

Keywords: Ecological economy · Environmental assets
Environmental monitoring · Heterogeneous data · Internet purchases
Key word searchers · Land remote sensing data

© Springer International Publishing AG, part of Springer Nature 2019
R. Silhavy (Ed.): CSOC 2018, AISC 765, pp. 176–185, 2019.
https://doi.org/10.1007/978-3-319-91192-2_19

1 Introduction

The protection of the environment is the key to the quality of life of the present and future generations. Such protection should be based on the balanced use and conservation of environmental assets. Currently, developed countries pay serious attention to the protection of the environment. The costs of environmental protection in the European Union countries in recent years are about 2.43% of GDP [6]. In the coastal countries of the Baltic Sea, this indicator is usually higher: 3.99% in Germany, 3.65% in Latvia, 3.37% in Netherlands, 3.06% in Estonia [6]. In Russia, the costs of protecting the environment amounted to 0.7% of GDP in 2014–2016 [4].

Existing approaches to evaluation of environmental activities include the use of cost indicators for these activities, estimations of environmental damage and characteristics of nature protected areas. To date, the practice of assessing the socio-economic results of economic activities aimed at the use, conservation and development of environmental assets has not been widely disseminated, and there are no methods for constructing such assessments using integrated socio-economic and environmental monitoring data.

The main directions of research in the field of environmental economics were formed in the twentieth century. The by-products of economic activity and consumption are part of the economic system in Leontiev-Ford's industrial balance model [10, 11], which additionally takes into account environmental factors. Dependencies between the volumes of production of "desirable" and "undesirable" products are presented in the form of coefficients that were used in the classical model of Leontiev to describe the interrelations between the production and consumption sectors.

The problems of quantitative evaluation of the effectiveness of nature protection were considered in many works by Hoffmann, Motkin, Ryumin [7, 8, 12, 15], and others. Hoffmann formed the main direction of environmental and economic analysis - the construction of monetary estimates of natural resources [7, 8]. A significant practical result of the work performed was the methodology for assessing the economic efficiency of environmental activities and economic damage [21] in economic calculations. The methodology allowed to justify the economic feasibility of achieving specified environmental standards, and to estimate cost values for damages from certain types of pollution.

The current state program on environmental protection in Russia [22] is aimed at preserving and restoring natural ecosystems, ensuring environmental safety and economic efficiency. At the regional level, environmental indicators are monitored, and the effectiveness of executive authorities in the environmental field is assessed [14].

The International Society of Ecological Economics (ISEE) [9] coordinates scientific research, conducts an international scientific conference [23] and publishes an interdisciplinary journal [5].

The United Nations Statistical Commission in 2012 approved an international standard for the System of Environmental and Economic Accounting [20]. The system integrates economic and environmental indicators to describe the dynamics of environmental assets and determine the impact of the economy on the environment. Russia has not yet joined this system.

Modern theoretical studies on the ecological economy consider the problems of interrelated description of ecological, economic and social efficiency at micro- and macro-levels. A typical example of such studies is [19]. Approaches to the complex analysis and synthesis of ecological, socio-economic, and medical statistical data of were developed in our works [2, 18, 26–29].

The drawbacks of existing approaches to environmental and economic monitoring are as follows. Data on the ecological state are fragmentary in time and localization of observations. There is no unified system of interrelated ecological and socio-economic indicators of eco-economic objects. The drawback, which is a consequence of the two previous ones, is the impossibility of establishing cause-effect relationships between the characteristics of anthropogenic environmental impacts and socio-economic indicators of economic objects.

The current level of digital technologies makes it possible to characterize functioning of eco-economic objects using heterogeneous data including statistical indicators describing dynamics of ecological assets and results of monitoring that involves ground-based measurements and observations through remote sensing of the earth which compensate fragmentation of terrestrial observations.

The purpose of the study was to develop methods for measuring and analyzing the dynamics of environmental assets using a set of indicators constructed from environmental, social and economic monitoring data.

2   Methods

2.1   Remote Sensing

Remote sensing of environment provides actual spatial and temporal ecological data [3, 13, 16, 30].

To overcome abovementioned drawbacks of existing approaches to environmental and economic monitoring, we facilitate an interdisciplinary approach to multi-criteria assessment of the state of eco-economic objects, with the use of remote sensing data of the Earth for given time points and given coordinates. These data can include not only ecological characteristics but also measurements of on-going activities using and changing environmental assets.

Remote sensing data should be coordinated in place and time with the calculated values of socio-economic indicators.

From a methodical point of view, integration of heterogeneous data obtained as a result of measurements (monitoring data), analytical calculations (statistical indices) and modeling is expedient due to the formation of a unified system of socio-economic indicators for eco-economic objects and definition of requirements for the monitoring system providing these indicators.

2.2   The System of Environmental and Economic Accounting

Methodological approaches to the construction of estimates of social and economic efficiency of activities leading to a change in environmental assets should reflect the relationship between the level of consumption of natural resources and production volumes and the value added by types of economic activity. Environmental assets include land, as a space for activities, natural resources (biological, water and soil resources). The complexity of assessing the value of environmental assets is due to the fact that they do not always have a market price. Over time, the cost may decrease because of depletion of resources and increase with the restoration and natural compensation of resources. The costs associated with the depletion of environmental assets can not be regarded as costs compared to the revenues of an enterprise that uses environmental assets.

In the formation and analysis of interrelated socio-economic and environmental indicators, it is advisable to use methodological approaches to data integration implemented in the System of Environmental and Economic Accounting [20]:

- delimitation of agreed ecological and economic observations;
- determination of the value of environmental assets;
- representation of all data in the form of stocks and flows in physical and monetary units.

The boundaries of observations are determined for the coordinated organization of information in space and time. Borders can be defined for countries, regions, territories, economic sectors, specified groups of organizations and individual environmental and economic facilities, such as ports, shoreline areas, coastal recreational zones, coastal urban areas, coastal agricultural lands, etc.

To assess the impact of eco-economic objects on environmental assets, it we propose to distinguish data from three levels: the first level - primary signs (primary parameters); second level - indicators; the third level is the results of modeling. The first-level data are determined as a result of statistical observations and direct measurements on the ground and using methods of remote sensing of the Earth. Direct measurements make it possible to determine not only ecological features (for example, water pollution levels, the state of the shoreline), but also characteristics of economic activity (for example, the intensity of recreational areas use). The indicators of the second level are directly calculated for the values of the primary characteristics. The results of the simulation are a description of trends and patterns of dynamics and mutual influence of indicators.

In the indicators characterizing the activity of enterprises using environmental assets, we propose [25] to use standard economic indicators of reserves and changes in assets, stocks and flows of products and services in monetary and physical units.

2.3 Online Socio-ecologic Activity

As part of the socio-demographic description we propose to consider the indicators of labor and employment, the natural movement of the population, morbidity, migration, satisfaction with the accessibility and quality of ecosystem services. Given the global world trends in the development of digital technologies and their diffusion into the economic and social space, it is expedient to study the online activity of the population concerned with the environment factors. The productivity of this approach is demonstrated, in particular, in [1, 17, 24].

To ground a general framework for integration of heterogeneous data in multi-criteria assessment of the state of eco-economic objects we introduce requirements to indicators to be used for the assessment, approaches for linking indicators to each other and an example illustrating a relationship between environmental indicators and indicators of ecology-related online activity in population.

3   Results

3.1   The System of Indicators

Determination of the flow of changes in environmental assets in monetary terms should be based on the use of three types of indicators:

- the cost of eliminating the consequences of adverse impacts on the environmental asset of anthropogenic and natural factors;
- the cost of efforts aimed at preventing environmental damage;
- costs associated with the depletion of environmental assets.

The latter indicator, unlike the first two types of indicators, can not be considered as an expense of an enterprise using environmental assets.

Socio-economic indicators describing the state of eco-economic objects and taking into account the impact of unfavorable factors are as follows.

Indicators of natural resource flows (land, recreational, soil, forest, fish, etc.) in value terms over the observation period should include: intermediate and final consumption of natural resources; gross added value created using natural resources.

Indicators of natural resource flows in physical terms over the observation period should include: physical amounts of ecosystem services, particularly recreational services; quantities of agricultural products, and amounts of waste.

Indicators of environmental assets at the end of the observation period in monetary units should include the value of environmental assets.

The indicators of the stocks of environmental assets at the end of the observation period in physical units should include the areas of recreational zones, agricultural lands, forests, etc.

Indicators of flows of changes in environmental assets should include the depletion of natural resources in monetary and physical units, the costs of replenishment and conservation of natural resources.

In constructing social indicators, it is reasonable to consider the social consequences of changing the state of environmental assets in terms of the dynamics of environmental interest indicators (environmental responsibility and environmental concerns) of the population.

3.2   Linking Indicators to Each Other

For multi-criteria estimation of the state of eco-economic objects, it is necessary to use models of three levels.

The first level includes the models for assessing adverse environmental factors from remote sensing data of the Earth:

- concentrations of substances in wastewater from observations in different spectral ranges,
- the concentration of filamentous blue-green algae from the temperature of water,
- surface contamination of oil products from the area of the spot and thickness of the film,
- dynamics of non-localized solid waste.

The composition of the indicators obtained from remote sensing data of the land in the port area can be determined as a source of adverse impact on environmental assets, expressed as flows in physical terms:

- discharges into water resources;
- separate indicators for types of discharged substances in wastewater: suspended solids, heavy metals, phosphorus, nitrogen, tons/month;
- volume of pollution by oil and oil products, tons/month;
- consequences of discharges into water resources on the state of the ecosystem of the reservoir (eutrophication);
- concentration and area of distribution of filamentous, blue-green water-trees;
- the area of reed thickets on the shoreline;
- the volume of solid waste;
- mixed domestic and commercial wastes (exported, imported, stored on the port territory, not localized in the port territory), tons/month.

The second level includes economic models for analyzing the economic activity of an enterprise using environmental assets to estimate the flows of changes in environmental assets in monetary and physical terms (produced agricultural products, ecosystem services); models for assessing the cost of environmental damage (the cost of activities aimed at preventing environmental damage and costs associated with the depletion of environmental assets).

The third level includes statistical, econometric and ecological models for assessing the elasticity of environmental and socio-economic indicators (sensitivity of indicators to changes in other indicators) and identifying the risks to loose equilibrium in the ecological and economic system.

3.3 An Example of an Online Response to Air Pollution

Taking into account the global world trends in the development of digital technologies and their diffusion into the economic and social space, we propose to use indicators of environmental interest (environmental responsibility and environmental concerns) of the population.

To estimate these indicators we can use online activity of the population.

The number, prevalence and dynamics of search queries for keywords related to recreational services, environmental pollution, cleanliness of water and air, the state of recreational resources reflect the degree of responsibility and concern of the population with the environmental situation in the region and are an indirect indicator of environmental ill-being. For example, we can examine the popularity of search queries for the keywords "dump", "waste", "water bodies", "waste water" among the population of territories associated with a coastal zone.

To find out a relationship between topical search queries and current environmental conditions we jointly used data on the number of regional online searches and the duration of unfavorable meteorological conditions rising concentrations of harmful impurities in the air (Fig. 1).

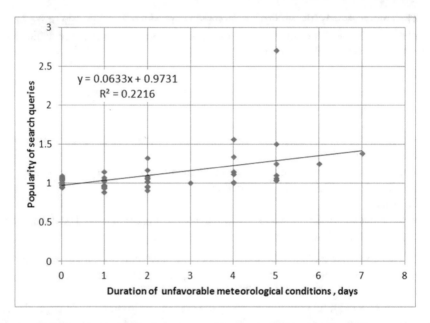

Fig. 1. Regression for the duration of unfavorable meteorological conditions (according to the Ministry of Ecology of Chelyabinsk region, http://www.mineco174.ru/, 20.02.2017–12.02.2018) and weekly popularity of search queries with the word "smog" (ratio of relative number of queries in Chelyabinsk and Russia, same weeks, according to the Yandex WordStat Service, https://wordstat.yandex.ru)

The regression obtained confirms the perspective of using data on regional online activity in multi-criteria assessment of environmental assets.

4   Discussion

Integration of social, economic and environmental data within the boundaries of the ecological and economic site, providing an interactive regime for environmental and economic monitoring (increasing the detail of observations and measurements, changing the composition of the observed parameters, etc.) will serve as a basis for planning and clarifying the measures to ensure environmental safety.

The system of ecological, economic, and social indicators for assessing activities on the use and conservation of environmental assets, as compared with classical environmental-economic accounting, firstly, directly links common environmental goals with strategic and tactical goals, facing the region where the assets are located, secondly, lets determine and select the sequence of tasks to be solved and the actions to be taken to control eco-economic object, and thirdly, justifies compromise multi-criteria decision in the allocation of limited resources in eco-economic objects.

Real-time acquisition of remote sensing results allows to build comprehensive assessments of the current values of the environmental parameters of ecological objects and the intensity of economic activity (recreational areas, the population using recreational services, recreational density of recreational areas, the intensity of traffic flows, the area of land used for agriculture, etc.).

The further steps for research should be aimed at detailing the general framework for integration of social, economic, and environmental data in evaluation of environmental assets. There are a lack of models describing mass ecology-related behavior.

Acknowledgments. The research described in this paper is partially supported by the Russian Foundation for Basic Research (grants 15-08-08459, 16-07-00779, 17-08-00797, 17-06-00108), Russian Humanitarian Found (grants 15-04-00400), state research 0073–2018–0003.

References

1. Brodovskaya, E., Dombrovskaya, A., Synyakov, A., Batanina, I.: Online social activity in Latin America (Mexico, Chile and Uruguay) and Russia: cross-national research. In: Chugunov, A.V., Bulgov, R., Kabanov, Y., Kampis, G., Wimmer, M. (eds.) Digital Transformation and Global Society First International Conference, DTGS 2016, St. Petersburg, Russia, 22–24 June 2016, pp. 26–34 (2016). Revised Selected Papers
2. Chereshnev, V.A., Verzilin, D.N., Maksimova, T.G., Versilin, S.D.: Ecologization and socioeconomic development of the regions: assessment of regional differentiation. Econ. Reg. 1(33), 33–46 (2013). (in Russian)
3. Choi, J.-K., Park, Y.J., Lee, B.R., Eom, J., Ryu, J.-H.: Application of the Geostationary Ocean Color Imager (GOCI) to mapping the temporal dynamics of coastal water turbidity. Remote Sens. Environ. **146**, 24–35 (2014)
4. Costs for environmental protection as a percentage of GDP. EMISS. https://fedstat.ru/indicator/57339
5. Ecological Economics: The Transdisciplinary Journal of the International Society for Ecological Economics (ISEE). http://www.journals.elsevier.com/ecological-economics
6. Environmental tax revenues. Eurostat Data Database. http://appsso.eurostat.ec.europa.eu/nui/submitViewTableAction.do
7. Gofman, K.G.: Economic mechanism of nature use in the transition to a market economy. Econ. Math. Meth. **27**(2), 315–321 (1991). (in Russian)
8. Gofman, K.G.: Transition to the market and the ecologization of the tax system in Russia. Econ. Math. Meth. **T.30**(4) (1994). (in Russian)
9. International Society of Ecological Economics (ISEE). http://www.isecoeco.org
10. Leontiev, V., Ford, D.: Interbranch analysis of the impact of the structure of the economy on the environment. Economics and mathematical methods, vol. VIII (1972). vol. 3, pp. 370–400 (in Russian)

11. Leontiev, V.: Impact on the environment and economic structure: the "input-output" approach. Economic essays, pp. 318–339. Politizdat, Moscow (1990). (in Russian)
12. Motkin, G.A.: Fundamentals of environmental insurance, 192 p. Nauka, Moscow (1996). (in Russian)
13. Potryasaev, S.A., Sokolov, B.V., Yusupov, R.M.: Conceptual and formal description of structure-function alsynthesis and development control problems of space and ground based date monitoring information systems. SPIIRAS Proc. **5**(28), 82–106 (2013)
14. Report on the environmental situation in St. Petersburg in 2014. Ed. I.A. Serebritsky. SPb.: Diton LLC, 180 p. (2015). (in Russian). http://gov.spb.ru/static/writable/ckeditor/uploads/2015/06/19/doklad_2014_SWipmNU.pdf
15. Ryumina, E.V.: Ecological factor in economic-mathematical models, 166 p. Nauka, Moscow (1980). (in Russian)
16. Santini, F., Alberotanza, L., Cavalli, R.M., Pignatti, S.: A two-step optimization procedure for assessing water constituent concentrations by hyperspectral remote sensing techniques: an application to the highly turbid Venice lagoon waters. Remote Sens. Environ. **114**(4), 887–898 (2010)
17. Sokolov, B., Verzilin, D., Maximova, T., Sokolova, I.: Dynamic models of self-organization through mass behavior in society. In: Advances in Intelligent Systems and Computing (AISC, vol. 679), vol. 1, pp. 114–123. Springer, Cham (2018). Part of the Advances in Intelligent Systems and Computing book series. Print ISBN 978-3-319-68320-1, Online ISBN 978-3-319-68321-80. https://doi.org/10.1007/978-3-319-68321-8, https://link.springer.com/chapter/10.1007/978-3-319-68321-8_12
18. Sokolov, B., Yusupov, R., Verzilin, D., Sokolova, I., Ignatjev, M.: Methodological basis of socio-cyber-physical systems structure-dynamics control and management. In: Chugunov, A.V., Bulgov, R., Kabanov, Y., Kampis, G., Wimmer, M. (eds.) Digital Transformation and Global Society First International Conference, DTGS 2016, St. Petersburg, Russia, 22–24 June 2016, pp. 610–618 (2016). Revised Selected Papers
19. Sukharev, O.S.: The general economic and "ecological" efficiency: theoretical statement. The Bulletin of the Perm university. Series: The Economy, No. 3 (2014). (in Russian)
20. System of Environmental-Economic Accounting 2012 – Central Framework. New York: United Nations, 378 p. (2014). ISBN: 987-92-1-161563-0. eISBN: 978-92-1-055926-3
21. Temporary sample methodology for determining the economic efficiency of implementing environmental measures and assessing economic damage caused to the national economy by environmental pollution. Approved by the decision of the State Planning Committee of the USSR, Gosstroy USSR and the Presidium of the USSR Academy of Sciences on 21 October 1983, No. 254/284/134. Economics (1986). (in Russian)
22. The State Program of the Russian Federation "Environmental Protection" for 2012–2020. Approved by Resolution of the Government of the Russian Federation of 15 April 2014, No. 326. (in Russian). http://programs.gov.ru/Portal/programs/passport/12
23. Transforming the Economy: Sustaining Food, Water, Energy and Justice. 2016 ISEE Conference/International Society of Ecological Economics (ISEE) (2016). http://www.isecoeco.org/isee-2016-conference/
24. Verzilin, D., Maximova, T., Sokolova, I.: Online socioeconomic activity in Russia: patterns of dynamics and regional diversity. Digital Transformation and Global Society: Second International Conference, DTGS 2017, St. Petersburg, Russia, 21–23 June 2017, vol. 745, pp. 55–69. Springer, Heidelberg (2017). Revised Selected Papers. Book Series: Communications in Computer and Information Science. Print ISBN: 978-3-319-69783-3. Electronic ISBN: 978-3-319-69784-0. https://www.springerprofessional.de/en/online-socioeconomic-activity-in-russia-patterns-of-dynamics-and/15206136

25. Verzilin, D.N., Maksimova, T.G., Antokhin, Y.N.: Socio-economic and ecological indicators of the state of ecological and economic objects: genesis and development. Soc. Polit. Econ. Law **12** (2017). (in Russian). http://www.dom-hors.ru/vipusk-12-2017-obshchestvo-politika-ekonomika-pravo/

26. Verzilin, D.N., Mamonov, S.A., Corbunova, I.R.: Modelling coherent and self-organization behaviour of social and economic system. In: XVI International Conference "Dynamical System Modeling and Stability Investigations" (DSMSI-2013), Taras Shevchenko National University of Kiev, Ukraine, 29–31 May 2013, 422 p. (2013)

27. Verzilin, D.N., Maximova, T.G.: Models of social actors' reaction on external impacts. St. Petersburg State Polytechnical Univ. J. Comput. Sci. Telecommun. Control Syst. **2**(120), 140–145 (2011)

28. Verzilin, D.N., Potapychev, S.N., Ryzhkov, N.A.: The use of simulation to evaluate the response of social and economic systems to external influences. In: V Conference "Simulation. Theory and Practice", IMMOD 2011, St. Petersburg, 19–21 October 2011, pp. 115–119 (2011). (in Russian)

29. Verzilin, D.N., Shanygin, S.I., Chereshneva, A.V.: Conceptual bases of modelling of social systems and estimation of stability of social processes. St. Petersburg State Polytechnical Univ. J. Comput. Sci. Telecommun. Control Syst. **5**(133), 123–128 (2011). (in Russian)

30. Zelentsov, V.A., Krylenko, I.N., Pimanov, I.Y., Postryasayev, S.A., Sokolov, B.V., Ahtman, J.: Principles of Earth Remote Sensing Processing Data System Design on the Base of Service Oriented Architecture. Izv. Universities. Instrument making, vol. 58, No. 3, pp. 241–243 (2015). (in Russian). https://cyberleninka.ru/article/v/osnovy-postroeniya-sistemy-obrabotki-dannyh-distantsionnogo-zondirovaniya-zemli-na-baze-servisorientirovannoy-arhitektury

Hidden Asymmetry in Shape
of Biological Patterns

Sergey G. Baranov[✉]

Vladimir State University, Vladimir 600000, Russia
bar.serg58@gmail.com

Abstract. Various bilaterally symmetrical traits have not the same variability in the magnitude of the fluctuating asymmetry. Directional asymmetry (DA) is the second type of asymmetry with a clear predominance of either right or left structures. Since the FA is a considered indicator of instability, traits with DA are not used in the integral environmental monitoring. In presented paper the geometric morphometrics method is considered. This takes into account the labels that are placed on the bilaterally symmetric structures. The centroid points of consensus figure are drawn by the averaging of landmarks in Cartesian coordinates and the value of the FA shape of lamina is evaluated. In present study the MorphoJ1.06d package was used. The sampling procedure resulted in a nested dataset design. The increase in the accuracy of the measurement indicated a large fraction of the directional asymmetry. 90% of population studied possessed this type asymmetry. 10% of samples were characterized by clear fluctuating asymmetry. The results conclude the importance fine compute approach to testing of stability of development in natural biosystem.

Keywords: Directional asymmetry · Method of geometric morphometrics
· Cartesian coordinates

1 Introduction

Fluctuating asymmetry (FA) is one of the types of bilateral asymmetry. It is characterized by a slight and statistically insignificant deviation from zero of the difference between the values of the right and the left parts of the homologous bilaterally symmetric trait, with the normal distribution of this difference [1, 2]. According to modern concepts, FA refers to a certain type of variability – random, or fluctuating [3].

Various bilaterally symmetrical traits have not the same variability in the magnitude of the FA. Directional asymmetry (DA) is the second type of asymmetry with a clear predominance of either right or left structures. Since the FA is a considered indicator of instability, traits with DA are not used in the integral environmental monitoring. Other mean – DA can serve as indicator of destabilizing of development. However, the presence of asymmetry in the mix of DA and FA and the ability of directional asymmetry to inherit raise the interest in this kind of asymmetry. In many studies was shown the importance of phenotypic plasticity and its contribution to the total amount of measured FA. The sides or

© Springer International Publishing AG, part of Springer Nature 2019
R. Silhavy (Ed.): CSOC 2018, AISC 765, pp. 186–195, 2019.
https://doi.org/10.1007/978-3-319-91192-2_20

repeated body parts of many sessiled organisms can be consistently exposed to differential environmental conditions, and therefore inflate the amount of FA [1, 4–8].

The most widely used method is the method of normalizing difference, when differences in the magnitude of the metric traits are related to the sum of the values of these traits. An alternative method is considered to be the geometric morphometric methods (GMMs) [9–11]. This takes into account the labels that are placed on the bilaterally symmetric structures. Research rejects these labels from the centroid points of consensus figure, which is drawn by the averaging of landmarks in Cartesian coordinates and the value of the FA shape of an organ (or the whole organism) evaluated. The value of FA is determined in a mixed two-factor analysis of variance on the magnitude of the mean square variance residuals of two factors interaction: "sample" (random) and "side" (fixed). The first factor is denoted by the code values corresponding to the level of population variability, to the individual or to the organ (leaf blade). Factor "side" is denoted only by two code values ("right" and "left"). The magnitude of variance of residues in the model, i.e., the deviation from the consensus symmetrical shape, is calculated. The value of the mean square of the factor "side" indicates the presence of directional asymmetry including its genotypic effect. According to Leamy et al. and other many sources, the fluctuating asymmetry is encoded by several genes on the principle of epistatic effect [12–14]. In plants DA appears approximately in 10% of some dimensional traits, for example in the leaf blades of woody plants, as silver birch [15]. In English oak with heterogeneous structure of leaf blade the amount of such traits was higher and DA tested in GMMs revealed in all populations observed [16].

Phenogenetics can be characterized as a study of homologous phenes and their individual combinations in evolutionary-ecological aspects. Currently, population ecology (in the field of phenogenetic monitoring) is actively developing by employing phenotypic traits, like phenogenetically markers [17, 18]. It is known that in ANOVA directional asymmetry mixed with FA gives an undesirable bias in the value of the FA. This fact does not prevent testing the value of DA, but the phenotypic effect of fluctuating asymmetry is not available or awkward for testing [1, 2], although the some approaches are used [6, 7, 19].

In previous studies the correlation obtained between the magnitude of the FA index, found by linear measurements and the value of the FA index, produced by the method of geometric morphometrics. Such a correlation cannot be regarded as mandatory, and depends on the magnitude of linear traits, making a greater contribution to the shape of bilaterally symmetrical halves. The magnitude of the FA and the stability/instability of development depended on a combination of factors, among which the following were significant: the value of vehicle's and industrial emissions and the height of the relief. The climatic factor was significant in the follow-up observational time [16].

The study of the relation between genotypic and phenotypic effect in populations of the plants was carried out only indirectly, depending on the habitat and climatic peculiarities, with the use of traditional linear methods for FA and DA testing [19, 21]. The apparent simplicity of the methodology very often led to a distortion of the results [22]. Most of the work on phenogenetics employed the geometric morphometric method in the study on populations of rodents and insects [23, 24]. Recent studies of the fluctuating asymmetry of birch leaves indicated the presence of a paradoxical non-monotonous effect and hormesis if the relatively small toxic dose increases the value of asymmetry and contrary, high dose decreases asymmetry [25].

Some studies appeal the size of leaf not the FA indicates stress level and the shape of leaves margin can play a serious role in variation of reproducibility of fluctuating asymmetry [26, 27]. The unbiased estimation of asymmetry, both at the population and individual level also is in focus [28]. Nevertheless the index of FA as an index of developmental stability remains the tool of environmental stress [29, 30].

The aim of this study was to test the level of phenotypic and genotypic variability of silver birch, or warty/weeping birch (*Betula pendula* Roth) leaf blade's shape at relatively normal, baseline environmental conditions. To test the phenotypic variability, its environmental variance component was used, as a value of leaf blade fluctuating asymmetry. To test the genotypic variability several components were used, like the value of shape leaf directional asymmetry. The working hypothesis was the following: asymmetry as an element of the form (shape) leaf blade includes genotypic and environmental components of variation, detected by geometric morphometric methods.

2   Materials and Methods

The silver birch has a very wide area; in Russia it is bounded in the north of the Arctic Circle, and in the south it is bounded of the 50th parallel of north latitude. Vladimirskay oblast (29 084 km^2; 56°5′0″ N; 40°37′0″ E) was chosen for study. The sites included as very close (2–5 km in limit of cities) as well remote ones 70–90 (km). So "populations" considered as relatively conditional ones and pronounced better as cenopopulations, as they included other species and forms of plants. The collection of sheet plates was carried out in 2016, September. 50 leaf blades from each population of 10 trees were selected using trees of the same age and the same generative stage of development. In a whole the laves were sampled from trees of age 15–25 years according to the method developed by Kryazheva *et al.* [29] and adopted by Gelashvili *et al.* for populations under different environmental stress level [31]. The trees were located on distance 2–3 m of each other. From each tree 5 plates were taken from the shortened shoots (brachioblasts) randomly on the height 1.5–2 m under conditions of relatively the same sun lightening. To reduce the allometric measurement error, leaves with a maximum leaf half-width equal to 3–3.5 cm were selected. The leaf collection sites varied in altitude elevation that assumed the physicochemical properties of soil.

The leaves are harvested in regular intervals, from the lower part of the tree crown and storage dry under the paper press for two weeks. The images of leaves were taken using a Panasonic DMC-FZ100 camera, and JPG file format was used. Files for data manipulation and digitization were created using the TPS software package [32, 33]. Every plate was photographed twice to calculate measurement error. The 12 landmarks were labeled in the same order on each picture, after setting a scale factor. The landmarks were digitized twice and were classified as homological landmarks type I, as represented by pair labels on the endpoints of the lateral vessels (Fig. 1).

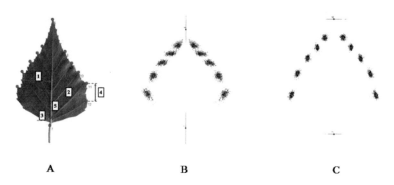

A   **B**   **C**

Fig. 1. A – 5 pair of landmarks for testing FA. 6–7th landmarks lie on the asymmetry axis. 1–5 are the traits used in traditional linear method. B – the landmarks represented on a Procrustes fit in a symmetric matrix and in the matrix of asymmetry (C). Black dots show the landmarks after aliment by superimposition. The symmetric matrix possesses more variance due to variety along the axis of asymmetry. The matrix of asymmetry reflects the variation of landmarks on the left and the right sides

For testing of both kinds of asymmetry the method of Procrustean analyses (Procrustes ANOVA), as analogue of 2WAY mixed model ANOVA (individual × side) which is used for FA value test in metric and in meristic traits. Procrustes fit is a space in limit of centroid size. Accordingly this method, the right and left point were aligned along with the mirror-reflected landmarks. The Procrustes alignment included the original and the mirrored configurations of a sample combined, and superimposes all of them simultaneously. For averaging consensus formation the method of least squares was used. In detail this method is observed in many basic studies, for example [9, 33].

In present study the MorphoJ1.06d package was used, available on web site www.morphometrics.org [34]. The total TPS file for population sample consisted of 200 TPS files (50 × leaf blades × 4 times repeats, as two photo × 2 measurements). The sampling procedure resulted in a nested dataset, with leaves nested within trees and trees nested within populations. Thus the plan consisted of table 200 × 4, first column included coordinates and served as identificator, others columns included coded values for factors: population, tree, leaf and measurement. After creating a Procrustean space (Procrustes fit) the Procrustean analysis was conducted on each individual biosystem level.

The magnitude of the fluctuating asymmetry was determined by the mean square MS and the value of the F-Goodall criterion evaluating the interaction of one of the random factors: "population", "tree", "leaf" or "measurement" with a fixed factor "side". The magnitude of DA was tested on the value of MS of factor "side" and on the value of F-Goodall criterion. To test antisymmetry, the third type of bilateral asymmetry, which has a bimodal distribution of the histogram of the frequencies of the difference between the values of the right and the left attributes and the negative value of the kurtosis, the program MorphoJ provides permutational multiplication of samples normalizing their distribution. In previous studies conducted by the traditional linear method, such properties were not met in metric traits [16]. The directional asymmetry of the linear characteristics was verified by the t-test with the verification of the null hypothesis H_0, on the equality of the right and left attributes. Auxiliary programs were the packages PAST 3.03 and STATISTICA 10.

3 Results

3.1 Measurement Error

The leaf blade represented a true replication, because each plate was measured twice. The author takes into account the point of view Klingenberg [36] about two types of errors: digitization and measurement (labeling). Therefore, a repeated survey of each plate was carried out, with a double marking on each image. The additional random factor "measure" took into account the measurement error, as the sum of the errors in photographing and labeling. The imaging and digitizing were performed as separate effects in the model with one nested into the other [37], (Table 1):

Table 1. Measurement error

Source	SS	MS	df	F	p
Population					
Error imaging	17,40	0,001	15820	9,43	<.0001
Error digitizing	19,33	0,001	31780		
Tree					
Error imaging	12,09	0,001	14380	7,21	<.0001
Error digitizing	14,02	0,000	30340		
Leaf					
Error imaging	2,69	0,000	7980	2,89	<.0001
Error digitizing	4,62	0,000	23940		

Procrustes analysis of variance of the amounts of shape variation attributable to population, tree and leaf blade that was photographed and digitized twice. Sums of squares and mean squares are in units of squared Procrustes distance.

The measurement error was up to 16.8% of the value of the sum square of total SS variation in Procrustes ANOVA for shape leaf variation. Error of imaging and digitizing were less on leaf level in comparison to levels of population and tree.

3.2 Allometric Effect

The centroid-sized (the square root of the sum of squares of distances from the landmarks to the image centre) was statistically significant, what indicated the allometric effect of the magnitude of the leaf blade on their shape variety (Table 2).

Table 2. Centroid size variation

Source	Explained SS	MS	df	F
Centroid size of pull population	30,48%	9476786	9	1200***
Centroid size tree	34,56%	1074397	90	675.7***
Centroid size leaves	34,96%	217379	450	184.2***
Error for leaves	8,63%	1180	1347	

*** – p < 0.0001

As can be seen no one biosystem level showed the absent of effect size-variance to shape asymmetry.

3.3   Phenotypic and Genotypic Components of Population Variability

General Procrustes analysis was performed for testing DA and FA for the pool of populations. The "pull population" source showed a high statistical significance, which indicated a difference in the asymmetry of leaf blades in the population pool ($F = 15.24$, $p < 0.0001$; Table 3).

Table 3. Generalized procrustes ANOVA results

Source	SS	MS	df	F
Pull population	5.237	0.058	90	15.24***
Side	0.048	0.005	10	1.25^{ns}
Population × side	0.344	0.004	90	6.09***
Error	22.414	0.001	35760	
Tree	10.457	0.012	900	8.18***
Side	0.048	0.005	10	3.35***
Tree × side	1.278	0.001	900	2.98***
Error	16.259	0.000	34140	
Leaf	18.448	0.004	4500	4.52***
Side	0.048	0.005	10	5.24***
Leaf × side	4.085	0.001	4500	4.48***

ns – not significant; *** – p < 0.0001

3.4   Directional Asymmetry in Populations

The increase in the accuracy of the measurement indicates a large fraction of the magnitude of the F-Goodall criterion (quotient effects MS "side" on MS "individual × side"). The DA value variance was coming higher at the order from "tree" to "leaf". The most statistically significant difference was in the population at high DA values; $F = 7.52$ for "tree"; $F = 15.09$ for "leaf". Thus, with a change in the level of the biosystem in the direction from the population to a lower level, there was an increase in the magnitude of the directional asymmetry ($df = 10$).

Only the first population, showed a 'clear' fluctuating asymmetry of the leaf blades. Population #1 was not special (in town Gus with 150 thousand dwellers, 131 m above sea level). Only weak negative correlation (Pearson's $r = -0.32$) was detected between values of FA (F-Goodall criterion) and altitude elevation. No difference found between remote and closest population sites in discriminate analysis or in principal component analysis.

The traditional method of normalizing difference showed a statistically significant presence of directional asymmetry in two populations #8 and #10 (Table 4). Population #1 had not indicated DA in linear traits as well.

Table 4. Results of directional asymmetry testing (t-test and F-Goodall Test)

Population, N	DA linear (t-test)	DA("side"; F-Goodall)		
	Trait, N	Tree	Leaf blade	Error (digitizing + imaging)
1	ns	0.22^{ns}	0.29^{ns}	0.56^{ns}
2	ns	1.62^{ns}	2.69**	3.56****
3	ns	1.72^{ns}	1.9*	2.67**
4	ns	1.81^{ns}	1.86*	2.95**
5	ns	1.86^{ns}	2.11*	3.53****
6	ns	2.01^{ns}	1.79*	3.28***
7	ns	3.09**	6.42****	6.07****
8	2*	3.13**	3.89****	6.39****
9	ns	7.52****	15.09****	24.15****
10	4*, 5*	9.78****	11.75****	20.98****

Note: $* - p < 0.05$; $** - p < 0.01$; $*** - p < 0.001$; $**** - p < 0.0001$

Traits with directional asymmetry were traits number 2, 4 and 5 (see Fig. 1). Thus, these features, including the angle between the middle and second veins (5th), demonstrated a relationship between the linear method and the geometric morphometry method, which assesses the asymmetry of the shape. No high kurtosis values (more than 4) were found in the samples (R – L) that indicated the absence of antisymmetry, as the third possible kind of bilateral asymmetry.

4   Discussion

As a species *Betula pendula* shows a high phenotypic plasticity from the absence of DA (pure FA, population #1) to a highly significant level of DA. The results showed that the presence of directional asymmetry in the context of the plate shape is a common type of asymmetry in the birch leaf blades. So studies showed that in the same region, similar study conducted in 2015, showed the presence of DA in populations with a level of statistical significance $p = 0.007$ (F-Goodall = 3.19). Accordingly, the "tree" factor showed a higher statistical significance ($p < 0.001$; F-Goodall = 4.97). The leaf blade also showed higher significance (F-Goodall = 12.17; $p < 0.001$), that confirmed the hypothesis of the presence of directional asymmetry in leaf blades a year earlier, under relatively similar environmental conditions. Only one population from seven showed "clear" fluctuating asymmetry.

The study shows that the variability of linear characteristics affects the asymmetry of the shape leaf blade. Variation of unemployed linear traits can significantly change the shape of the leaf blades. Thus, total asymmetry of shape contained joined components of the genotypic and phenotypic variability. The dependence value mixed FA from elevation altitude is corresponding to the study of *Betula pubescence* [38] as well as to previous study conducted in 2015 [16].

The method of geometric morphometry could be seemed as more preferable for evaluating fluctuating asymmetry and stability of development. This method takes into account the shape of the organ, i.e. lamina. The method of the normalizing difference takes into account only the sum of the values of the FA of the individual bilateral traits, anyone of which cannot be free of directional asymmetry or antisymmetry. If populations are genetically differ (real populations with genotype) we can say they carry individual genetic component highly heterogeneous for *Betula pendula*.

There is a controversial means on technogenic stress factors influencing developmental stability of birch species. In spite of a big body studies two opposite ideas arise: from direct dependence on value of FA and developmental instability for example [30, 39] to reducing FA value in response to stress [40].

The presented study confirmed the heterogeneity DA contributes the biasing in FA value in inter individual and intra individual levels. Mixed FA as a rule corresponds to high error of measurement and heterogeneity in variance of left and right homological traits [19, 41]. Study showed a high heterogeneity in DA value and possible confounding effect on FA value. It means a high individual and among individual variability in asymmetry including the genotype effect of DA.

Thus, a study confirmed hypothesis about the joint presence of both components of variability: genotypic and phenotypic in the asymmetry of the shape of the leaf blade.

Fluctuating asymmetry in its pure form was met only in 1 out of 10 cases population studied, that should be taken into account in assessing the stability of development of the birch and possibly other woody plants.

References

1. Palmer, A.R., Strobeck, C.: Fluctuating asymmetry analyses revisited. In: Polak, M. (ed.) Developmental Instability: Causes and Consequences, pp. 279–319. Oxford University Press, New York (2003)
2. Graham, J.H., Raz, S., Hel-Or, H., Nevo, E.: Fluctuating asymmetry: methods, theory, and applications. Symmetry **2**, 466–540 (2010)
3. Tikhodeyev, O.N.: Classification of variability forms based on phenotype determining factors: traditional views and their revision. Ecol. Genet. **11**(3), 79–92 (2013). https://doi.org/10.17816/ecogen11379-92
4. Viscosi, V., Cardini, A.: Leaf morphology, taxonomy and geometric morphometrics: a simplified protocol for beginners. PLoS ONE **6**(10), e25630 (2011). https://doi.org/10.1371/journal.pone.0025630

5. Savriama, Y., Gómez, J.M., Perfectti, F., Klingenberg, C.P.: Geometric morphometrics of corolla shape: dissecting components of symmetric and asymmetric variation in *Erysimum mediohispanicum* (Brassicaceae). New Phytol. **196**, 945–954 (2012)

6. Graham, J.H., Emlen, J.M., Freeman, D.C., Leamy, L.J., Kieser, J.: Directional asymmetry and the measurement of developmental instability. Biol. J. Linn. Soc. **64**, 1–16 (1998)

7. Van Dongen, S., Lens, L., Molenberghs, G.: Mixture analysis of asymmetry: modelling directional asymmetry, antisymmetry and heterogeneity in fluctuating asymmetry. Ecol. Lett. **2**, 387–396 (1999)

8. Savriama, Y., Klingenberg, C.P.: Beyond bilateral symmetry: geometric morphometric methods for any type of symmetry. BMC Evol. Biol. **11**, 280 (2011)

9. Rohlf, F.J.: Shape statistics: procrustes superimpositions and tangent spaces. J. Classif. **16**, 197–223 (1999)

10. Mardia, K.V., Bookstein, F.L., Moreton, I.J.: Statistical assessment of bilateral symmetry of shapes. Biometrika **87**(2), 285–300 (2000). https://doi.org/10.1093/biomet/87.2.285

11. Klingenberg, C.P., Barluenga, M., Meyer, A.: Shape analysis of symmetric structures: quantifying variation among individuals and asymmetry. Evolution **56**, 1909–1920 (2002)

12. Leamy, L.J., Routman, E.J., Cheverud, J.M.: An epistatic genetic basis for fluctuating asymmetry of mandible size in mice. Evolution **56**(3), 642–653 (2002)

13. Hochwender, C.G., Robert, S.: Fritz. Fluctuating asymmetry in a Salix hybrid system: the importance of genetic versus environmental causes. Evolution **53**(2), 408–416 (1999)

14. Leamy, L.J., Klingenberg, C.P.: The genetics and evolution of fluctuating asymmetry. Annu. Rev. Ecol. Evol. Syst. **36**, 1–21 (2005)

15. Zorina, A.A., Korosov, A.V.: Variability of asymmetry measures and indexes of leaf characters in the crown of *Betula pendula* (Betulaceae). Botanicheskiĭ Zhurnal **94**(8), 1172–1192 (2009)

16. Baranov, S.G.: Geometric morphometric methods for Testing Developmental Stability of *Betula pendula* Roth. Biol. Bull. **5**, 567–572 (2017)

17. Baranov, S.G.: Fenogenetic aspect of asymmetry of leaf blade *Betula pendula* Roth. Nauchnye vedomosti BelGU. Seriya: Estestvennye nauki **11**(232), 10–17 (2016)

18. Vasilyev, A.G., Vasilyeva, I.A., Bol'shakov, V.N.: Evolutionary-ecological analysis of trends in phenogenetic variation of homologous morphological structures: from populations to ecological series of species. Russ. J. Ecol. **41**(5), 365–371 (2010)

19. Stige, L.C., David, B., Alibert, P.: On hidden heterogeneity in directional asymmetry – can systematic bias be avoided? J. Evol. Biol. **19**, 492–499 (2006). https://doi.org/10.1111/j.1420-9101.2005.01011.x

20. Fair, J.M., Breshears, D.D.: Drought stress and fluctuating asymmetry in *Quercus undulata* leaves: confounding effects of absolute and relative amounts of stress? J. Arid. Enviro. **62**(2), 235–249 (2005)

21. Fei, X., Weihua, G., Weihong, X., Renqing, W.: Habitat effects on leaf morphological plasticity in *Quercus acutissima*. Acta Biologica Cracoviensia Series Botanica **50**(2), 19–26 (2008)

22. Kozlov, M.V.: Plant studies on fluctuating asymmetry in Russia: mythology and methodology. Russ. J. Ecol. **48**, 1 (2017). https://doi.org/10.1134/S1067413617010106

23. Bol'shakov, V.N., Vasil'ev, A.G., Vasil'eva, I.A., Gorodilova, Y.V., Chibiryak, M.V.: Coupled biotopic variation in populations of sympatric rodent species in the Southern Urals. Russ. J. Ecol. **46**(4), 339–344 (2015)

24. Voita, L.L., Omelko, V.E., Petroava, E.A.: Analysis of the morphometrics variability and intraspecific structure of *Sorex Minutissimus* Zimmermann 1780 (Lipotyphla: Soricidae) in Russia. Acta Theriol. **8**, 167–179 (2013). https://doi.org/10.4098/at.arch.64-10

25. Erofeeva, E.A.: Hormesis and paradoxical effects of Drooping Birch (*Betula pendula* Roth) parameters under motor traffic pollution. Dose-Response **13**(2), 1559325815588508 (2015). https://doi.org/10.1177/1559325815588508

26. Koroteeva, E.V., Veselkin, D.V., Kuyantseva, N.B., Chashchina, O.E.: The size, but not the fluctuating asymmetry of the leaf, of silver birch changes under the gradient influence of emissions of the Karabash Copper Smelter Plant. Dokl. Biol. Sci. **460**, 36–39 (2015). Springer

27. Kozlov, M.V., Cornelissen, T., Gavrikov, D.E., Kunavin, M.A., Lamma, A.D., Milligan, J.R., Zverev, V., Zvereva, E.L.: Reproducibility of fluctuating asymmetry measurements in plants: sources of variation and implications for study design. Ecol. Indic. **73**, 733–740 (2017)

28. Dongen, V.S.: Variation in measurement error in asymmetry studies: a new model, simulations and application. Symmetry **7**(2), 284–293 (2015)

29. Kryazheva, N.G., Chistyakova, E.K., Zakharov, V.M.: Analysis of development stability of *Betula pendula* under conditions of chemical pollution. Russ. J. Ecol. **27**(6), 422–424 (1996)

30. Ivanov, V.P., Ivanov, Y.V., Marchenko, S.I., Kuznetsov, V.V.: Application of fluctuating asymmetry indexes of silver birch leaves for diagnostics of plant communities under technogenic pollution. Russ. J. Plant Physiol. **62**(3), 340–348 (2015)

31. Gelashvili, D.B., Cheprunov, E.V., Iudin, D.I.: Structural and bioindicative aspects of fluctuated asymmetry of bilateral organisms. Zhurnal obshchei biologii **65**(5), 433–441 (2003)

32. Rohlf, F.J.: Tps Series. Department of Ecology and Evolution, State University of New York, Stony Brook, New York (2010). http://life.bio.sunysb.edu/morph/. Accessed 8 June 2011

33. Rohlf, F.J.: The tps series of software. Hystrix (Ital. J. Mammal.) **26**(1), 9–12 (2015)

34. Klingenberg, C.P.: MorphoJ: an integrated software package for geometric morphometrics. Mol. Ecol. Resour. **11**, 353–357 (2011). https://doi.org/10.1111/j.1755-0998.2010.02924.x

35. Pavlinov, L.Y., Mikeshina, N.G.: Principles and methods of geometric morphometrics. Russ. J. Ecol. **3**(6), 473–493 (2002)

36. Klingenberg, C.P.: Analyzing fluctuating asymmetry with geometric morphometrics: concepts, methods, and applications. Symmetry, 7843–934 (2015). https://doi.org/10.3390/sym7020843

37. Breuker, C.J., Patterson, J.S., Klingenberg, C.P.: A single basis for developmental buffering of Drosophila wing shape. PLoS ONE **1**(1), e7 (2006)

38. Hagen, S.B., Ims, R.A., Yoccoz, N.G., Sørlibråten, O.: Fluctuating asymmetry as an indicator of elevation stress and distribution limits in mountain birch (*Betula pubescens*). Plant Ecol. **195**(2), 157–163 (2008)

39. Vasil'ev, A.G., Vasil'eva, I.A., Marin, Y.F.: Phenogenetic monitoring of the weeping birch (*Betula pendula* Roth.) in the Middle Urals: testing a new method for assessing developmental instability in higher plants. Russ. J. Ecol. **39**(7), 483–489 (2008)

40. Franiel, I.: Fluctuating asymmetry of *Betula pendula* Roth. leaves – an index of environment quality. Biodiv. Res. Conserv. **9–10**, 7–10 (2008)

41. Van Dongen, S.: Unbiased estimation of individual asymmetry. J. Evol. Biol. **13**(1), 107–112 (2000)

Classification, Clustering and Association Rule Mining in Educational Datasets Using Data Mining Tools: A Case Study

Sadiq Hussain[1]([✉]), Rasha Atallah[2], Amirrudin Kamsin[3],
and Jiten Hazarika[4]

[1] Dibrugarh University, Assam, India
sadiqdu@rediffmail.com
[2] Faculty of Computer Science and IT,
University of Malaya, Kuala Lumpur, Malaysia
[3] Department of Computer System and Technology,
University of Malaya, Kuala Lumpur, Malaysia
[4] Department of Statistics, Dibrugarh University, Assam, India

Abstract. Educational Data Mining is an emerging field in the data mining domain. In this competitive world scenario, the quality of education needs to improve. Unfortunately most of the students' data are becoming data tombs for not analyzing the hidden knowledge. The educational data mining tries to uncover the hidden knowledge by discovering relationships between student learning characteristics and behavior. With this educational data modeling, the educators may plan for future learning pedagogy to support the student's learning style. This knowledge may be applied by the academic planners to improve the quality of education and decrease the failure rate. In this paper, we had collected real dataset containing 666 instances with 11 attributes. The data is from the Common Entrance Examination (CEE) data of a particular year for admission to medical colleges of Assam, India conducted by Dibrugarh University. We tried to find out the association rules using the data. Various clustering and classification methods were also used to compare the suitable one for the dataset. The data mining tools applied in the educational data were Orange, Weka and R Studio.

Keywords: Classification · Clustering · Association rule mining
Educational data mining · Data mining tools

1 Introduction

Educational Data Mining is a sub area of Data Mining Domain. This new area has immense potential to mine various aspects for betterment of students as well as in decision making by the authorities of educational institutes. There is a collection of student data year after year without much utility. These data may be classified or clustered to distinguish between excellent, good and academically poor students. The excellent and good students may be encouraged to perform better, while the academically poor students may be given remedial classes and extra attention to prevent

© Springer International Publishing AG, part of Springer Nature 2019
R. Silhavy (Ed.): CSOC 2018, AISC 765, pp. 196–211, 2019.
https://doi.org/10.1007/978-3-319-91192-2_21

dropouts and to enrich the quality of education across higher educational institutes across the globe. So, the aim of Educational Data Mining is to extract some hidden yet useful information from the large educational institutes' datasets which varies from the schools to university levels [1]. In case of unsupervised learning or descriptive analysis, the clustering may be applied. For supervised learning where the class labels are known, classification techniques may be used. Using the student datasets, various aspects may be predicted. The authorities may use these predictions for quality enhancements of the institutes. The association rule mining may be applied to discover some of the interesting relation among the attributes of the datasets [2].

There are various clustering techniques used in the field of knowledge extraction. The clustering techniques used by in this paper are K-means, Hierarchical and Partitioning Around Medoids. For classification, neural networks, naïve bayes and decision tree methods were used. For association rule mining, Apriori algorithm was used. A comparison was also made among these techniques. The data mining tools used for this work are Weka, Orange and R. The reasons for selecting these tools are that all are open source, easy to use and platform independent data mining tools. All have scripting interface. They are good at data visualization and analysis [3]. The Government of Assam authorized Dibrugarh University which is the easternmost University of India, conducted common entrance examinations (CEE) for a particular period for admission of students to medical colleges of Assam. The collected data were of students who came for counseling cum admission into medical colleges of Assam in the year 2013. The data were collected as per our requirement by one of the authors of this paper through direct personal interview at the time of admission. Altogether, data with 12 attributes were collected. The CEE rank and CEE percentage is ambiguous data, as if the CEE percentage is high so is the rank. So, only the CEE percentage termed as performance of the candidate is used as response variable whereas the rank of the candidate was not used. The performance is converted to categorical data from the real data. Not only performance but all the explanatory variables data were converted to categorical data in case of classification and association. For Clustering, the data was not converted and only selected numerical attributes were used to find some meaningful clusters.

The rest of the paper is organized as follows: Sect. 2 present Literature Review, Sect. 3 describes Methodology, Sect. 4 presents Experiments and Results and the Sect. 5 describes the Conclusion and Future Work.

2   Literature Review

This section explains various works done by the researchers on educational datasets using clustering and classification techniques.

2.1   Clustering Techniques Used by the Researchers on Educational Datasets

DeFreitas et al. [4] made a comparative analysis of clustering algorithms. The dataset used was Learning Management System log data. They compared K-means, DBSCAN and BIRCH methods to select the most suited clustering methods for educational data sets.

Based on the silhouette coefficient score, it was found that K-means performs better than the other two algorithms. It is also better from the point of view of distribution of clusters.

Dutt et al. [5] reviews how the large data that is generated by the educational institutes may be utilized properly. Most of such data remains unused and the useful information is not extracted. This application of Clustering, Classification, prediction modeling is very low in case of educational setup. The authors surveys different clustering methods applied in the field of Educational Data Mining (EDM).

Nagy et al. [6] proposed a student advisory framework which offers consultation to first year students of Cairo Higher Institute for Engineering. This intelligent system is based on clustering and classification techniques. This framework aims for decreasing the rate of failures which was very high otherwise.

Oyelade et al. [7] proposed a deterministic educational model to evaluate the performance of student academic achievement. The data was collected from a private institute based at Nigeria. The students' results were clustered by using K-means algorithm. The authors claimed that these mining results would be helpful in effective decision of higher authorities of the institutes.

2.2   Classification Techniques Used by the Researchers on Educational Datasets

Almarabeh [8] used Weka tool to compare five different classifier available in the toolkit for the University student dataset. The classifiers were Bayes Network, J48, ID3, Neural Network and Naïve Bayes. It was found that Bayes Network classifier performs better than the others by using different evaluation measures.

Bhardwaj et al. [1] used predictive model for classification of good learners and slow learners. The dataset used from different affiliating institutes under Dr. R.M.L. Awadh University, Faizabad, India and it contains 300 instances. They had used 17attributes for the classification task. The high attributes were obtained from the dataset to predict the academic performance of the students.

Yadav et al. [2] applied decision tree algorithms for prediction of engineering students' results. The algorithms used were C4.5, ID3 and CART. The results of the analysis helped the academically weaker students to perform better. The data was collected from VBS Purbanchal University with 90 records for the session 2010. C4.5 decision tree algorithm performs better than other two with 67.78% accuracy.

3   Methodology

This section describes the about various techniques used by the data miners for educational datasets, e.g. classification, clustering, association rule mining, classification errors, evaluation methods, and data mining open source tools.

3.1   Unsupervised Learning and Clustering Methods

Clustering is grouping or partitioning data into some subsets. The clustering is made in such a way that the object belong to one cluster similar to other objects in that cluster and the dissimilar clusters belong to other clusters. Clustering, an unsupervised learning technique may be classified depending on the data types used, similarity measures and the theory involved for defining the cluster [9]. There are two types of clustering [10], hard and soft. In the hard clustering, each data point belongs to one cluster or not, but in case of soft clustering each data point may belong to one or more clusters.

In this paper, we used K-means clustering, Hierarchical Clustering, Partitioning Around Medoids. K-means algorithm tries to group n items into some user defined subgroups k where k should be less than or equal to n. The grouping is done in iterative manner by minimizing the sum of squared distances and centroids of the items until there are no longer any changes in the structure or a threshold is reached. So, Partitioning Around Medoids (PAM) technique needs to define the number of clusters in advance like k-means algorithm. PAM is based on medoids that are centrally located in clusters [11]. PAM is another partitioning algorithm like k-means whereas the later performs better in wide variety of datasets [12]. Hierarchical Clustering technique is used to group items into a tree of clusters. This method has two types; divisive and agglomerative. In divisive clustering, the single cluster is iteratively splitted to make more clusters where as the clusters are merged into a single cluster in case of agglomerative. The merging or splitting is stopped until a certain condition is reached [13].

We used one neural reduction unsupervised technique called Self Organizing Map (SOM) for visualizing its data. SOM is useful for visualizing high dimensional data for low dimensional view and is also useful for dimensionality reduction [14]. The clustering methods were performed by Orange and R Studio.

3.2 Supervised Learning and Classification Methods

In this paper, we used three classification techniques; decision tree (J48 in Weka), neural network (Multilayer Perceptron in Weka) and Naïve Bayes. Various researchers have used these classification techniques to classify educational datasets [1, 15, 16]. All the classifications were performed by the WEKA software. The decision tree is a predictive supervised modeling technique where all the class labels are known. It is used for predicting and analyzing the data. This technique builds a tree like top down model. The C4.5 algorithm (J48 in case of WEKA) implements decision tree algorithm by using pruning which means that the node may be deleted if it does not add any significance. In WEKA, the neural network classifier is implemented by using Multi-layer Perceptron algorithm. Neural network is emulation of human neuron system. In pattern recognition scenario, neural network is a very popular classifier [17]. It maps the input data to acceptable output, and feeds forward in nature. It has many layers including hidden layers which generally use sigmoid activation functions. Naïve Bayes is predictive classification technique which is based on bayes' probability theory with strong assumptions among the fields. Naïve Bayes classifier is suitable for small datasets and it is simple and easy to interpret. The model can represent only linear class boundaries with categorical data, so the representation is difficult compared to decision tree classification algorithm in such cases [18].

3.3   Association Rule Mining

The occurrence of an item may be predicted by using the occurrence of other items in the transactions. The rules that define such transactions in the form X → Y are called association rules [19]. Support is termed as frequency of occurrence of set of items or itemset, while confidence is fraction of transaction that contains the itemset. The frequent itemset is the itemset whose support is greater than the minimum support threshold. While generating association rule, minimum support threshold, size, dimension and average transaction width are the factors that affects complexity in mining. Agrawal et al. [20] proposed the Apriori Algorithm which is based on the principle that any subset of a frequent itemset is frequent, and any superset of infrequent itemset should not tested or generated.

3.4   Classification Errors

Mean Absolute Error (MAE) is the average of absolute differences of the predicted and observed samples [21]. Root Mean Squared Error (RMSE) is square root of the average of the difference between the forecast and observed values. Both MAE and RMSE vary from zero to infinity. The lower the value of both, the better is the results. Since MAE do not use the square, so it is more robust to outliers. RMSE is more useful when the large errors are not expected.

$$\text{MAE} = \frac{1}{n} \sum\nolimits_{i=1}^{n} |X_i - X| \tag{1}$$

where n = the number of errors, $|x_i - x|$ = the absolute errors.

$$\text{RMSE} = \sqrt{\frac{\sum_{i=1}^{n} (X_{obs,i} - X_{model,i})^2}{n}} \tag{2}$$

where X_{obs} is observed values and X_{model} is modeled values at time/place i.

3.5   Accuracy, Confusion Matrix, Silhouette and Multidimensional Scaling

Accuracy is the proportion of the true positives and true negatives to the total number of cases [8]

$$\text{Accuracy} = \frac{(\text{True Positives} + \text{True Negatives})}{(\text{True Positives} + \text{True Negatives} + \text{False Negatives} + \text{False Positives})} \tag{3}$$

To view the performance of the machine learning algorithms especially for classifier, a confusion matrix or error matrix is obtained where each row represents the instances of predicted values and column the actual ones [22]. It is a special contingency table with predicted and actual classes.

Silhouette value is a measure to find out how close each point to its own cluster than to its neighboring clusters [23]. This value ranges from −1 to +1. The value close +1 means that the clustering has some impact and the samples are close to its own cluster where as the value close to −1 indicates the cluster has no meaning or objects are in wrong clusters. Silhouette Plot may be used to visualize the clusters.

Multidimensional Scaling (MDS) visualizes the level of similarity among the objects in the datasets [24]. This visualization is made with the help of distance matrix.

3.6 Data Mining Tools

The following data mining tools had been used for the analysis.

3.6.1 Orange

Orange is python based open source machine learning data mining tool [25]. It uses various visualizations for the mined data. It is interactive in nature. So, orange is visual programming environment for data analysis and mining. It runs on Linux, Windows and Mac. The current version is 3.7. Data fusion, text mining and bioinformatics packages may be installed as add-ons.

3.6.2 Weka

Waikato Environment for Knowledge Analysis (Weka) is another machine learning open source under GNU Genera Public License for classification [26], regression, association, visualization. Weka is written in Java. It was developed at University of Waikato, New Zealand. This tool runs on all the computing platforms and may be used easily because of its GUI.

3.6.3 R Studio

R Studio is a freeware for integrated development environment of R programming language [27]. It is available for two editions. One is R Studio Desktop and another is R Studio Server. R Studio is written in C++ and Qt framework is for its graphical programming. R packages are implemented in the R Studio. R programming is machine learning programming language used extensively in the field of data mining.

4 Experiments and Results

4.1 Data Cleaning and Feature Selection

The collected data has 666 instances. Some of the attributes are ambiguous. The rank of the candidate is removed from the dataset in the data cleaning phase of the data pre-processing. Performance is the response variable and others are explanatory variables. Performance attribute is taken as numerical attribute in case of k-means clustering whereas in other cases it is termed as categorical attribute. The following is the data table used for the data analysis (Table 1).

The attributes were ranked using InfogainAttribute Eval ranker search method of Weka. InfoGainAttributeEval method is used for feature selection. Entropy measures the degree of impurity. It can be assumed that the dataset is less impure if the degree is

Table 1. Attribute Description with their values

Attribute	Description	Values
Performance	Performance in Common Entrance Examination (CEE)	{'Excellent','Vg','Good','Average'} If the percentage is top 100, then Excellent If the percentage is next 200, then Very Good (Vg) If the percentage is next 200, then Good The rest is termed as Average Here the percentage means the percentage in the CEE Examination
Gender	Sex of the Candidate	{'male','female'}
Caste	Caste of the Candidate	{'General','OBC','SC','ST'} OBC – Other Backward Caste SC – Schedule Caste ST – Schedule Tribes
coaching	Whether the candidate attended any coaching classes within Assam, outside Assam or not	{'NO','WA','OA'} No – No Coaching WA – Within Assam OA – Outside Assam
Class_ten_education	Name of the board where the candidate studied at Class X level	'SEBA','OTHERS','CBSE'
twelve_education	Name of the board where the candidate studied at Class XII level	'AHSEC','CBSE','OTHERS'
medium	Medium of instructions for the study at Class XII level	'ENGLISH','OTHERS','ASSAMESE'
Class_X_Percentage	The percentage secured by the candidate at Class X standard	'Excellent','Vg','Good','Average' If the percentage is above 80%, then Excellent If the percentage is less than 80% but more than or equal to 70%, then Very Good (Vg) If the percentage is less than 70% but more than or equal to 60%, then Good The rest are termed as Average
Class_XII_Percentage	The percentage secured by the candidate at Class XII standard	'Excellent','Vg','Good','Average' If the percentage is above 80%, then Excellent If the percentage is less than 80% but more than or equal to 70%, then Very Good (Vg) If the percentage is less than 70% but more than or equal to 60%, then Good The rest are termed as Average
Father_occupation	The occupation of the father of the candidate	'DOCTOR','SCHOOL_TEACHER','BUSINESS','COLLEGE_TEACHER','OTHERS','BANK_OFFICIAL','ENGINEER','CULTIVATOR'
Mother_occupation	The occupation of the mother of the candidate	'OTHERS','HOUSE_WIFE','SCHOOL_TEACHER','DOCTOR','COLLEGE_TEACHER','BANK_OFFICIAL','BUSINESS','CULTIVATOR','ENGINEER'

close to zero. A good attribute is an attribute that reduces the most entropy and are highly ranked [28]. The caste was the highest rank attribute followed by Class_XII_ Percentage, Father_Occupation, Mother_Occupation, Class_X_Percentage and Medium. The other high ranked attributes are obvious, but the Caste attribute rank is quite surprising with the ranked value 0.51393.

4.2   Association Rule Mining

The authors tried to find the association rules on the datasets using Orange. The orange canvas for the associations with widgets looks as in the Fig. 1.

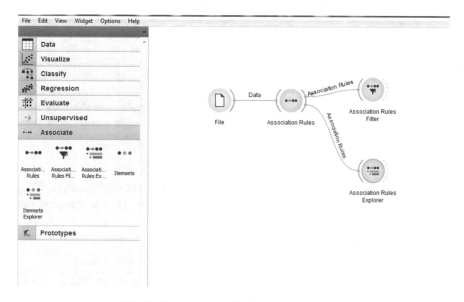

Fig. 1. Orange canvas for the association rules

With the threshold for support and confidence set as 50% and 60% respectively, the followings are the rule statistics, rule matrix and the extracted rules (Tables 2 and 3).

Table 2. Rule statistics for the datasets

Association Rules Filter Fri Nov 03 17, 15:51:26
Rules statistics
Total number of rules: 14
Support: 51%–65%
Confidence: 63%–95%
Number of rules in the graph: 14
Support: 51%–65%
Confidence: 63%–95%
Selected rules: 14
Support: 52%–65%
Confidence: 64%–96%

Table 3. Selected rules obtained by Orange

Supp	Conf	Antecedent	→	Consequent
0.527	0.794	Mother_occupation=HOUSE_WIFE		medium=ENGLISH
71	0.744	Class_X_Percentage=Excellent		Class_XII_Percentage=Excellent
0.527	0.655	medium=ENGLISH		Mother_occupation=HOUSE_WIFE
0.652	0.849	Class_X_Percentage=Excellent		medium=ENGLISH
0.517	0.673	Class_X_Percentage=Excellent		coaching=WA
0.515	0.862	Class_XII_Percentage=Excellent		medium=ENGLISH
0.517	0.766	coaching=WA		Class_X_Percentage=Excellent
0.529	0.889	Class_ten_education=SEBA		twelve_education=AHSEC
0.524	0.651	medium=ENGLISH		coaching=WA
0.652	0.810	medium=ENGLISH		Class_X_Percentage=Excellent
0.515	0.640	medium=ENGLISH		Class_XII_Percentage=Excellent
0.529	0.957	twelve_education=AHSEC		Class_ten_education=SEBA
0.524	0.777	coaching=WA		medium=ENGLISH
0.571	0.955	Class_XII_Percentage=Excellent		Class_X_Percentage=Excellent

4.3   Classification

Three Classification Algorithms were used in the datasets viz. Naïve Bayes, Neural Network and Decision Tree using Weka. The Table 4 shows the classification summary of the three classifiers.

Table 4. The comparison of classifiers on the educational datasets

Classifier	Correctly classified instances	Incorrectly classified instances	Accuracy	MAE	RMSE
Decision Tree (J48)	431	235	64.71%	0.2296	0.3388
Neural Network (MLP)	605	61	90.84%	0.0658	0.1861
Naïve Bayes (NB)	385	281	57.81%	0.2567	0.3625

The following figures compare the correctly classified instances with incorrectly classified instances and MAE and RMSE errors (Figs. 2 and 3).

The Table 5 shows the confusion matrix of the Neural Network (Multilayer Perceptron).

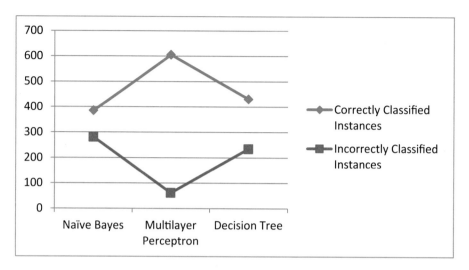

Fig. 2. Comparison of three classifiers

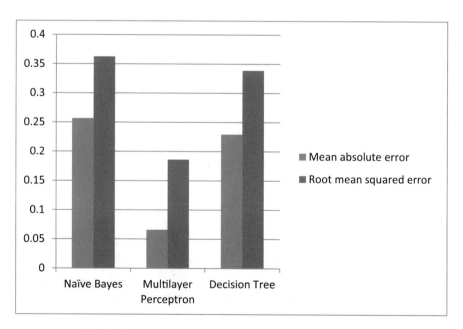

Fig. 3. Comparison of three classifiers using MAE and RMSE

Table 5. Confusion matrix for multilayer perceptron classification

a	b	c	d	Classified as
84	12	4	1	a = Excellent
4	**182**	11	1	b = Vg
5	12	**185**	8	c = Good
0	1	2	**154**	d = Average

4.4   Clustering

The clustering was applied in the datasets using Orange and R Studio. The Clustering methods applied was K-means clustering, Hierarchical Clustering and Partitioning Around Medoids (PAM). K-means and PAM performs better than the Hierarchical Clustering in one considers the Silhouette score. The following figure shows the dendrogram of Hierarchical Clustering using R Studio. For PAM and K-means, the silhouette score is 0.54 which means that a moderate score is achieved. The experiment is done with varying number of clusters. But it was found that as the number of clusters increase the silhouette score decreases. So, the performance is good with number of clusters i.e. k = 3. The following figures depict the cluster analysis results (Figs. 4, 5, 6, 7, 8, 9 and 10).

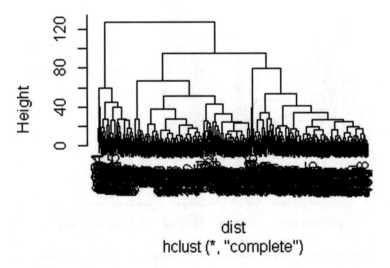

Fig. 4. Dendrogram of Hierarchical Clustering using R Studio

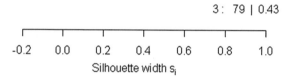

Silhouette plot of pam(x = dist, k = 3)

n = 666

3 clusters C_j

j : $n_j \mid ave_{i \in C_j}\ s_i$

1 :   291 | 0.68

2 :   296 | 0.43

3 :   79 | 0.43

-0.2   0.0   0.2   0.4   0.6   0.8   1.0

Silhouette width s_i

Average silhouette width : 0.54

Fig. 5. Silhouette Plot of Pam with 3 Clusters using R Studio

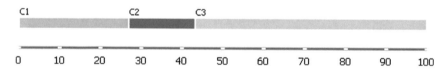

C1   C2   C3

0   10   20   30   40   50   60   70   80   90   100

Fig. 6. Box Plot of 3 Clusters using K-means Algorithm using Orange

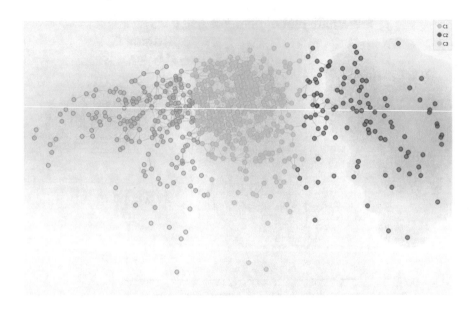

Fig. 7. MDS of K-means Clustering using Orange

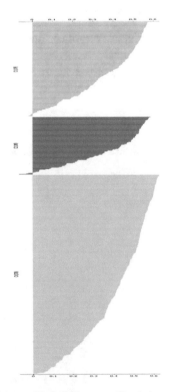

Fig. 8. Silhouette Plot of K-means Clustering using Orange

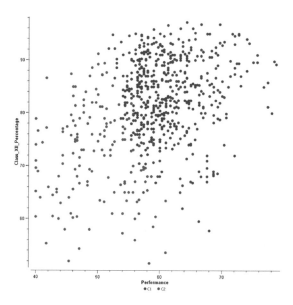

Fig. 9. Scatter Plot of two Clusters using K-means Clustering (Performance against Class XII Percentage)

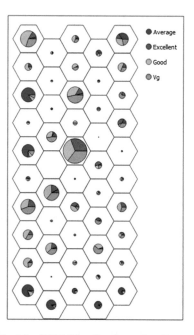

Fig. 10. SOM Visualization using Orange

5 Conclusion and Future Work

The authors tried different data mining techniques using different visualization tools. In case of association rule mining, apriori algorithm was used to mine some of the interesting rules. In case of Classification, three classifiers were used and based on its accuracy and classification errors; it was found that the neural network was the best classifier. Although neural network performs well on big dataset and was believed that it is not the best classifier for educational datasets, but in our case, the neural network classifier outperforms the other classifier with 90.84% accuracy. At the end, the authors tried out for any meaningful clustering structure. It was observed that PAM and K-means clustering performs better than hierarchical clustering with silhouette with 0.54 and number of clusters is three.

We may try to find out the trends of CEE data using time series methods as future work. The classification and clustering may be applied to the students with different sets of relevant attributes. The highly influence factors for cracking the competitive examination like CEE may also explored as future work.

References

1. Bhardwaj, B.K., Pal, S.: Data mining: a prediction for performance improvement using classification. Int. J. Comput. Sci. Inf. Secur. (IJCSIS) **9**(4), 136–140 (2012)
2. Yadav, S.K., Pal, S.: Data mining: a prediction for performance improvement of engineering students using classification. World Comput. Sci. Inf. Technol. J. **2**(2), 51–56 (2012)
3. Kukasvadiya, M.S., Divecha, N.H.: Analysis of Data Using Data Mining tool Orange. Int. J. Eng. Develop. Res. **5**(2), 1836–1840 (2017)
4. DeFreitas, K., Bernard, M.: Comparative performance analysis of clustering techniques in educational data mining. IADIS Int. J. Comput. Sci. Inf. Syst. **10**(2), 65–78 (2015)
5. Dutt, A., Aghabozrgi, S., Ismail, M.A.B., Mahroein, H.: Clustering algorithms applied in educational data mining. Int. J. Inf. Electron. Eng. **5**(2), 112–116 (2015)
6. Nagy, H.M., Aly, W.M., Hegazy, O.F.: An educational data mining system for advising higher education students. Int. J. Comput. Inf. Eng. **7**(10), 1226–1270 (2013)
7. Oyelade, O.J., Oladipupo, O.O., Obagbuwa, I.C.: Application of K-means clustering algorithm for prediction of students' academic performance. Int. J. Comput. Sci. Inf. Secur. **7**(1), 292–295 (2010)
8. Almarabeh, H.: Analysis of students' performance by using different data mining classifiers. Int. J. Mod. Educ. Comput. Sci. **9**(8), 9–15 (2017)
9. Sivogolovko, E., Novikov, B.: Validating cluster structures in data mining tasks. In: Proceedings of the 2012 Joint EDBT/ICDT Workshops on - EDBT-ICDT 2012, p. 245. ACM, New York (2012)
10. Everitt, B.: Cluster Analysis. Wiley, Chichester (2011). ISBN 9780470749913
11. Park, H.S., Jun, C.H.: A simple and fast algorithm for K-medoids clustering. Exp. Syst. Appl. **36**(2), 3336–3341 (2009)
12. Maulik, U., Bandyopadhyay, S.: Performance evaluation of some clustering algorithms and validity indices. IEEE Trans. Patt. Anal. Mach. Intel. **24**(12), 1650–1654 (2002)
13. Berkhin, P.P.: A Survey of Clustering Data Mining Techniques. Springer, Heidelberg (2006)

14. Chang, W.L., Pang, L.M., Tay, K.M.: Application of self-organizing map to failure modes and effects analysis methodology. Neurocomputing (2017). https://doi.org/10.1016/j.neucom.2016.04.073
15. Ahmed, A.B.E.D., Elaraby, I.S.: Data mining: a prediction for student's performance using classification method. World J. Comput. Appl. Technol. **2**(2), 43–47 (2014)
16. Pandey, U.K., Pal, S.: Data mining: a prediction of performer or underperformer using classification. Int. J. Comput. Sci. Inf. Technol. (IJCSIT) **2**(2), 686–690 (2011)
17. Duda, R.O., Hart, P.E., Stork, D.G.: Pattern Classification, 2nd edn. Wiley-Interscience Publication, New York (2000)
18. Domingos, P., Pazzani, M.: On the optimality of the simple Bayesian classifier under zero-one loss. Mach. Learn. **29**, 103–130 (1997)
19. Jiawei, H., Micheline, K.: Data Mining: Concepts and Techniques. Elsevier Book Series (2000)
20. Rakesh, A., Ramakrishnan, S.: Fast algorithms for mining association rules. In: Proceedings of the 20th International Conference on Very Large Data Bases, VLDB, pp. 487–499 (1994)
21. Willmott, C.J., Matsuura, K.: Advantages of the mean absolute error (MAE) over the root mean square error (RMSE) in assessing average model performance. Clim. Res. **30**, 79–82 (2005)
22. Powers, D.M.W.: Evaluation: from precision, recall and f-measure to roc., informedness, markedness & correlation. J. Mach. Learn. Technol. **2**(1), 37–63 (2011)
23. de Amorim, R.C., Hennig, C.: Recovering the number of clusters in data sets with noise features using feature rescaling factors. Inf. Sci. **324**, 126–145 (2015). https://doi.org/10.1016/j.ins.2015.06.039
24. Borg, I., Groenen, P.: Modern Multidimensional Scaling: Theory and Applications, pp. 207–212, 2nd edn. Springer, New York (2005). ISBN 0-387-94845-7
25. Demšar, J., Curk, T., Erjavec, A., Gorup, Č., Hočevar, T., Milutinovič, M., Možina, M., Polajnar, M., Toplak, M., Starič, A., Stajdohar, M., Umek, L., Žagar, L., Žbontar, J., Žitnik, M., Zupan, B.: Orange: data mining toolbox in Python. JMLR. **14**(1), 2349–2353 (2013)
26. Witten, I.H., Frank, E., Hall, M.A.: Data Mining: Practical Machine Learning Tools and Techniques, 3rd edn. Morgan Kaufmann, San Francisco (2011)
27. Verzani, J.: Getting Started with RStudio, p. 4. O'Reilly Media, Inc. (2011). ISBN 9781449309039
28. Sharma, A., Dey, S.: Performance investigation of feature selection methods and sentiment lexicons for sentiment analysis. IJCA **3**, 15–20 (2012). Special Issue on Advanced Computing and Communication Technologies for HPC Applications ACCTHPCA

Review of Research Progress, Trends and Gap in Occupancy Sensing for Sophisticated Sensory Operation

Preethi K. Mane[(⊠)] and K. Narasimha Rao

Department of Electronics and Instrumentation Engineering,
BMS College of Engineering, Bangalore, India
pkmph2008@gmail.com

Abstract. With the adoption of increasing number of occupancy sensor in building premises, there is a growing concern about the inclusion of the smarter features for catering up sophisticated demands of information processing in Internet-of-Things (IoT). Although, there are various commercially available occupancy sensors, but there is a bigger deal of trade-off between the existing offered featured and actual demands of the user that is quite dynamic. Therefore, we reviewed the most potential research work carried out towards incorporating various features of occupancy sensor in present times in order to investigate the degree of effectiveness in existing research contribution with respect to problems, techniques, advantages, and limitation. This is the first reported review manuscript in occupancy sensing that offers a quick view of existing research trends as well as brief of potential research gap with respect to open-end problems that are yet to be solved in future studies.

Keywords: Occupancy sensor · PIR sensor · Occupation estimation
Object detection · Energy · Accuracy

1 Introduction

Occupation sensing is one of the developing technologies that are used within a building or any specific infrastructure in order to sense the inhabitants within the premises. There are different forms of sensors with different capabilities that are used for this purpose [1]. Both similar as well as heterogeneous forms of sensors are used for sensing the environmental parameters of the subject in order to perform the specific level of controlling [2]. Usage of visual sensors, Pyro-electric Infrared sensor (PIR), CO_2 sensors, etc. are used for counting, motion sensing, and capturing concentration readings respectively [3]. Majority of the utilization of sensors can be seen at present in bigger infrastructure, e.g., shopping malls, hospitals, bigger corporate office, etc. But all of them use very simple forms of sensors as well as sensing technologies that don't demand much dependency. However, with the proliferation of Internet-of-Things (IoT), there is a growing need for performing more number of smarter sensing that is capable of

© Springer International Publishing AG, part of Springer Nature 2019
R. Silhavy (Ed.): CSOC 2018, AISC 765, pp. 212–222, 2019.
https://doi.org/10.1007/978-3-319-91192-2_22

sustaining under variable conditions [4]. Unfortunately, for the occupancy sensing in order to grow in this line, there are some definitive challenges, e.g., (i) although the sensor comes at the lower price but the software or its internal algorithm will become much expensive with the rise of sophisticated user demand. To cater up such demands, the conventional microcontroller will not be eligible enough to maintain the low cost of operation. At the same time, none of the existing commercial products or the research-based techniques have ever spoken about the reduced computational complexity that finally leads to degradation of accuracy as well as expensive maintenance, (ii) there is a higher cost involved in automating the complete process of sensing as well as incorporating learning capabilities in occupancy sensing, (iii) in order to make the sensing more smarter, the existing research is also found to include more intelligent learning techniques, e.g., neural network, support vector machine, fuzzy logic, etc. However, all of them are still in very nascent stages of development, and such smarter tools are yet to be come out in commercial market, (iv) although products before launching undergoes series of test, but they are relatively more comfortable as feature offered by existing occupancy sensors are not much (not even more than 2–3 features). However, it is potentially challenging task to subject the algorithm (under research) for appropriate testing and hence many of the existing algorithms at present is not found to be benchmarked or are found to be compliant under any globally recognized standard, (v) with more research direction on small-scale integration of such sensors, the number of constraints will only keep on increasing where maintaining a well balance between sensing accuracy and all other parameters (e.g. energy efficiency, algorithm complexity) is near to impossible challenge). Therefore, this is the first of the kind of review work towards occupancy sensor where the effectiveness of existing research work has been discussed. The proposed manuscript has briefly discussed all the existing research techniques associated with the occupancy sensing. Section 1.1 discusses the existing literature that offers information about the study background followed by the discussion of research problems in Sect. 1.2 and proposed solution in Sect. 1.3. Section 2 discusses existing research work and its contribution with respect to advantages and limitation followed by the discussion of research trend in Sect. 3 and research gap highlighting all the potential unsolved and unaddressed problems connected to occupancy sensing in Sect. 4. Finally, the conclusive remarks are provided in Sect. 5.

1.1 Background

At present, there is no significant presence of any review manuscript related to the occupancy sensing mechanism. Hence, we only discuss some manuscripts which offer good amount of information about a functional operation associated with occupancy sensor. The work carried out by Pritoni et al. [5] have discussed about the contribution of occupancy sensor with respect to energy preservation. Similarly, the base discussion of energy-related utilization and the respective corresponding approaches were discussed by Rafsanjani et al. [6]. Kjaergaard et al. [7] have presented a discussion of the fundamental methods involved in occupancy sensing. The work carried out by Kleiminger et al. [8] has discussed a unique mechanism of occupancy sensing on the basis of the electricity bill. We find that such similar technique has been used by many researchers. Zhang et al. [9] have presented a detailed report about the occupancy sensing technologies that offers comprehensive information about the various components as well as the cumulative technology involved in the design process. All these studies provided significant information about the potential features of different technologies involved in the construction of different applications for improving the performance of the occupancy sensing system.

1.2   Research Problem

The significant research problems are as follows:

- Existing research work for implementation of occupancy sensing is highly scattered that renders impediment towards clear visualization of a research problem in the specified area of radar.
- There is few research work towards reviewing the existing system of occupancy sensing for which reason it is difficult to have the true picture of progress in the area of sensing.
- Existing implementation doesn't offer much comparative analysis with other equivalent research-based approaches that don't provide much information about the effectiveness of the techniques.
- There are many problems associated with occupancy sensing that are yet to be addressed in existing or upcoming research work, which is quite unclear in the existing system.

Therefore, the problem statement of the proposed study can be stated as "*To understand the research effectiveness pertaining to elite outcomes in occupancy sensing and extract a true picture of research trend and potential research outcome.*"

1.3   Proposed Solution

The core aim of this manuscript is to offer a snapshot of study effectiveness of approaches associated with occupancy sensing presented during the years 2010–2017. The study also aimed for exploring different forms of problems that have been solved using different techniques. The technique also discusses about the advantages as well as limitations associated with the existing problems. Finally, the present manuscript discusses about the existing research trend as well as research gap, and hence this acts as a significant contribution to the proposed study. This solution can be used by the future researchers in order to understand the effectiveness of the contribution of the existing researcher towards occupancy sensing. The next section discusses about existing research contribution towards occupancy sensing.

2 Existing Research on OS

At present, there are various forms of research contribution towards enhancing the applicability of the occupancy sensing. A closer look into the existing approaches shows that there are evolutions of multiple techniques for facilitating sensing of an object with the aid of WiFi data [10, 31], patterns of usage based on battery [21], illumination factor [14, 15], etc. There are also various forms of research-based technique in order to facilitate the process where the dominance of methodology is quite higher for an experimental approach. There are also certain degree of soft-computational-based approach e.g. usage of Markov approach [12, 13], fuzzy logic [18], stochastic approach [38], clustering [39], evidence theory [40], etc. It was also noticed that existing approaches of problem identification is mainly governed by application side and therefore existing approaches using experimental methods are more application specific and doesn't work effectively if the working environment is altered. Some of the problems addressed are occupation estimation [10, 12, 31, 35], energy efficiency [15, 22, 29, 34, 38], infrastructure monitoring and management [11, 16, 21, 28, 33, 37] etc.

Figure 1 highlights the frequently used scheme of Occupancy Sensing, e.g., (i) exploring the presence of an object; (ii) localizing the presence of an object; (iii) counting the number of objects; (iv) monitoring the activity of a user; (v) identification of the precise user, and (vi) tracking the user synching with its mobility. Existing research approaches are evident to apply CO_2 as one of the physiological factors to perform detection. It also makes use of infrared sensors (PIR Sensor), ultrasonic sensors, etc. Different forms of thermal sensors are also reported to capture the thermal imageries for proper detection of the subject. Adoption of acoustic-based, as well as electromagnetic-based approach, was also seen in existing methods of detection. At the same time, there are adoptions of the different form of experimental test-bed, where different types of wireless sensors or transceivers were used for capturing the presence of any object within the monitoring area. However, the bigger challenge lies in the usage of different forms of hardware for experimenting, as well as different software for performing the simulation. The brief outline of existing research contribution is as shown in Table 1.

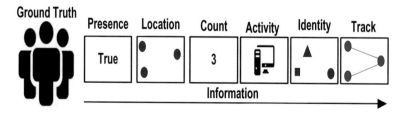

Fig. 1. Scheme of existing research on occupancy sensing

Table 1. Existing research contribution in occupancy sensing

Authors	Problems	Techniques	Advantages	Limitation
Eedera et al. [10]	Occupancy estimation	Experimental, WiFi metadata	Accurate estimation	No benchmarking
Iyer et al. [11]	Position monitoring	Experimental, dual-band sensor	Reduced false alarm	No benchmarking
Liu et al. [12]	Occupancy inference	Experimental, Hidden Markov Model	Minimize user annoyance	Doesn't emphasize on complexity
Li et al. [13]	Deep sensing	Markov chain, Bernoulli filter	Offers unified sensing	Involves system complexity
Avestruz et al. [14]	Capacitative sensing	Experimental, lightning factor (Ballast)	Works both in light and dark	Application specific to light
Cooley et al. [15]	Energy conservation	Experimental, fluorescent lamp, capacitative model	Simplified implementation	No classification capability
George et al. [16]	Seat occupancy detection	Experimental, integrated IC sensing	Cost-effective	Less numerical analysis
Hossain [17]	Wideband spectrum sensing	Analytical, Markov	Multi-user detection	No benchmarking
Reena et al. [18]	Power Scheduling	Genetic algorithm, fuzzy logic	Enhanced stability of system	No benchmarking
Vidal et al. [19]	Autonomic networks	Model-driven development	Supports open source	No supportability of dynamicity
Hua and Chen [20]	Track occupancy	Chaotic neural network	Maximum accuracy	Dependency of training
Man et al. [21]	Security in Occupancy sensing	Battery-based scheme	Simplistic solution	Not energy efficient
Agarwal et al. [22]	Energy management	Simulation	10–15% of energy reduction	Not focused on accuracy
Hammoud et al. [23]	Self-calibration	Unsupervised learning	Highly adaptable	Low scope of sensing
Shih et al. [24]	Low-power sensing	Semi-Supervised, Weighted PCA	Benchmarked recall outcome	Introduces harmonics
Schoofs et al. [25]	Constructing time strategies	Correlation of power consumption	10% reduction in delay	Less emphasize on accuracy
Chen et al. [26]	Energy and memory	Convolution network, deep structure	Enhanced accuracy estimation	Doesn't emphasize on complexity

(continued)

Table 1. (*continued*)

Authors	Problems	Techniques	Advantages	Limitation
Depatla et al. [27]	Walking people count	Experimental, Scattering effect, blocking NLOS, RSSI	Good accuracy	No benchmarking
Ebadat et al. [28]	Room occupancy	Fused lasso estimator, regularization	Better accuracy	Application specific
Lam et al. [29]	Energy+thermal comfort	Correlation model, field experiment	Optimizes setpoint	Less emphasize on accuracy
Forouzanfar et al. [30]	Contactless Event recognition	Feature-based, Bhattacharyya distance, Bayesian classifier	Superior accuracy detection and classification	Only for stationary objects
Mikkelsen et al. [31]	Occupancy estimation	WiFi probe	Reduces overestimation	Issues in accuracy scores
Munir et al. [32]	Real-time occupancy sensing	Usage of depth sensor, ARM platform, histogram of oriented gradient	Good accuracy	Involve higher sensor cost
Nagarathinam et al. [33]	Spatial data of occupancy	Control strategy	18% of energy saving	Less focus on accuracy
Lu et al. [34]	Energy saving in occupancy sensing	Simulation, Smart thermostat	Benchmarked outcome	Less focus on accuracy
Tyndall et al. [35]	Occupancy estimation	Thermal image-based pixel count, classification using entropy, machine learning	Good accuracy performance	No benchmarking
Scott et al. [36]	Occupancy prediction	Experimental, RFID	Good accuracy performance	Associated with complexity, application specific
Yang et al. [37]	Building smart sensing	Experimental, LED sensing	High accuracy	Less improved performance
Bakker et al. [38]	Energy issues	Stochastic modeling,	Results shows better correlation and convergence	Elite outcome depends on higher iteration
Steyer et al. [39]	Object tracking	Clustering, velocity variance analysis	Ensure robustness	No dynamic estimations
Nesa et al. [40]	Data fusion for occupancy sensing	Dempster Shafer	Enhanced precision & specificity	No focus on classification

3   Research Trend

We carry out an explicit investigation of existing research trends from all the major research manuscript publishers e.g. IEEE, ACM, InderScience, Elsevier, Springer, etc. Only the manuscripts published between 2010 to till date is considered for this analysis.

Figure 2 shows that there have been approximately 100 or less number of journals published towards addressing occupancy sensing, occupancy estimation, and occupancy prediction problems. The trend on occupation sensing and estimation is more than prediction problems. This figure is evidence to claim that there is more scope of significant research work towards sense. Figure 3 showcases the different forms of sensors used in recent times which shows that majority of the existing studies have used sensor fusion and Image sensors instead of electromagnetic sensors for capturing the information about the subject's occupancy. Although, there are various studies towards the adoption of other forms of the sensor, but they are less relatively repeated or chosen in different forms of other experiments. A closer look will also show that adoption of image sensors is quite higher, but it is not witnessed in any significant journals. Similarly, adoption of the ultrasonic sensor has gained slight momentum as well as PIR sensors, but utilization of them are quite less reported in the existing system. We also find that there are 161 manuscripts dedicated for addressing energy problems and eight research manuscripts for counting/estimation problems. Therefore, the existing system is found to emphasize more on occupancy sensing problem with more adoption of sensor fusion. The inclination of problems is more on energy efficiency and not much on accuracy.

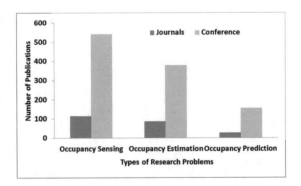

Fig. 2. Research trends towards problem addressed

4   Research Gap

Existing research work towards occupancy sensing has made a lot of progress and technological advancement that has caused them to make fast entries into the global commercial market. At present, various commercial brands offer occupancy sensors at a very competitive price. However, all the existing commercial products are also

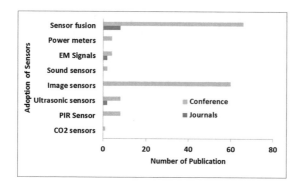

Fig. 3. Research trends towards adoption of sensor

closely associated with challenges, e.g., cross-platform interoperability, easy commissioning, reduced error rates, etc.

- **Deviation in research focus:** There is splitted research work where the research-based community has not mentioned the criticality of research problems. For an example, by observing research trend, it can be just understood that energy efficiency is a critical problem that has attracted the researchers in this, but on the other hand there are various applications where accuracy is highly demanded (may be at the cost of energy also), e.g., healthcare application, elderly people monitoring system, etc. Even the work carried out in addressing energy problems has not received more than 40% of energy saving.
- **Lack of focus on complexity:** Majority of the existing studies have used recursive principle as well as a rule-based mechanism based on which the occupancy sensors are operated. Unfortunately, such schemes potentially suffer from a reduction in computational complexity that is even not computed. There is not a single study which has reported the efficiency of an algorithm.
- **Less Adoption of Doppler Radar:** It is now proven from the research trend that there are less number of studies that use ultrasonic sensors, which is known for using the Doppler shift methods for tracking the objects. Utilization of such autonomous sensors is gaining pace at a very slow rate. The second generation of waves that are reported to be used in IoT application suffers from lack of cost-effectiveness problems as well as privacy problems.
- **Non-inclusion of granularity in research contribution:** Existing research work is too much scattered on addressing the problems using different forms of research methodology, where experimental methods are more dominant. Usage of experimental methods gives good reliability measure, but at the same time, it also makes the research either much application specific or too narrow. Because of such methodology adoption, the current occupancy sensor is quite capable of identifying the specific physiological factor but fails if the patterns of physiological factors differ from usual patterns. Hence, they are incapable for sensing the complex biological signal pattern.

- **Lack of capability of performing classification:** Usage of existing occupancy sensing mechanism performs frequency modulation in order to control the translational motion of the subject. Such forms of the signals are represented using time and frequency in order to generate a spectrum. The stationary spectrums are quite easier to be studied and are also addressed in the majority of existing studies. However, dynamic spectrums are quite difficult to be categorized or classified that poses a bigger impediment towards occupancy prediction, which is next line of research in IoT.

5   Conclusion

This manuscript presents a discussion of existing research approaches in occupancy sensing where the prime motive is to understand the effectiveness of research techniques. After reviewing the existing research contribution, we found that majority of them have similar kinds of problems, e.g., no benchmarking, no focus on complexity analysis, less balance between accuracy and other features, etc. As a research trend, we find that there is less work towards usage of ultrasonic sensors and more experimental work has been carried out towards occupancy sensing/estimation problems. Therefore, our research work will continue in the direction of overcoming the research gap with respect to incorporating of granularity in occupancy sensing framework and further work for classification-based problems that are yet to be addressed in the existing system. Also, the occupancy sensing is often made by HD-cameras and HPC-powered image processing/machine learning algorithms.

References

1. Krarti, M.: Energy Audit of Building Systems: An Engineering Approach, 2nd edn. CRC Press, Boca Raton (2016)
2. Benya, J.R., Leban, D.J.: Lighting Retrofit, and Relighting: A Guide to Energy Efficient Lighting. Wiley, Hoboken (2011)
3. Fraden, J.: Handbook of Modern Sensors: Physics, Designs, and Applications. Springer, Heidelberg (2015)
4. Yasuura, H., Kyung, C.-M., Liu, Y., Lin, Y.-L.: Smart Sensors at the IoT Frontier. Springer, Heidelberg (2017)
5. Pritoni, M., Wooley, J.M., Modera, M.P.: Do occupancy-responsive learning thermostats save energy? A field study in university residence halls. Elsevier J. Energy Buildings **127**, 469–478 (2016)
6. Rafsanjani, H.N., Ahn, C.R., Alahmad, M.: A review of approaches for sensing, understanding, and improving occupancy-related energy-use behaviors in commercial buildings. J. Energies **8**, 10996–11029 (2015)
7. Kjærgaard, M.B., Lazarova-Molnar, S., Jradi, M.: Poster abstract: towards a categorization framework for occupancy sensing systems. In: Proceedings of the Sixth ACM International Conference on Future Energy Systems (e-Energy), pp. 215–216. Association for Computing Machinery (2015). https://doi.org/10.1145/2768510.2770947

8. Kleiminger, W., Staake, T., Santini, S.: Occupancy Detection from Electricity Consumption Data. ACM, New York (2013)
9. Zhang, J., Liu, G., Dasu, A.: Review of literature on terminal box control, occupancy sensing technology and multi-zone Demand Control Ventilation (DCV). Technical report of U.S. Department of Energy (2012)
10. Eedara, P., Li, H., Janakiraman, N., Tungala, N.R.A., Chamberland, J.F., Huff, G.H.: Occupancy estimation with wireless monitoring devices and application-specific antennas. IEEE Trans. Sig. Process. 65(8), 2123–2135 (2017)
11. Iyer, B., Pathak, N.P., Ghosh, D.: Dual-Input Dual-Output RF sensor for indoor human occupancy and position monitoring. IEEE Sens. J. 15(7), 3959–3966 (2015)
12. Liu, P., Nguang, S.K., Partridge, A.: Occupancy inference using pyroelectric infrared sensors through hidden markov models. IEEE Sens. J. 16(4), 1062–1068 (2016)
13. Li, B., Li, S., Nallanathan, A., Nan, Y., Zhao, C., Zhou, Z.: Deep sensing for next-generation dynamic spectrum sharing: more than detecting the occupancy state of primary spectrum. IEEE Trans. Commun. 63(7), 2442–2457 (2015)
14. Avestruz, A.T., Cooley, J.J., Vickery, D., Paris, J., Leeb, S.B.: Dimmable solid state ballast with integral capacitive occupancy sensor. IEEE Trans. Ind. Electronics 59(4), 1739–1750 (2012)
15. Cooley, J.J., Avestruz, A.T., Leeb, S.B.: A retrofit capacitive sensing occupancy detector using fluorescent lamps. IEEE Trans. Industr. Electronics 59(4), 1898–1911 (2012)
16. George, B., Zangl, H., Bretterklieber, T., Brasseur, G.: A combined inductive-capacitive proximity sensor for seat occupancy detection. IEEE Trans. Instrum. Meas. 59(5), 1463–1470 (2010)
17. Hossain, K., Champagne, B.: Wideband spectrum sensing for cognitive radios with correlated subband occupancy. IEEE Sig. Process. Lett. 18(1), 35–38 (2011)
18. Mary Reena, K.E., Mathew, A.T., Jacob, L.: An occupancy based cyber-physical system design for intelligent building automation. Math. Prob. Eng. 2015, 15 (2015)
19. Vidal, C., F-Sánchez, C., Díaz, J., Pérez, J.: A model-driven engineering process for autonomic sensor-actuator networks. Int. J. Distrib. Sens. Netw. 11(3), 684892 (2015)
20. Hua, Z.-X., Chen, X.: Multisensor track occupancy detection model based on chaotic neural networks. Int. J. Distrib. Sens. Netw. 11(7), 896340 (2015)
21. Man, D., Yang, W., Xuan, S., Du, X.: Thwarting nonintrusive occupancy detection attacks from smart meters. Secur. Commun. Netw. 2017, 9 (2017)
22. Agarwal, Y., Balaji, B., Gupta, R., Lyles, J., Wei, M., Weng, T.: Occupancy-driven energy management for smart building automation. In: Proceedings of the 2nd ACM Workshop on Embedded Sensing Systems for Energy-Efficiency in Building, pp. 1–6 (2010)
23. Hammoud, A., Deriaz, M., Konstantas, D.: UltraSense: a self-calibrating ultrasound-based room occupancy sensing system. Procedia Comput. Sci. 109, 75–83 (2017)
24. Shih, O., Lazik, P., Rowe, A.: AURES: a wide-band ultrasonic occupancy sensing platform. In: Proceedings of the 3rd ACM International Conference on Systems for Energy-Efficient Built Environments, pp. 157–166 (2016)
25. Schoofs, A., Delaney, D.T., MP O'Hare, G., Ruzzelli, A.G.: COPOLAN: non-invasive occupancy profiling for preliminary assessment of HVAC fixed timing strategies. In: Proceedings of the Third ACM Workshop on Embedded Sensing Systems for Energy-Efficiency in Buildings, pp. 25–30 (2011)
26. Chen, Z., Zhao, R., Zhu, Q., Masood, M.K., Soh, Y.C., Mao, K.: Building occupancy estimation with environmental sensors via CDBLSTM. IEEE Trans. Ind. Electronics PP(99), 1 (2017)
27. Depatla, S., Muralidharan, A., Mostofi, Y.: Occupancy estimation using only WiFi power measurements. IEEE J. Sel. Areas Commun. 33(7), 1381–1393 (2015)

28. Ebadat, A., Bottegal, G., Varagnolo, D., Wahlberg, B., Johansson, K.H.: Regularized deconvolution-based approaches for estimating room occupancies. IEEE Trans. Autom. Sci. Eng. **12**(4), 1157–1168 (2015)

29. Lam, A.H., Yuan, Y., Wang, D.: An occupant-participatory approach for thermal comfort enhancement and energy conservation in buildings. In: Proceedings of the 5th International Conference on Future Energy Systems, pp. 133–143 (2014)

30. Forouzanfar, M., Mabrouk, M., Rajan, S., Bolic, M., Dajani, H.R., Groza, V.Z.: Event recognition for contactless activity monitoring using phase-modulated continuous wave Radar. IEEE Trans. Biomed. Eng. **64**(2), 479–491 (2017)

31. Mikkelsen, L., Buchakchiev, R., Madsen, T., Schwefel, H.P.: Public transport occupancy estimation using WLAN probing. In: 2016 8th International Workshop on Resilient Networks Design and Modeling (RNDM), Halmstad, pp. 302–308 (2016)

32. Munir, S., et al.: Real-time fine grained occupancy estimation using depth sensors on ARM embedded platforms. In: 2017 IEEE Real-Time and Embedded Technology and Applications Symposium (RTAS), Pittsburgh, PA, pp. 295–306 (2017)

33. Nagarathinam, S., Iyer, S.R., Vasan, A., Sarangan, V., Sivasubramaniam, A.: On the utility of occupancy sensing for managing HVAC energy in large zones. In: Proceedings of the ACM Sixth International Conference on Future Energy Systems, pp. 219–220 (2015)

34. Lu, J., Sookoor, T., Srinivasan, V., Gao, G., Holben, B., Stankovic, J., Field, E., Whitehouse, K.: The smart thermostat: using occupancy sensors to save energy in homes. In: Proceedings of the 8th ACM Conference on Embedded Networked Sensor Systems, pp. 211–224. ACM (2010)

35. Tyndall, A., Cardell-Oliver, R., Keating, A.: Occupancy estimation using a low-pixel count thermal imager. IEEE Sens. J. **1**(10), 3784–3791 (2016)

36. Scott, J., Brush, A.J.B., Krumm, J., Meyers, B., Hazas, M., Hodges, S., Villar, N.: PreHeat: controlling home heating using occupancy prediction. In: Proceedings of the 13th International Conference on Ubiquitous Computing, pp. 281–290. ACM (2011)

37. Yang, Y., Hao, J., Luo, J., Pan, S.J.: CeilingSee: device-free occupancy inference through lighting infrastructure based LED sensing. In: 2017 IEEE International Conference on Pervasive Computing and Communications (PerCom), Kona, HI, pp. 247–256 (2017)

38. de Bakker, C., van de Voort, T., van Duijhoven, J., Rosemann, A.: Assessing the energy use of occupancy-based lighting control strategies in open-plan offices. In: 2017 IEEE 14th International Conference on Networking, Sensing and Control (ICNSC), Calabria, Italy, pp. 476–481 (2017)

39. Steyer, S., Tanzmeister, G., Wollherr, D.: Object tracking based on evidential dynamic occupancy grids in urban environments. In: 2017 IEEE Intelligent Vehicles Symposium (IV), Los Angeles, CA, USA, pp. 1064–1070 (2017)

40. Nesa, N., Banerjee, I.: IoT-based sensor data fusion for occupancy sensing using dempster-shafer evidence theory for smart buildings. IEEE Internet of Things J. **PP**(99), 1 (2017)

A 3D Visualization Application
of Zlín in the Eighteen-Nineties

Pavel Pokorný[✉] and Pavla Dočkalová

Department of Computer and Communication Systems, Faculty of Applied Informatics,
Tomas Bata University in Zlín, Nad Stráněmi 4511, 760 05 Zlín, Czech Republic
{pokorny,p2_dockalova}@fai.utb.cz

Abstract. This paper describes a 3D interactive web application which visualizes the appearance of the town of Zlín in the Eighteen-nineties. The Blender software suite was used to create a complex 3D model of Zlín and its municipality in previous years. Now, this model has been revisited, improved and extended. This was followed by work in order to convert it into a 3D web application. The Blend4Web - which is an external software module that extends Blender was used to do this. In order to obtain an application with the correct level of interactivity and rendered image outputs, it is necessary to set all of the corresponding parameters and to optimize them.

Keywords: Modeling · Visualization · Application · History

1 Introduction

Visualization is the process of representing data, information, or knowledge in a visual form in order to support the exploration, confirmation, presentation, and understanding of these tasks. It is used as a communication mechanism. A single picture can contain a wealth of information, and can be processed much more quickly than a comparable page of words. Pictures can also be independent of local language, just as a graph or a map may be understood by a group of people with no common tongue [1].

With the development and performance of computer technologies, the possibilities and limits of computer graphics continue to increase. The consequences of this are 3D visualizations, which are used more frequently, image and animation outputs now have better quality, and the number of configurable parameters is rising. These visualizations are used in many scientific and other areas of human interest.

This paper describes a 3D interactive web application, which visualizes the appearance of the town of Zlín in the Eighteen-nineties. The main goal was to be able to show Zlín's past to the lay and professional publics, and to breathe life into the atmosphere of the late 19th Century.

© Springer International Publishing AG, part of Springer Nature 2019
R. Silhavy (Ed.): CSOC 2018, AISC 765, pp. 223–232, 2019.
https://doi.org/10.1007/978-3-319-91192-2_23

1.1   Zlín's History

The first written records of Zlín date back to 1322, when it was the center of an independent feudal estate. Zlín became a town in 1397. Later, in the 16th Century, Zlín's Lords brought additional benefits to the town. In this period, the population grew, the town expanded, and was one of the most important settlements in this region. The main and largest building was the Zlín Zámek (Chateau), which served the Tetour family as their residence. There were also other important buildings. The church, which was embellished with a tower (1566), and the Town Hall on the Main Square, which was adapted in Renaissance Style, (1569 and 1586) [2].

During the Thirty Year Wars (1618–1648), the residents of Zlín - along with people from the whole Wallachian Region, led an uprising against the Habsburg Monarchy. The town was heavily damaged and economically ruined. It took almost a century before the number of houses and the population returned to their earlier levels. In the 18th and 19th Centuries, the town developed gradually and local crafts and markets had a good reputation. There was an attempt to make a manufacturing production company (for bleaching linen 1779); and later, a factory - for producing matches (1850), and another for shoes (1870).

One of the oldest images of Zlín and its vicinity is shown in Fig. 1. This contains a painted image dating from 1846. In the oldest photos - taken around 1890, one can see the rural town with one-story and two-story houses. This was just at the time that Zlín began to grow. Table 1 contains statements relating to the number of inhabitants in Zlín

Fig. 1.   Zlín in 1846 (Painted image) [3]

in different years. All of this information has been included in order to make the visualization application based on data from the Eighteen-nineties.

Table 1. Zlín population in the years [2]

Year	Population
1771	1622
1834	2630
1869	2823
1900	2975

2   3D Model Analysis and Its Improvements

This work follows the 3D model that was created in previous years, which was also published in [4]. This model was developed completely in Blender [5] and contains a relatively complex 3D scene of Zlín. It includes models of the terrain, water resources, houses, buildings, bridges, trees, simple shrubs and trees. Technically, the whole scene contains over three hundred thousand vertices, almost 3000 objects, and 59 different textures in the 128 × 128 pixel resolution.

A high-resolution rendered image of this model is shown in Fig. 2. The classic render algorithm – Blender Render was used for this purpose.

Fig. 2. Rendered image of the previous model of Zlín [4]

2.1   Model Improvements

In order to create an interactive web application, the simplest way is to install the Blend4Web module [6] and - in Blender, to switch the rendering algorithm from Blender

Render to Blend4Web. Since Blend4Web uses a different rendering algorithm, the output images showed mistakes in the models, their textures and where the mapping was wrong.

The most common mistake of these 3D models was their inaccurate location and an inability to model the terrain. So these models were moved to their correct position. Some models also contained unnecessary faces with textures which could not be rendered (faces inside objects); these were deleted. The most common mistake was faulty texture mapping. Blender Render can render textures on both sides of faces, but Blend4Web can usually render texture on only one side – the side where its normal vector is set. In simpler cases, it was enough to change the normal vector orientation. In other cases, it was necessary to perform new texture mapping.

3   An Interactive Web Application

As described in the previous chapter, the interactive web application was created in Blender using the Blend4Web module.

Blender is a free, open source 3D creation suite. It supports the entirety of the 3D pipeline — modeling, rigging, animation, simulation, rendering, composition as well as motion tracking; and even, video editing and game creation. Blender has a wide variety of tools which make it suitable for almost any sort of media production. People and studios around the world use it for hobby projects, commercials, feature films, games and other interactive applications - like kiosks, games and scientific research [7].

Blend4Web is a web-oriented 3D engine - a software framework for authoring and interactive rendering three-dimensional graphics and audio in browsers. The platform is intended for visualizations, presentations, online-shops, games and other rich internet applications. The Blend4Web framework is closely integrated with Blender - a 3D modeling and animation tool (hence the name). The content is rendered by means of WebGL [8], and other browser technologies, without the use of plugins. Technically speaking, Blend4Web is a library for web pages, a Blender add-on and some debugging and optimization tools [6].

Blend4Web is very closely integrated with Blender and can create applications that use most of the settings and properties from Blender, including advanced material and texture options, complex models and animations, a powerful physics system, sounds, logic editor settings, etc.

Blend4Web can generate outputs of the whole Blender scene in two ways. Both outputs are available in the File–Export menu in Blender. The first output is given by a classic HTML format, using the HTML5 standard [9], where all of the resources are packed into one HTML file that can then be easily deployed in a web-page. The interactive logic is then performed by the JavaScript commands [10], and these commands are also automatically generated. The benefit of this option is, that everything is created without any great knowledge of programming being necessary.

The JSON format is the second output. JSON - (JavaScript Object Notation), is an open standard for representing data as attributes with values and thus provides an ideal means to encapsulate data between the client and server [11]. The benefit of this option is that the user obtains only the complete data structure of the 3D scene, so they can create their application with its own interaction.

3.1   Realization

An interaction application, in this case, means the possibility of rendering a scene in line with user inputs via a keyboard and mouse. The user controls a character object in the scene environment, which can be to look and move in all directions, just like the real movements of a human.

Any physical object can be used like a character. For the application presented herein, a cube object with Object Physics settings was inserted into the scene in order to get a dynamic object, which can be moved by user commands. A camera was linked to this object, because this camera renders images from its position and the direction one looks. So interaction with the object makes the same interaction with the camera. With the camera setting "Eye Move Style", rendered images are like in 1st-person computer games. The camera and cube settings are shown in Fig. 3. Camera manipulation limits settings are also visible in this picture.

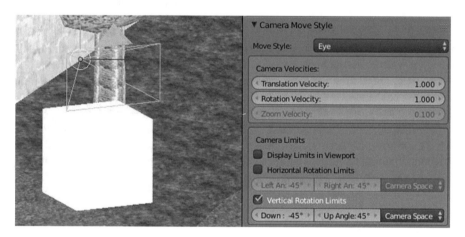

Fig. 3. The character object with camera and its settings

The next step is to define static objects, which will collide with the character object. Mainly, these static objects include the terrain and whole houses and buildings. But this was a problem since the 3D scene is very complex and contains a few hundred thousand vertices. Therefore, the calculation of collision would be too time-consuming and would slow down the whole interaction and rendering process. For the continuous running of the application, it was necessary to create simplified objects, which would not be rendered, but would be used for the calculation of collisions. This technology is very often used in applications that contain large 3D scenes, e.g. 3D computer games.

The object simplifications for collision detection described above, were mainly used for houses and buildings. The terrain object was not modified (this would only be very difficult - with unclear results), and so new collision objects were created in the border of water surfaces, in order that the character cannot move there.

Settings for collision were very simple. It is only necessary set the same material for collision faces and to set the Special Collision parameter for this. In this project, red color was set for this special material and the collision examples around the water with a bridge object are shown in Fig. 4.

Fig. 4. Collision objects around the water and bridge

When the collision objects were created, the application was tested using the Fast Preview function. This showed up some shortcomings – which were subsequently removed. The main problem was the height from which the image was rendered. The view was too high for a man, and would be unnatural. Blend4Web does not solve this problem - so the whole scene was scaled up in order that the rendered image would correspond to the view-point of an adult.

The inequality of the terrain model was another problem. In some places, where there was a large slope for example, the character object could not move further. This can be solved with extended settings of the character object, where the parameters for moving could be set, including "Step High" or "Walk Speed".

4   Application Optimization and Extension

After successful implementation of interactions with the 3D model of Zlín, the next step was to improve the quality of the graphics in order to obtain better visual outputs. The whole model is very large and complex; so, because of this, it was possible to implement some details - otherwise, a loss of flow of the running application would occur. The main attention was focused on surroundings – sky, trees and shadows.

The settings of the sky in the previous version of the 3D model did not work with the Blend4Web module. So it was necessary to implement a new sky and, in order to achieve greater reality, this was animated. The animation in the Blend4Web module is performed by the use of time nodes, which can be run anytime, and can change any material or texture parameters.

In the case of the sky, the gradient from dark blue to light blue in the background was set. To make animated clouds, a grid object was inserted into the scene. This object was then scaled up to a bigger dimension than the whole model. The following step was to shape it to the part of the sphere object. The cloud texture - with transparency, was mapped on this object (Fig. 5). What is important is that this texture is seamless, since the animation process was performed by moving (rolling) it. This texture was downloaded from [12] and improved, (cropped, scaled and make seamless), in GIMP [13].

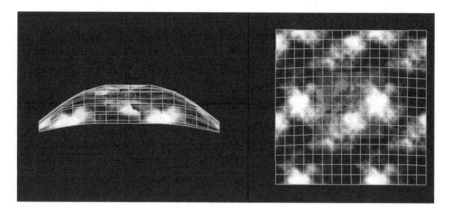

Fig. 5. The object for a sky with a cloudy texture

The previous scene contained simple tree models. These models are shown in Fig. 6. They were simply created by combining deformed sphere and cylinder objects – each model was composed from about 70 vertices. This solution offers good resolution, when the scene is rendered from a greater distance. But if the image is rendered from up close, their appearance is disturbing.

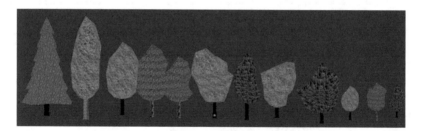

Fig. 6. Old 3D models of trees with their textures

In order to achieve better quality trees, these models need to have more vertices. The problem is that the whole scene contains over 2200 models of trees, so the total number of vertices can be increased too much - and the running application could be slowed down even more.

The solution was to make models of trees based on a cylinder and some transformed planes, on which the texture of branches with leaves or needles were mapped. These textures were downloaded from [12], and - when necessary, improved in GIMP [13]. The resolution for textures was set at 256 × 256 pixels.

The shapes of the trees were modified to better correspond to the appearance of real trees (Fig. 7). The average number of vertices was about 350 vertices, but the total number of them - which were inserted into the 3D scene, was reduced to about 1300. Their position was set according to the historical cadastral map and photos, mainly those of gardens and along the roads.

Fig. 7. Examples of new 3D models of trees with textures

Shadows can also add greater reality to the scenes. Blend4Web supports them for most types of light sources in Blender. The environmental lighting setting was selected to represent the global light. But this light does not support the generation of shadows. So, the next light source to be added to the scene was the Sun object - which is able to generate shadows. It was placed in the west part of the scene of the above-mentioned models. Further, it was necessary to allow shadows on the surfaces of each object that can receive them. This is the object properties and it is very simple to check or uncheck it in the Blender environment. Advanced settings for shadows is available in the Render menu; there, it is possible to set shadow resolution, softness, offset, blurring, fade-out and calculation algorithms.

When these tasks were finished, the complete scene with set necessary parameters was exported to the HTML5 web application. Interactivity with the user is performed by means of a mouse (direction of view), and keyboard - (keys W, S, A, D for moving). The C key can activate/deactivate gravity.

An example of a rendered image from the interactive application is shown in Fig. 8. This image was created from the same position and direction of view as in Fig. 2 in order to compare the differences between these outputs.

Fig. 8. The rendered image of the interactive application

5   Conclusion

This paper describes 3D interactive web application design and realization using the Blender software and Blend4Web module. The content of this application was a 3D model of the town of Zlín and its municipality in the Eighteen-nineties.

This model was created in previous years; but, in order to use it as an interactive application, it was necessary to make some modifications and improvements. Thus, the article describes this process and the text also includes problems and their solutions during the application realization process. The application is fully functional and is based on the HTML5 and JavaScript technologies, without the use of third-party libraries. This means it is able to run on any platform that supports these technologies.

More ideas exist for future work on this project. The main task is to ensure that the visualization is more realistic in order to approximate the atmosphere of the end of the 19th Century in Zlín, for the general public. Therefore, the graphics should be improved and limits found, where the application can be still continuously run on the ordinary personal computers. Another plan is to try to use other JavaScript libraries for rendering – for instance, Three.js or Babylon.js for the comparison of optimizations and rendering results and then to use the best of them.

References

1. Ward, M.: Interactive Data Visualization: Foundations, Techniques, and Applications, 1st edn. A K Peters Ltd., India (2010). ISBN 978-1568814735
2. Pokluda, Z., Nováček, S.: Zlín ve fotografii (1890-1950). 1st edn. ESPRINT a Nadace Tomáše Bati, Zlín (2008). ISBN 978-80-254-3144-3

3. Estranky Zlin Homepage. http://www.zlin.estranky.cz/. Accessed 05 Dec 2017
4. Pokorný, P., Dorazínová, I.: The 3D Visualization of the town Zlín at the turn of the 19th/20th century. In: Proceedings of the 5th WSEAS International Conference on Visualization, Imaging and Simulation, pp. 154–158. Sliema, Malta. Wseas Press (2012). ISBN 978-1-61804-119-7, ISSN 1790-5117
5. Blender – Free and Open Source 3D Creation Suite. https://www.blender.org/. Accessed 05 Dec 2017
6. Blend4Web – Unleashing the Power of 3D Internet. https://www.blend4web.com/en/. Accessed 05 Dec 2017
7. Blender manual. https://docs.blender.org/manual/en/dev/. Accessed 05 Dec 2017
8. Parisi, T.: WebGL: Up and Running, 1st edn. O'Reilly, Sebastopol (2012). ISBN 978-1-449-32357-8
9. Pilgrim, M.: HTML5: Up and Running, 1st edn. O'Reilly, Sebastopol (2010). ISBN 978-0-596-80602-6
10. Daggett, M.E.: Expert JavaScript, 1st edn. Apress, New York (2013). ISBN 978-1-4302-6097-4
11. Rischapter, R.: JavaScript JSON Cookbook, 1st edn. Packt Publishing, Birmingham (2015). ISBN 978-1-78528-690-2
12. Textures for 3D, graphic design and Photoshop! https://www.textures.com/. Accessed 05 Dec 2017
13. GIMP – GNU Impage Manipulation Program. https://www.gimp.org/. Accessed 05 Dec 2017

Multi-agent Systems Interacting (Addressing Scopes, Control Resources)

Mohamad Kadi, Said Krayem, Roman Jasek[(✉)], Petr Zacek,
and Bronislav Chramcov

Faculty of Applied Informatics, Tomas Bata University in Zlin,
Zlin, Czech Republic
mohamad76kadi@gmail.com, drsaid@seznam.cz,
{jasek,zacek,chramcov}@fai.utb.cz

Abstract. Multi-agent systems consist of agents and their environment. the agents in a multi-agent system could equally well be robots, humans or human teams. And may contain combined human-agent.

Multi-agent systems can be used to solve problems that are difficult or impossible for an individual agent or a monolithic system to solve. Intelligence may include some methodic, functional, procedural approach, algorithmic search or reinforcement learning.

In a system with agents that have their own objectives and schedules, when tasks are dependent on one another or when resources are to be shared, it can be important to add the function of coordination to the system, otherwise there is a risk of redundancy or even of a "locked" situation occurring.

With the modeling in Event-B we are now ready to make precise what we mean by a "faultless" system, which represents our ultimate goal as the title of this prologue indicates.

In this paper and with the abstract machine, we are going to present a formal approach to develop the addressing and the relation between Multi-Agents and its convenient scope achieving the allocated missions.

On the refinement machine the technique of adding auxiliary resources is considered during the mission life-cycle.

Keywords: Multi-agent systems · Addressing scopes
Control resources Event-B · Rodin

1 Introduction

An agent is a collection of a small amount of features in software capable to solve complex problems. Agents do not simply act in response to their environment. Agents are able to exhibit purposive behavior and performance by taking the initiative, which means that an agent is Proactive.

Agents can be competent of flexible and autonomous in environment to convene its design objectives. Technically, have roles, behaviors, functionalities, and goals. makes a decision about what action to perform based on the history of the system that it has witnessed to date.

© Springer International Publishing AG, part of Springer Nature 2019
R. Silhavy (Ed.): CSOC 2018, AISC 765, pp. 233–245, 2019.
https://doi.org/10.1007/978-3-319-91192-2_24

Actually, if each agent is connected; after that, everything has to be a part of a particular system. In addition, if each agent has to take decisions on its own with its capacity, then such a system is a Multi agent system (MAS). A system is a pair containing an agent and an environment: to create a system that is pervasively embedded in the environment, completely connected, intuitive, portable, and constantly available. Each node of the system should be able to initiate its tasks and actions based on what it learned, with its interaction with other nodes [1].

Figure 1 shows the essential concepts of agent-based computing are agents, high-level interactions and organizational relationships. It can be seen that there can be numerous agents wherein specific agents can interact amongst themselves and/or have the same sphere of visibility and influence. There can also be agents who act independently without any interaction of other agents and unique sphere of influence [2].

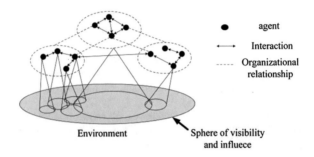

Fig. 1. Canonical view of an agent-based system (source [2])

Hence, Multi-agent systems consist of agents and their environment. However, the agents in a multi-agent system could equally well be robots, humans or human teams. A multi-agent system may contain combined human-agent [3].

Agents can be divided into different types ranging from simple to complex. Some categories suggested to define these types include:

- Passive agents: or agent without goals (like obstacle, apple or key in any simple simulation)
- Active agents: with simple goals (like birds in flocking, or wolf–sheep in prey-predator model)
- Cognitive agents (which contain complex calculations)

Agent environments can be divided into:

- Virtual Environment
- Discrete Environment
- Continuous Environment [4]

The main application of multi-agent systems at the moment can be listed as follows:

- Problem Solving: As an alternative to centralized problem solving, either because problems are themselves distributed, or because the distribution of problem solving between different agents reveals itself to be more efficient way to organize the problem solving - it can be flexible and allow failures in the system - or because, in some cases, it is the only way to solve the problem.

- Multi-Agent Simulation: Simulation is widely used to enhance knowledge in biology or in social science and MAS gives us the possibility to make artificial universes that are small laboratories for the testing of theories about local behaviors. Examples include Simdelta (Cambier and Bousquet) and in SimPop (Bura).
- Construction of Synthetic Worlds: These artificial universes can be used to describe specific interaction mechanisms and analyses their impact at a global level in the system. The entities that are represented are usually called animates, since they are mainly inspired by animal behaviors (hunting, searching or gathering habits). The aim of this research is to have societies of agents that are very flexible and can adapt even in cases of individual failure. (For example, when robots are sent on an expedition and they are required to be very independent from the instructions they could receive.)
- Collective Robotics: Defining the robots as a MAS where each subsystem has a specific goal and deals with that goal only. Once all the small tasks are accomplished the big task is too. MAS approaches can also be used in the co-ordination of different mobile robots in a common space.
- Kenetic Program Design: MAS can also be seen as a very efficient modular way to program.

In a system with agents that have their own objectives and schedules, when tasks are dependent on one another or when resources are to be shared, it can be important to add the function of coordination to the system. Otherwise there is a risk of redundancy or even of a "locked" situation occurring.

Optimization in the resolution of these problems is often very difficult when one treats them in a centralized way and implies that one must make choices about which agents should make more effort to adapt to the needs and actions of others [5].

2   Event-B and the Early Identification of Errors Through Formal Modelling

This paper addresses the key role played by formal modelling and verification in software systems engineering. Modelling may be used at all stages of the development process from requirements analysis to system acceptance testing. Formal modelling and verification lead to deeper understanding and higher consistency of specification and design than informal or semi-formal methods. In order to manage system complexity, abstraction and refinement are key methods for structuring the formal modelling effort since they support separation of concerns and layered reasoning. A refinement approach means that models represent different abstraction levels of system design; consistency between abstraction levels are ensured by formal verification. We use the Event-B formal modelling language developed by Abrial [6] and the associated Rodin1 toolset for Event-B [7].

We outline the general structure of an Event-B specification. A specification consists of a static part, specified in a context, and a dynamic part, specified in a machine.

An Event-B context contains the following elements:

- Sets: Abstract types used in specification to distinguish various entities
- Constants: Logical variables whose value remain constant
- Axioms: Predicates that specify assumptions about the constants.

An Event-B machine contains the following elements:

- Variables: State variables whose values can change.
- Invariants: Predicates that specify properties about the variable that should always, remain true.
- Initialization: Initial values for the abstract variables.
- Events: Guarded actions specifying ways in which the variables can change.

Events may have parameters. A machine may see the static elements defined in a context meaning that these elements are visible within the machine.

An Event-B machine M1 may be declared to be a refinement of some other Event-B machine M0. In this case, we refer to M0 as the abstract machine and M1 as the refined machine. Machine M1 is said to be a correct refinement of M0 if any behavior that may be exhibited by M1 is also a possible behavior of M0. Refinement represents our expectation that the behavior of M1 should conform to the behavior of M0. Of course, the declaring that M1 refines M0 does not on its own guarantee the correctness of a refinement. Rather the declaration gives rise to proof obligations that need to be discharged in order to guarantee the correctness of a refinement.

When refining a machine, it is common to specify new types and constants to be used in the refinement. This is achieved by specifying a new context for the refined machine. If the specification of any new types and constants depend on the types and constants used by the abstract machine, the new context is declared to be an extension of the context of the abstract model. A refined context C1 is declared as an extension of the abstract context C0 meaning context C1 may refer to types and constants specified in context C0. The dashed line from machine M1 to context C0 indicates that M1 implicitly see definitions in C0 (via C1) [8].

The main fundamental characteristics of Event-B which is providing a rich modelling language based on set theory thus this allows precise descriptions of intended system behavior (models) to be written in an abstract way. It provides a mathematical notion of consistency together with techniques for identifying inconsistencies or verifying consistency within a model. It also provides a notion of refinement of models together with a notion of consistency between a model and its refinement.

By abstracting and modelling system behavior in Event-B, it is possible to identify and fix requirements ambiguities and inconsistencies at the specification phase, much earlier in the development cycle than system testing. Figure 2 shows, that by this way, rather than having an error-discovery profile in which most errors are discovered during system testing.

Following situation is illustrated by Fig. 3. We would arrive at an ideal profile in which more errors are discovered as soon as they are introduced [9].

Formal modelling and reasoning help to increase understanding and reduce defects in requirements specification. Sets and relations play a key role in modelling as do operators on these structures.

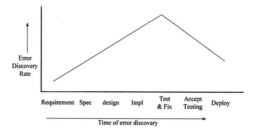

Fig. 2. Typical error discovery rate at different stage of development (source [9])

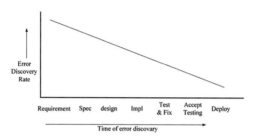

Fig. 3. Idealized error discovery rate through early stage formal modelling and analysis (source [9])

Precise definitions and rules are provided in order to help the reader gain a strong understanding of the mathematical operators for sets and relations. While the emphasis is on mathematical reasoning, particularly through invariant proofs [10].

Hence, modelling in Event-B we are now ready to make precise what we mean by a "faultless" system, which represents our ultimate goal as the title of this prologue indicates [11].

3 Controlling Multi-agent System Interacting with the Mission Environment

In this paper, we are going to present a formal approach to develop the addressing and the relation between Multi-Agents and its convenient scope achieving to handle the allocated missions as can be seen in Fig. 4.

As a case study and in the abstract idea we will assume the following. *AGENT* is the set of microprocessor or functionality robots likewise, automated cars or vehicles and even company employees.

agent is set of trusted and qualified agents, who are evaluated and approved to get *readiness* relation accessing the scope,

$$\text{inv1} : \text{agent} \subseteq \text{AGENTS}$$

missions are sets of roles and tasks can be carried out individually or parallel by the qualified agents,

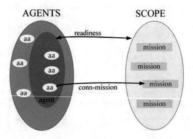

Fig. 4. Addressing the SCOPE relations (own source)

readiness is the relation between the qualified agent and the compatible mission

$$\text{inv2} : \text{readiness} \in \text{AGENTS} \leftrightarrow \text{SCOPE}$$
$$\text{inv3} : \text{dom(readiness)} \subseteq \text{agent}$$

We assume that, each qualified agent can be connected with on mission at one time and just once be ready to carry out the mission,

$$\text{inv4} : \text{conn\_mission} \in \text{AGENTS} \nrightarrow \text{SCOPE}$$
$$\text{inv5} : \text{conn\_mission} \subseteq \text{readiness}$$

So accordingly, the abstract context, variable and invariant are presented as follows:

```
CONTEXT
        C0
SETS
        AGENTS
        SCOPE
AXIOMS
        axm1    :    AGENTS ≠ ∅
        axm2    :    SCOPE ≠ ∅
END

MACHINE
        M0
SEES
        C0
VARIABLES
        agent
        readiness
        conn_mission
INVARIANTS
        inv1    :    agent ⊆ AGENTS
        inv2    :    readiness ∈ AGENTS ↔ SCOPE
        inv3    :    dom(readiness)⊆ agent
        inv4    :    conn_mission ∈ AGENTS ⇸ SCOPE
    inv5    :    conn_mission ⊆ readiness
```

In Event-B we also specification a special initialization event for a model that de-fines how the variables are to be initialized before other events are performed

```
EVENTS
INITIALISATION   ≙
BEGIN
          act1    :    agent ≔ ∅
          act2    :    readiness ≔ ∅
          act3    :    conn_mission ≔ ∅
END
```

The missions that are combined together presenting *SCOPE* should match the ability of all members of *AGNET*.

Means the robots should have an ability to work with the proper SCOPE and handle the certain kind of mission or the automated cars or vehicles should be compatible with the surrounded environment to do certain duties.

Even in company structure the employee expertise should be elected and improved before addressing them handling special scope-of-work missions related to their, we assume here that the qualified agents have allocated to *agent* set to be ready handling the mission.

```
EngageAgent   ≙
ANY
          aa
WHERE
          grd1    :    aa ∈ AGENTS
          grd2    :    aa ∉ agent
THEN
          act1    :    agent ≔ agent ∪ {aa}
END
```

Then the *agent* members and the *SCOPE* of mission have a pre-concluded relation named *readiness*.

The *readiness* will be assign as a relation between *agents* and SCOPE just when the agent's members are evaluated and absolutely confirm its ability, reliability and compatibility to match this kind of SCOPE to handle the related missions.

```
AssignReadiness   ≙
ANY
      aa
      mission
WHERE
      grd1    :    aa ∈ agent
      grd2    :    mission ∈ SCOPE
THEN
      act1    :    readiness ≔ readiness ∪ {aa↦mission}
END
```

The certain mission of the *SCOPE* will be allocated to the stand-by agent so the *conn_mission* relation will take place and the involved mission is triggered.

```
AllocatMission    ≙
ANY
            aa
            mission
WHERE
            grd1   :    aa↦mission ∈ readiness
            grd2   :    aa ∉ dom(conn_mission)
THEN
            act1   :    conn_mission(aa) ≔ mission
END
```

Once the agent completely achieves the allocated mission, it should be released and leaves the *conn_mission* relation.

```
CloseMission    ≙
ANY
            aa
WHERE
            grd1   :    aa ∈ dom(conn_mission)
THEN
            act1   :    conn_mission ≔ {aa} ⩤ conn_mission
END
```

Even the agent should be got out from the readiness status.

```
ReleaseReadiness    ≙
ANY
            aa
            mission
WHERE
            grd1   :    aa↦mission ∈ readiness
            grd2   :    aa↦mission ∉ conn_mission
THEN
            act1   :    readiness ≔ readiness \ {aa↦mission}
END
```

The revised process will get the agent back to the initial status as a stand-by situation to be ready for the next relation and mission.

```
DisEngageAgent    ≙
ANY
            aa
WHERE
            grd1   :    aa ∈ agent
            grd2   :    aa ∉ dom(conn_mission)
THEN
            act1   :    agent ≔ agent \ {aa}
            act2   :    readiness ≔ {aa}⩤ readiness
END
```

4 Controlling Multi-agent System Interacting with Shared Resources

The refinement machine will guarantee the efficiency, reliability and the productivity of the multi-agents system, the technique of adding auxiliary resources should be considered during the mission life-cycle as can be seen in Fig. 5.

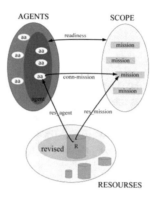

Fig. 5. Addressing Resources (own source)

The resources could be allocated based on the importance and the priority of the requested mission, hence it need more efficient resources to be assigned and reserved.

Suppose the zero-fail down mission will be take place like the Air navigation services which is the term applied to the bundle of services provided to aircraft to enable safe and efficient flight from one destination to another.

Likewise, the ICU and the surgery services are also another type of critical mission requiring some special unique resources to be involved along the operation, by their concept the resources here like an addition facility added to mission scope.

The other factor is related to the agent by itself, for an instance, the ability to improve its force by adding extra resources, let's assume that the agent here as a virtual machine so it has adjustable performance and can be improved dependently so the allocated resource here like a pool of processing, memory and storage so can be offered to improve the agent performance during the process.

Accordingly, the allocating of the recourses may have related to both the performance of the agent and the status of the mission.

So, the set of *recourses* will be presented as a pool and the resource-agent (*res_agent*) along with the resource-mission (*res_mission*) will present the resource relations with agents and mission respectively.

```
CONTEXT
        C1
EXTENDS
        C0
SETS
        RESOURCES
CONSTANTS
        res_agent
        res_mission
AXIOMS
        axm1    :    res_agent∈RESOURCES→AGENTS
        axm2    :    res_mission∈RESOURCES→SCOPE
END
```

The *reserved* action should be taken place first, this distinctive approach will indeed the two goals, the first one is that the reserved resources should NOT be involved again to any other relation until it get release back, the second critical goal is giving an indicator for the accurate capacity of the un-allocated recourse and whether it is enough to be engaged with another relation or not, so the ReserveResource event will handle this approach.

```
ReserveResource    ≜
ANY
        aa
        mission
        R
WHERE
        grd1    :    R∉reserved
        grd2    :    res_agent(R)=aa
        grd3    :    res_mission(R)=mission
        grd4    :    aa↦mission ∈ readiness
THEN
        act1    :    reserved:=reserved∪{R}
END
```

And defiantly the resources should be released again by *ReleaseResource* event when the mission is get closing successfully to be addressable again for the other agent and mission.

```
ReleaseResource    ≜
ANY
        R
WHERE
        grd1    :    R∈reserved
THEN
        act1    :    reserved:=reserved\{R}
END
```

The conclusion of the total refinement machine is presented as follows:

```
EngageAgent    ≜                          ReleaseReadiness    ≜
REFINES                                   REFINES
         EngageAgent                               ReleaseReadiness
ANY                                       ANY
      aa                                        aa
WHERE                                           mission
      grd1   :   aa ∈ AGENTS             WHERE
      grd2   :   aa ∉ agent                     grd1   :   aa↦mission ∈ readiness
THEN                                            grd2   :   aa↦mission ∉ conn_mission
      act1   :   agent  ≔  agent  ∪             grd3   :   ∀R·R∈reserved              ⇒
   {aa}                                         res_agent(R)↦res_mission(R)≠aa↦mission
END                                       THEN
                                                act1   :      readiness  ≔  readiness  \
AssignReadiness    ≜                         {aa↦mission}
REFINES                                   END
         AssignReadiness
ANY                                       ReserveResource    ≜
      aa                                  ANY
      mission                                   aa
WHERE                                           mission
      grd1   :   aa ∈ agent                     R
      grd2   :   mission ∈ SCOPE         WHERE
THEN                                            grd1   :   R∉reserved
      act1   :   readiness  ≔  readi-          grd2   :   res_agent(R)=aa
   ness ∪ {aa↦mission}                         grd3   :   res_mission(R)=mission
END                                             grd4   :   aa↦mission ∈ readiness
                                          THEN
AllocatMission    ≜                             act1   :      reserved≔reserved∪{R}
REFINES                                   END
         AllocatMission
ANY                                       RleaseResource    ≜
      aa                                  ANY
      mission                                   R
      R                                   WHERE
WHERE                                           grd1   :   R∈reserved
      grd1   :   aa↦mission ∈ readiness   THEN
      grd2   :   aa ∉ dom(conn_mission)         act1   :      reserved≔reserved\{R}
      grd3   :   R ∈ reserved             END
      grd4   :   res_agent(R)=aa
      grd5   :   res_mission(R)=mission   DisEngageAgent    ≜
THEN                                      REFINES
      act1   :      conn_mission(aa)  ≔            DisEngageAgent
   mission                                ANY
END                                             aa
CloseMission    ≜                         WHERE
REFINES                                          grd1   :   aa ∈ agent
         CloseMission                            grd2   :   aa ∉ dom(conn_mission)
ANY                                              grd3   :   ∀R·R∈reserved              ⇒
      aa                                       res_agent(R)≠aa
WHERE                                            THEN act1   :      agent ≔ agent \ {aa}
      grd1   :   aa ∈ dom(conn_mission)          act2   :      readiness ≔ {aa}◁ readiness
THEN                                      END
      act1   :   conn_mission ≔ {aa}
   ◁ conn_mission
END                                       END
```

5 Proof Obligations Statistics

In Table 1, there can be seen the proof of statistics for our model using the Rodin3.2 platform, the statistics give us the proof obligations generated and discharged by the Rodin, The finial development of our model results in 20 POs (Proof obligations), around (95%) of them have been proved automatically by the Rodin platform and the rest have been proved manually in the Rodin interactive proving environment.

Table 1. The summary of the proof obligation statistics

Element name	Total	Auto	Manual	Reviewed	Undischarged
MAS_SCOPE	20	19	1	0	0
C0	0	0	0	0	0
C1	0	0	0	0	0
M0	14	13	1	0	0
M1	6	6	0	0	0

6   Conclusion

In this paper, we have presented a rigorous approach to developing interactive relations for agent systems.

Our approach has proposed the relations how the addressing will be going on between the agents and convention scope to kick-out the requested missions, further more we presented how to improve this approach and addressing auxiliary resources to meet the mission requirements.

Indeed, by starting the development from a high level of abstraction and progressively introducing representation of implementation details, we gradually increased complexity of a system model. Since each refinement step was formally verified, the final system model was proved to be correct by construction.

Our final model is translated into the Event-B notation to verify required properties, so we can say, event-B allows us to define a kind of modeling methodology by write the correct mathematical notions; wherefore we can apply event-B in modeling Multi-agent systems, but we should choose carefully invariants and variables to ease effort of proof.

In the future, we are planning to extend the proposed approach in two more directions. One of them to investigate the use of fault tolerance concept shifting the mission from the fault agent to another.

And the other one, improving the allocation technic of the Recourses by addressing it dynamically during the process life-cycle achieving the mission deadline.

Acknowledgement. This work was supported by the Ministry of Education, Youth and Sports of the Czech Republic within the National Sustainability Programme project No. LO1303 (MSMT-7778/2014) and by the European Regional Development Fund under the project CEBIA-Tech No. CZ.1.05/2.1.00/03.0089. Also supported by grant No. IGA/CebiaTech/2017/ 007 from IGA (Internal Grant Agency) of Tomas Bata University in Zlin.

References

1. Brahnam, S., Jain, L.C.: Dhu'l-H. 17, 1435 AH – Computers, 291 p. Springer, Heidelberg
2. Cellular Automata and Agent - based Models - A Theoritical Framework. http://wgbis.ces. iisc.ernet.in/energy/paper/TR100/tr100_cel.htm
3. Kaminka, G.A.: Robots are Agents, Too! AgentLink News, pp. 16–17, December 2004

4. Kubera, Y., Mathieu, P., Picault, S.: Everything can be Agent! (PDF). In: Proceedings of the Ninth International Joint Conference on Autonomous Agents and Multi-Agent Systems (AAMAS 2010), Toronto, Canada, pp. 1547–1548 (2010)
5. Ferber, J.: Multi-Agent System: An Introduction to Distributed Artificial Intelligence. Addison Wesley Longman, Harlow (1999). Paper: ISBN 0-201-36048-9
6. Abrial, J-R.: The B book: Assigning Programs to Meanings. Cambridge University Press, Cambridge (1996)
7. Abrial, J-R.: Extending B without changing it (for developing distributed systems). In: First B Conference Putting Into Practice Methods and Tools for Information System Design, Nantes, pp. 169–190 (1996)
8. Metayer, C., Abrial, J., Voisin, L.: Event-B Language. Technical report D7, z RODIN Project Deliverable (2005)
9. Butler, M.: Mastering System Analysis and Design through Abstraction and Refinement. Deploy (2012)
10. Butler, M.: Reasoned Modelling with Event-B, 30 January 2017
11. Abrial, J.-R.: Modeling in Event-B: System and Software Engineering. Cambridge University Press, Cambridge (2010)

Improved Adaptive Fault Tolerance Model for Increasing Reliability in Cloud Computing Using Event-B

Ammar Alhaj Ali, Roman Jasek$^{(\boxtimes)}$, Said Krayem,
Bronislav Chramcov, and Petr Zacek

Faculty of Applied Informatics,
Tomas Bata University in Zlin, Zlin, Czech Republic
ammar282n@hotmail.com,
{jasek,chramcov,zacek}@fai.utb.cz, drsaid@seznam.cz

Abstract. Cloud computing provide services to many users at the same time by providing virtual resources via internet, General example of cloud services is Google apps, provided by Google. In most of cloud applications, processing is done on remote cloud computing nodes. So, there are more chances of errors, due to the undetermined latency and lose control over remote nodes, so it is very important apply the techniques for fault tolerance in cloud computing.

In this paper, a fault tolerance in real time cloud computing is proposed. In our model, the system tolerates the faults and makes the decision on the basis of reliability of the processing nodes.

And we will present an event-B as formal method that can be used in the development of reactive distributed systems and we propose using the Rodin modeling tool for Event-B that integrates modeling and proving.

Keywords: Cloud computing · Fault tolerance · Reliability · Virtual machines
Event B · Rodin

1 Introduction

Recently, with the increasing demand and benefits of cloud computing and real time systems, we started seeing large number of organizations and companies that dependent on cloud computing to delivery its services over the internet, through cloud computing, companies can use resources, such as virtual machine, applications or storage centers [1].

In other side, using of cloud infrastructure for real time applications increases the chances of errors, so these systems have to be reliable [2].

The reliability of a system is linked directly to its ability to operate correctly despite the presence of faults; from here we can know reason for increasing number of real-time applications that require fault-tolerance [3].

In this paper, we will offer algorithm for reliability assessment and will propose improving to this algorithm.

© Springer International Publishing AG, part of Springer Nature 2019
R. Silhavy (Ed.): CSOC 2018, AISC 765, pp. 246–258, 2019.
https://doi.org/10.1007/978-3-319-91192-2_25

2   Related Work

Fault tolerance is a major issue to guarantee the availability and reliability of computing cloud, where fault tolerance means that Cloud infrastructure should continue to offer services under fault presence.

Malik and Huet [4] proposed the model that deals with the fault tolerance mechanism, a model name fault tolerance model for cloud (FTMC). This model tolerates the faults on the basis of reliability of each computing node, i.e. virtual machine. A virtual machine is selected for computation on the basis of its reliability and can be removed, if does not perform well for real time applications, In this model choose the best VM depends on highest reliability in each cycle. If two nodes have the same highest reliability level then the output of the node with smaller IP address is selected as computing cycle output.

In our paper, we will improve reliability assessment algorithm, by using average of the response time of nodes instead of smaller IP address to assess reliability to select the output of the node which has highest reliability among all the competing nodes, Because when we use smaller IP, probably we won't get best node, where we suggested in case we have same highest reliability for many nodes we will select node that have best response time and with this additional factor certainly we will get best node in cloud, and we added new factors to calculate new reliability value of node based on response time.

3   Proposed Model

In this scheme, we have 'N' VM (Virtual machine), which run the 'N' variant algorithms. Algorithm 'A1' runs on VM1, 'A2' runs on VM2 up till 'An', which runs on VMn, shown in Fig. 1.

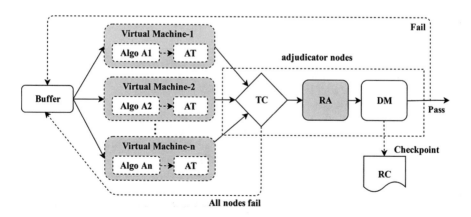

Fig. 1. Proposed system model (own source)

Then we have AT module which is responsible for the verification of output result of each node. The outputs are then passed to TC module which checks the timing of each result. On the basis of the timing, the RA module calculates and reassigns the reliability of each module.

Then all the results are forwarded to DM module which selects the output on the basis of best reliability. The output of a node with highest reliability is selected as the system cycle output [4–6].

In this model choose the best VM depends on highest reliability in each cycle. If two nodes have the same highest reliability level then the output of the node with smaller average of the response time is selected as computing cycle output

In this model, represented by Fig. 1 we have two main types of node:

1. **Virtual machines:** virtual machine contains the real-time application algorithm and an acceptance test for its logical validity [4].
2. **The adjudication node:** we have the time checker (TC), reliability assessor (RA) and decision mechanism (DM) modules, and the location of adjudication node can be a part of the cloud infrastructure or can be a part of the user infrastructure [4].

4   Virtual Machines

Each virtual machine contains:

- **The real time application algorithm:** Real time application algorithm is the application logic to perform the real time computations. These algorithms can be different from each other either by implementation language or software engineering process. The algorithm passes the result to the acceptance test (AT) for verification [4].
- **Acceptance test (AT):** That performs the acceptance test for each cycle output. It verifies the correctness of the result produced by an algorithm running on the same virtual machine [4].

5   Adjudication Node

We have follow nodes:

- **Time checker (TC):** module is a time checking sensor, TC module passes the results to RA (reliability assessor) module. It only passes the correct results of those nodes which produces the result before deadline time.
- **Reliability assessor (RA):** module assesses the reliability for each virtual machine. In the beginning the reliability of each virtual machine is 100%. If the algorithm in node produces a correct result within the time limit, its reliability increases, and if the algorithm in node fails to produce the correct result or result within time, its reliability decreases.

If reliability of any processing node falls below minimum reliability level, the RA stops that node to work further and removes it. It then adds a new node in place of the removed node. Handling of removal and addition of node is not the responsibility of RA. It only asks to some other responsible system module.

If reliability of all the nodes fall below minimum reliability level, the system either perform the backward recovery or stop the system to work further [4].

- **Decision mechanism (DM):** It selects the output of the node which has highest reliability among all the competing nodes. Competing nodes are those nodes which have produced the results within time. We have included only the competing nodes, because a node which fails to produce the result can have higher existing reliability [4]. If two nodes have the same highest reliability level then the output of the node with smaller average of the response time is selected as computing cycle output, and there is a SRL (System Reliability Level). It is the minimum reliability level to be achieved to pass the result. DM compares the best reliability with the SRL. Best reliability level should be greater than or equal to SRL, DM also request to some responsible module to remove one node with minimum reliability and add a new node [4].
- **Recovery cache (RC):** Recovery cache is a repository area to hold the checkpoints. At the end of each computing cycle DM makes checkpoint in it. In case of a complete failure, backward recovery is performed with the help of checkpoints maintained in this recovery cache [4].

6   Fault Tolerance Mechanism

In (FTMC) model, we have already reliability assessment algorithm for each node as it is shown in Fig. 2 as the representation of the source code [4, 5].

Initially Reliability of a node is set to 1(100%). There is an adaptability factor *n*, which controls the adaptability of reliability assessment. The value of n is always greater than 0.

The algorithm has three factors *RF*, *MinReliability* and *MaxReliability*:

- *RF:* is a reliability factor which increases or decreases the reliability of the node. It decreases the reliability of the node more quickly as compare to the increase in reliability. It is due to its multiplication with the adaptability factor *n*.
- *MinReliability:* is the minimum reliability level. If a node reaches to this level, it is stopped to perform further operations.
- *MaxReliability:* is the maximum reliability level. Node reliability cannot be more than this level. It is really important in a situation, where an initially produces correct results in consecutive cycles, but then fails again and again. So its reliability should not be high enough to make the reliability difficult to decrease and converge towards lower reliability [4, 5].

```
Begin
InitiallyReliability: =1, n: =1
Input RF, MaxReliability, MinReliability
Input nodestatus
If NodeStatus =Pass then
        Reliability: = Reliability + (Reliability * RF)
        If n > 1 Then
                    n: = n-1;
        End If
Else if nodeStatus = Fail then
        Reliability: = Reliability – (Reliability * RF * n)
        n: = n+1;
End If
If Reliability >= MaxReliability then
        Reliability: = MaxReliability
End If
If Reliability < MinReliability then
        NodeStatus: =Dead
        call_proc: Remove_this_node
        call_proc: Add_new_node
End If
End
```

Fig. 2. Reliability Assessment Algorithm (own source)

We propose improving the reliability assessment algorithm, in proposed algorithm we have **RF, RF1, RF2** where **RF1 < RF < RF2** and **2*RF = RF1 + RF2,** Example **(RF = 0.02, RF = 0.01, RF2 = 0.03),** and we depended on timeliness of each virtual machine result.

- **RF:** is a reliability factor which decreases the reliability of the node in 'Fail' status.
- **RF1:** is a reliability factor which increases the reliability of the node in 'Pass' status and average time of node is smaller from last time.
- **RF2**: is a reliability factor which increases the reliability of the node in 'Pass' status and average time of node is larger from last time, as can be seen in Fig. 3.

In DM (Decision mechanism) we apply decision mechanism algorithm, as can be seen in Fig. 4.

We have a **SRL** (system reliability level). It is the minimum reliability level to be achieved to pass the result. Best reliability level should be greater than or equal to system reliability level, in case best reliability was smaller from SRL In this case, backward recovery is performed. Also request to some responsible module (resource manager or scheduler) to remove one node with minimum reliability and add a new node) [4, 5].

7 Discussion of Results

We designed a web page to simulate a model [7]. We assume the existence of some virtual machines, and set environment variables exactly as can be seen in Table 1.

```
Begin
InitiallyReliability: = 1,
n: =1,
InitiallySumTime: = 0,
n_Pass: = 0,
InitiallyAvrTime:=0
Input RF, RF1, RF2, MaxReliability, MinReliability
Input nodestatus, nodetime
If NodeStatus =Pass then
        If nodetime > AvrTime Then
                Reliability: = Reliability + (Reliability * RF1)
        Else
                Reliability: = Reliability + (Reliability * RF2)
        End If
        n_Pass:= n_Pass+1
        SumTime: = SumTime+ nodetime
        AvrTime: = SumTime / n_Pass
        If n > 1 Then
            n: = n-1;
        End If
Else if nodeStatus = Fail then
        Reliability: = Reliability – (Reliability * RF * n)
        n: = n+1;
End If
If Reliability >= MaxReliability then
        Reliability: = MaxReliability
End If
If Reliability < MinReliability then
        NodeStatus: =Dead
        call_proc: Remove_this_node
        call_proc: Add_new_node
End If
End
```

Fig. 3. Proposed algorithm for reliability assessment (own source)

```
Begin
Input from configuration SRL
BestReliability: = find_Reliability of node with highest reliability
If BestReliability >= SRL
        Status: = success
Else
        perform_backward_recovery
        call_proc: remove_node_minReliability
        call_proc: add_new_node
End If
End
```

Fig. 4. Decision mechanism algorithm (own source)

After 10 cycles, we have got these results, which are represented by Table 2. There can be seen the comparison between virtual machines in each cycle and also status of VMs in modules (AT, TC and average time). If an AT fails to produce a correct result

Table 1. Environment variables

FTMC model – Reliability Assessment Algorithm (Alg1)	Proposed Reliability Assessment Algorithm (Alg2)
Reliability = 1, n = 1, RF = 0.02, SRL = 0.8, MaxReliability = 1.2, MinReliability = 0.7	Reliability = 1, n = 1, RF = 0.02, RF1 = 0.01, RF2 = 0.03, SRL = 0.8, MaxReliability = 1.2, MinReliability = 0.7

then TC status is also 'Fail', and TC is 'Pass' if AT was 'Pass' and Time is less than Real Time Limit, and we can see difference in results between FTMC model Algorithm (column **Alg1**) and Proposed Algorithm (column **Alg2**) for the results follow Table 3.

And in first cycle, all Virtual Machines have the same reliability, but the result is VM-1 that has been selected as it has a lower IP address.

8 Modeling and Refinement in Event B

Event-B is a mathematical approach for developing formal models of systems. An Event-B model is constructed from a collection of modeling elements. These elements include invariants, events, guards and actions.

The modeling elements have attributes that can be based on Set Theory and Predicate Logic. Set Theory is used to represent data-types and to manipulate the data. Logic is used to apply conditions to the data.

The development of an Event-B model goes through two stages; abstraction and refinement, the abstract machine specifies the initial requirements of the system. Refinement is carried out in several steps - with each step adding more detail to the system [8].

The Rodin Platform is an Eclipse-based IDE for Event-B that provides effective support for refinement and mathematical proof. The platform is open source, contributes to the Eclipse framework and is further extendable with plugins [9].

In Event-B we have two kinds of components.

1. **Context:** describes the static elements of a model.
2. **Machine:** describes the dynamic behavior of a model. Figure 5 shows relationships between machine and context [10].

8.1 Abstract Model

In abstract model (initial model) defined by the machine *Mac0*, we focus on following points:

- Add nodes to buffer by event *AddNotesToBuffer*.
- Perform the acceptance test for each node by event *CheckByAT*.
- Check nodes which produce the correct result before deadline time by event *CheckByTC_PASS*, in case, node produces the incorrect result or after deadline, we use event *CheckByTC_FAIL*.

Table 2. Results for our algorithms and FTMC Algorithm

Cycle	Real Time Limit (ms)	VM-1						VM-2						VM-3						Alg1	Alg2
		AT	TC	Time (ms)	AVR Time (ms)	Rel1	Rel2	AT	TC	Time (ms)	AVR Time (ms)	Rel1	Rel2	AT	TC	Time (ms)	AVR Time (ms)	Rel1	Rel2		
Start	-	-	-	-	0	1	1	-	-	-	0	1	1	-	-	-	0	1	1	-	-
1	4006	Pass	Pass	2922	2922	1.020	1.010	Pass	Pass	1366	1366	1.020	1.010	Pass	Pass	3562	3562	1.020	1.010	VM-1	VM-1
2	4862	**Fail**	**Fail**	**3453**	**2922**	**1.000**	**0.990**	Pass	Pass	2780	2073	1.040	1.020	Pass	Pass	3662	3612	1.040	1.020	VM-2	VM-2
3	4187	Pass	Pass	2221	2572	1.020	1.019	Pass	Pass	1973	2040	1.061	1.051	Pass	Pass	1445	2890	1.061	1.051	VM-2	VM-2
4	4859	Pass	Pass	2554	2566	1.040	1.050	Pass	Pass	2532	2163	1.082	1.061	Pass	Pass	1779	2612	1.082	1.082	VM-2	VM-3
5	4963	Pass	Pass	4685	3096	1.061	1.061	Pass	Pass	1383	2007	1.104	1.093	Pass	Pass	1272	2344	1.104	1.115	VM-2	VM-3
6	4250	Pass	Pass	2834	3043	1.082	1.092	Pass	Pass	2086	2020	1.126	1.104	Pass	Pass	1243	2160	1.126	1.148	VM-2	VM-3
7	4244	Pass	Pass	2596	2969	1.104	1.125	Pass	Pass	1544	1952	1.149	1.137	Pass	Pass	1650	2088	1.149	1.183	VM-2	VM-3
8	4116	Pass	Pass	1509	2760	1.126	1.159	Pass	Pass	2715	2047	1.172	1.148	Pass	Pass	2035	2081	1.172	1.218	VM-2	VM-3
9	4114	Pass	Pass	2554	2734	1.148	1.194	**Pass**	**Fail**	**4739**	**2047**	**1.148**	**1.125**	Pass	Pass	3638	2254	1.195	1.230	VM-3	VM-3
10	4351	Pass	Pass	4143	2891	1.171	1.206	**Pass**	**Fail**	**4894**	**2047**	**1.102**	**1.080**	Pass	Pass	2086	2237	1.219	1.267	VM-3	VM-3

Table 3. The result comparison

Cycle	VM-2				
	FTMC model Algorithm		Proposed Algorithm		
	Reliability (Rel1)	Change	Reliability (Rel2)	Change	
3	1.061	–	1.051	–	–
4	1.082	+0.02	1.061	+0.01	AVR Time(ms) < Time
5	1.104	+0.02	1.093	+0.03	AVR Time(ms) > Time

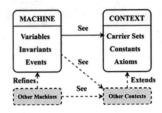

Fig. 5. Decision mechanism algorithm (own source)

- To get best virtual machine, by event *GetBestVmAcc,* we are selecting node that has smaller IP and *TC = Pass.*
- To remove node from cloud, we use event *RemoveNodes,* and change value of variable *RequestNew_VM* to True to request new node to cloud, The overall model structure is given on Fig. 6.

8.2   First Refinement

In first refinement, we added new variables to calculate average time and reliability of nodes, in event *AddNotesToBuffer;* we set 100(%) as initial value of reliability for each node. And in event *CheckByTC_PASS* we calculate total time of node and number of obtained PASS statuses to calculate average time for each node, as can be seen in Fig. 7.

According to proposed algorithm represented by Fig. 3, we will increase the reliability by RF1 or RF2, this depends on average time and current time of node, and we used *inv_RF1* rather from *RF1* where *(RF1 = 1/inv_RF1)* and changed the multiplication to division to avoid work with real number in Rodin, and by event *Ckeck-Max.* Node reliability cannot be more than this level as it is shown in Fig. 8.

And in event *RemoveNodes,* we added new guard, *Nodes_Reliability (VM) < MinReliability,* where we will remove each node with reliability smaller from *MinReliability*

To get virtual machine with best reliability we use event GetBestVmAcc1 but in status there are two nodes have same reliability we select machine that have smaller IP by event GetBestVmAcc2. See Fig. 9.

event AddNotesToBuffer
any VM, ip
where
  @grd1 VM∉buffer
  @grd2 VM∈cloud
  @grd3 ip∈Z
then
  @act1 buffer:=buffer∪{VM}
  @act2 Nodes_ips(VM):=ip
  @act3 Nodes_AT_Status(VM):=INIATIAL_AT
  @act4 Nodes_TC_Status(VM):=INIATIAL_TC
end
event CheckByAT
any VM , s
where
  @grd1 VM∈buffer
  @grd2 s∈AT_Status\{INIATIAL_AT}
then
  @act1 Nodes_AT_Status(VM):=s
End
event CheckByTC_PASS
any VM, t, RealTimeLmt
where
  @grd1 VM∈buffer
  @grd2 t∈Z
  @grd3 RealTimeLmt∈Z
  @grd4 Nodes_AT_Status(VM)=PASS_AT
  @grd5 t≤RealTimeLmt
then
  @act1 Nodes_TC_Status(VM):=PASS_TC
end

event CheckByTC_FAIL any VM, t ,RealTimeLmt
where
  @grd1 VM∈buffer
  @grd2 t∈N
  @grd3 RealTimeLmt∈N
  @grd4 Nodes_AT_Status(VM)=FAIL_AT∨
t>RealTimeLmt
then
  @act1 Nodes_TC_Status(VM):=FAIL_TC
end
event GetBestVmAcc any VM
where
  @grd1 VM∈buffer
  @grd2 Nodes_TC_Status(VM)=PASS_TC
  @grd3 ∀n:n ∈ buffer ⇒Nodes_ips(VM)< Nodes_ips(n)
then
  @act1 Best_VM:=Nodes_ips(VM)
end
event RemoveNodes
any VM
where
  @grd1 VM∈ buffer
  @grd2 Nodes_TC_Status(VM)=FAIL_TC
then
  @act1 Nodes_AT_Status:= {VM} ◁ Nodes_AT_Status
  @act2 Nodes_TC_Status:= {VM} ◁Nodes_TC_Status
  @act3 Nodes_ips:= {VM} ◁ Nodes_ips
  @act4 buffer:=buffer\{VM}
  @act5 cloud:=cloud\{VM}
  @act6 RequestNew_VM:=TRUE
end
end

Fig. 6. A specification of *Mac0* (own source)

event AddNotesToBuffer refines AddNotesToBuffer
any VM ,ip
where
  @grd1 VM∉buffer
  @grd2 VM∈cloud
  @grd3 ip∈Z
then
  @act1 buffer:=buffer∪{VM}
  @act2 Nodes_ips(VM):=ip
  @act3 Nodes_n(VM):=1
  @act4 Nodes_Reliability(VM):=100
  @act5 Nodes_AT_Status(VM):=INIATIAL_AT
  @act6 Nodes_TC_Status(VM):=INIATIAL_TC
  @act7 Nodes_SumTime(VM):=0
  @act8 Nodes_nPass(VM):=0
end

event CheckByTC_PASS refines CheckByTC_PASS
any VM, t, RealTimeLmt
where
  @grd1 VM∈buffer
  @grd2 t∈Z
  @grd3 RealTimeLmt∈Z
  @grd4 Nodes_AT_Status(VM)=PASS_AT
  @grd5 t≤RealTimeLmt
then
  @act1 Nodes_TC_Status(VM):=PASS_TC
  @act2 Nodes_nPass(VM):=Nodes_nPass(VM)+1
  @act3 Nodes_SumTime(VM):=Nodes_SumTime(VM)+t
End

Fig. 7. Variables, invariants and some events of *Mac1* (own source)

8.3   Second Refinement

In Second refinement we added new event to check if reliability of best node best greater than or equal to SRL (system reliability level), in case best reliability was smaller from SRL In this case, backward recovery is performed, see Fig. 10.

event Cal_ReliabilityInPass1
  any VM, t
  where
    @grd1 VM∈buffer
    @grd2 Nodes_TC_Status(VM)=PASS_TC
    @grd3 Nodes_n(VM)≤1
    @grd4 t∈Z
    @grd5 Nodes_nPass(VM)>0
    @grd6 Nodes_SumTime(VM)÷Nodes_nPass(VM)<t
  then
    @act1 Nodes_Reliability(VM):=Nodes_Reliability(VM) + (Nodes_Reliability(VM) ÷ inv_RF1)
  End

event Cal_ReliabilityInPass2
  any VM, t
  where
    @grd1 VM∈buffer
    @grd2 Nodes_TC_Status(VM)=PASS_TC
    @grd3 Nodes_n(VM)≤1
    @grd4 t∈Z
    @grd5 Nodes_nPass(VM)>0
    @grd6 Nodes_SumTime(VM)÷Nodes_nPass(VM)≥t
  then
    @act1 Nodes_Reliability(VM):=Nodes_Reliability(VM) + (Nodes_Reliability(VM) ÷ inv_RF2)
  end

event CkeckMax
  any VM
  where
    @grd1 VM∈buffer
    @grd2 Nodes_Reliability(VM) >MaxReliability
  then
    @act1 Nodes_Reliability(VM):=MaxReliability
  end

Fig. 8. A specification of Events *Cal_ReliabilityInPass1, Cal_ReliabilityInPass2 and CkeckMax* (own source)

event GetBestVmAcc1 refines GetBestVmAcc
  any VM
  where
    @grd1 VM∈buffer
    @grd2 Nodes_TC_Status(VM)=PASS_TC
    @grd3 ∀n·n ∈buffer ⇒Nodes_Reliability(VM)> Nodes_Reliability(n)
  then
    @act1 Best_VM:=Nodes_ips(VM)
  end

event GetBestVmAcc2 refines GetBestVmAcc
  any VM
  where
    @grd1 VM∈buffer
    @grd3 Nodes_TC_Status(VM)=PASS_TC
    @grd4 ∀n·n ∈buffer ⇒Nodes_Reliability(VM)≥ Nodes_Reliability(n)
    @grd5 ∀n·n ∈buffer ⇒ Nodes_ips(VM)<Nodes_ips(n)
  then
    @act1 Best_VM:=Nodes_ips(VM)
  end

Fig. 9. A specification of Events Cal_ GetBestVmAcc1 & GetBestVmAcc1 (own source)

Event CkeckBestVmSRL
  any *VM*
  where
    @grd1 *VM*∈buffer
    @grd2 Nodes_TC_Status(*VM*)=PASS_TC
    @grd3 ∀n·n ∈buffer ⇒Nodes_Reliability(*VM*)≥ Nodes_Reliability(n)
    @grd4 Nodes_Reliability(*VM*)<**SRL**
  then
    @act1 RequestBackRecovery:=TRUE
  end

Fig. 10. A specification of event CkeckBestVmSRL (own source)

8.4   Statistics and Results

In Table 4, we can see proof statistics for our model using the Rodin3.2 platform, the statistics give us the proof obligations generated and discharged by the Rodin, The finial development of our model results in 131 POs (Proof obligations), around (91%) of them have been proved automatically by the Rodin platform and the rest have been proved manually in the Rodin interactive proving environment.

Table 4. Proof obligations of our model

Reliability (Our model)	Total	Auto	Manual
	131 (100%)	119 (91%)	12 (9%)
Ctx0	0	0	0
Ctx1	3	3	0
Ctx2	0	0	0
Mac0	18	18	0
Mac1	67	61	6
Mac2	43	37	6

9 Conclusion

We are surrounded by a large number of clouds that depended on real-time systems and many real time systems are safety critical systems, so they require a higher level of fault tolerance. A lot of work has been done in the area of fault tolerance for standard real time systems and the use of cloud infrastructure for real time computing is quite new. Most of the real time applications require the fault tolerance capability to be provided.

In this paper, we have presented enhancing of reliability assessment algorithm to improve our model, by using average of the response time of nodes instead of smaller IP address to assess reliability to select the output of the node which has highest reliability among all the competing nodes.

Our final models are translated into the Event-B, so we can say following; event-B allows us to define a kind of modeling methodology by write the correct mathematical notions; wherefore we can apply event-B in modeling many different reactive distributed and real time systems.

As well as the Rodin tool offers reactive environment for constructing and analyzing models as do most modern integrated development environments, and provides integration between modeling and proving.

Acknowledgement. This work was supported by the Ministry of Education, Youth and Sports of the Czech Republic within the National Sustainability Programme project No. LO1303 (MSMT-7778/2014) and by the European Regional Development Fund under the project CEBIA-Tech No. CZ.1.05/2.1.00/03.0089. Also supported by grant No. IGA/CebiaTech/2017/ 007 from IGA (Internal Grant Agency) of Tomas Bata University in Zlin.

References

1. http://searchcloudcomputing.techtarget.com/definition/cloud-computing. Tec Target, Accessed 26 Apr 2012
2. Amin, Z., Sethi, N., Singh, H.: Review on fault tolerance techniques in cloud computing (2015). http://research.ijcaonline.org/volume116/number18/pxc3902768.pdf
3. Unsal, O.S., Koren, I., Mani Krishna, C.: Towards energy-aware software-based fault tolerance in real-time systems (2008). http://euler.ecs.umass.edu/research/ukk-islped-2002.pdf

4. Malik, S., Huet, F.: Adaptive Fault Tolerance in Real Time Cloud Computing, Research Team OASIS INRIA - Sophia Antipolis 06902 Sophia Antipolis, France, IEEE World Congress on Services (2011)
5. Sudha Lakshmi, S.: Fault Tolerance in Cloud Computing (2013). http://ijesr.in/wp-content/uploads/2013/12/ACICE-050.pdf
6. Kaur, P., Kaur, M.: Enhanced adaptive fault tolerance model for increasing reliability in cloud computing (2015). http://www.techrepublic.com/resource-library/whitepapers/enhanced-adaptive-fault-tolerance-model-for-increasing-reliability-in-cloud-computing/
7. http://tbu.ic-q8.net/
8. Krayem, S.: Modern theory of information - Structured and knowledge based system design, Faculty of Applied Informatics, Tomas Bata University in Zlín, Czech republic (2016)
9. http://www.event-b.org/
10. Damchoom, K., Butler, M., Abria, J.-R.: Modelling and Proof of a tree-structured file system in Event-B and Rodin (2008). http://www.ensiie.fr/~dubois/PR_2010/TreeFileSysICFEM2008.pdf

Adaptive Access Mechanism Based on Network State Detection in Multi-rate IEEE802.11 WLANs

Jianjun Lei[✉] [ID], Shengjie Peng [ID], and Yu Dai [ID]

School of Computer Science and Technology,
Chongqing University of Posts and Telecommunications, Chongqing 400065, China
{leijj,daiyu}@cqupt.edu.cn, 353351748@qq.com

Abstract. This paper addresses the problem of channel access in multi-rates IEEE802.11 WLANs (Wireless Local Area Networks). Unlike some current solutions for performance anomaly only aiming to the saturated network, we propose a fine-grain channel access mechanism that takes the saturated and unsaturated network into account simultaneously. Meanwhile, we present a model to detect the network state under different network density. This novel framework can detect the network state by collision rate for the dynamic network scenarios and perform more efficient channel access by combining the legacy IEEE802.11 DCF (Distributed Coordination Function) and airtime fairness algorithm. The simulations show that our algorithm can improve the aggregate throughput and reduce the collision rate during channel access compared to the legacy DCF mechanism.

Keywords: WLANs · Network state detection · Performance anomaly
Airtime fairness

1 Introduction

The IEEE 802.11 WLANs have been widely adopted because of its low cost and high flexibility. In the 802.11 protocol, the DCF mechanism is a basic radio channel access method for shared nodes. It exploits CSMA/CA (Carrier Sense Multiple Access with Collision Detection) so that each node almost has equal probability to access the channel. In addition, the 802.11 WLANs adopt the adaptive rate selection strategy that the nodes in the network have different transmission rates. And With the rapid increasement of wireless terminals, the density of WLANs becomes very large, multi-rate nodes usually coexist and prone to resulting in performance anomalies [1] that causes unfairness among clients to access channels and degrades the overall throughput of the system. Nowadays, the trending methods to solve the performance anomaly exploit fairness airtime shared strategy for the nodes by changing some parameters in the MAC (Media Access Control) layer. However, almost all airtime fairness strategies tend to prefer in saturation network, and do not consider that most of the actual networks are performing in unsaturation state. These airtime fairness algorithms will degrade the low-rate clients'

© Springer International Publishing AG, part of Springer Nature 2019
R. Silhavy (Ed.): CSOC 2018, AISC 765, pp. 259–271, 2019.
https://doi.org/10.1007/978-3-319-91192-2_26

throughput and affect their QoS (Quality of Service). Therefore, to solve the performance anomaly problem and obtain a better performance, a sophisticated method should take network state into account when performing channel access.

In this paper, we propose an algorithm that performs flexible channel access by combining the legacy DCF and the airtime fairness method. We use an innovative approach to detect network state and give the state detecting model. By analysis and simulation, this algorithm can achieve the better performance in terms of throughput and transmission delay in multi-rate and dynamic network environments. The remainder of the paper is organized as follows. Section 2 discusses the related work. In Sect. 3, we introduce the DCF network model in unsaturated environment and do some experiments to analyze the network characteristics for the state detection. The saturation network detection mechanism and hybrid channel access algorithm are described in Sect. 4. Section 5 gives the performance evaluation. Section 6 concludes this paper.

2   Related Work

Most literatures aim at achieving airtime fairness providing an equal length of long-term channel occupancy time for each competing node. The airtime fairness concept is proposed in [2, 3] and is achieved generally by changing the contention window (CW) in legacy IEEE802.11 contention-based binary-exponential backoff algorithm. In [4, 5], the authors use a centralized approach to change the CW and achieve time fairness between clients. In [6], the author proposes a distributed medium access control algorithm for fairly sharing network resources among contending nodes. Also, some articles use the queue management to offer airtime fairness. In [7], the author proposes an IAQM (Improved Active Queue Management) algorithm for fairly sharing network resources among all contending nodes. The above literatures can solve the problem of performance anomaly, but only performing in saturation network. However, most environment is not saturated in real network applications, and the bandwidth demand for each node is often different. If only the airtime fairness algorithm is exploited, the low-rate nodes will suffer in unnecessary latency. There also has been a significant effort to improve the performance in the case of unsaturated state. In [8], the author dynamically changes the value of CW according to the client density and PER (Packet Error Rate) to adapt to the changing of network topology and traffic. In [9], the authors present a distributed approach by tuning CW that can achieve maximum throughput in both unsaturated and saturated states. In [10], the author puts forward the analysis model for IEEE 802.11e EDCA in unsaturated states based on the frame transmission cycle method. In [11], the author uses the equilibrium point analysis model to represent the algorithm under various traffic load states and propose a dynamic controlling backoff time algorithm considering the distinction between high and low traffic loads to deal with unsaturated traffic load state. In [12], the author increases CW either exponentially or linearly based on collision rate to enhance the network utilization under high and low traffic conditions. Although these literatures have proposed many methods to provide more efficient channel utilization under unsaturated conditions. However, they all lack an efficient detection mechanism to distinguish the saturated and unsaturated state.

This paper presents a comprehensive method to perform channel access in saturated and unsaturated network environment. In particular, we propose a network state detection mechanism by constructing network model under different dense network. Through this detection, we obtain a trigger mechanism to make the alternation between airtime fairness access and traffic-based access. Therefore, each node can obtain fair airtime to increase the overall throughput and eliminate the performance anomaly problem in saturated state. And during the unsaturated state, they can obtain the same probability to access the channel so that the low-rate nodes can occupy more airtime to transmit burst traffic. This mechanism can achieve the better performance in multi-rate and dense network.

3 Network Model and Analysis

3.1 Motivation

A specific example is provided to demonstrate the performance loss caused by airtime fairness algorithm under unsaturated network state. In experiments, each group contains 4 STAs with the different rates such as 1 Mbps, 2 Mbps, 5.5 Mbps and 11 Mbps. And all STAs are associated with the same AP (Access Point). A 5 Mega file is transmitted to the server connecting the AP for each STA. We join STAs by group to the network exponentially during experiment, which indicates the traffic load varies from low to high. We exploit legacy DCF and DAFA (Distributed Airtime Fairness Algorithm) [13] to control channel access, and then compare the file transmission time needed by the STA with 1 Mbps rate.

Figure 1 depicts the completion time for all STAs with 1 Mbps rate that perform under DCF and DAFA scheme respectively. In the unsaturation stage, low-rate STAs can obtain more transmission opportunities for DCF algorithm, therefore they can complete the file transmission faster compared for DAFA algorithm. With the increasement of network load and density, DAFA allocates the airtime more fairly to each STA so that it can complete the file transmission faster. Contrast to DCF, transmission time changes longer and longer due to collisions occurring tensely under saturation state. Consequently, the airtime fairness algorithm can perform well under heavy network load, but they often over-operate in

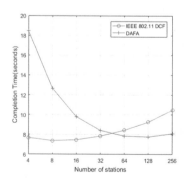

Fig. 1. Comparison of completion time for DCF and DAFA schemes

unsaturated network environment. Therefore, we often need to combine these two kinds of algorithms and perform fine-grain channel access controlling, as well as supported by an efficient network state detection mechanism.

3.2   Network Model

Most of the performance analysis of IEEE802.11 MAC is performing under a saturated network, where each STA always has data to send. To handle easily, they often ignore the dynamic queues and avoid to model the upper layer traffic for the saturated application. However, most of the network applications are performing in the unsaturated state, such as web browsing, e-mail, VoIP (Voice over Internet Protocol) and other Internet applications, showing bursty or switching traffic characteristics. Therefore, we propose a traffic model that is simple enough and still can give some key features in the unsaturated network, such as throughput, delay and collision probability. We model a typical multi-rate WLAN consisting of one AP and several associating STAs and sharing the same channel among them.

In this paper, to describe the upper-layer load more easily, we use a metric q to represent the probability of a STA that has at least one packet awaiting in transmission queue. Therefore, we can simply describe the load status as the unsaturated state when the counter is beginning to decrease. The probability that each distributed STA tries to transmit data in one slot is τ statistically, which can be described as follows.

$$\tau = b_{(0,0)_e} \frac{q^2}{1-q} \left(\frac{W_0}{(1-p)(1-(1-q)^{w_0})} - (1-p) \right) \tag{1}$$

Where, $b_{(0,0)_e}$ can be expressed by the following equation.

$$1/b_{(0,0)_e} = 1 - q + \frac{q^2 W_0 (W_0+1)}{2(1-(1-q)^{W_0})} + \frac{q(W_0+1)}{2(1-q)} \left(\frac{q^2 W_0}{1-(1-q)^{W_0}} + p(1-q) - q(1-p)^2 \right)$$
$$+ \frac{pq^2}{2(1-q)(1-p)} \left(\frac{W_0}{1-(1-q)^{W_0}} - (1-p)^2 \right) \left(2W_0 \frac{1-p-p(2p)^{m-1}}{1-2p} + 1 \right) \tag{2}$$

Where, W_0 is the CW in IEEE802.11 DCF, p is the probability of collision, n is the number of STA and m is the maximum retransmission number. Therefore, given q, W_0, m, n, we can compute p and τ by using the formulas: $1 - p = (1 - \tau)^{n-1}$.

According to τ, if we set i from 1 to n, the probability of successful transmission of STA can be expressed as follows:

$$P_s^i = \tau_i \prod_{j \neq 1} (1 - \tau_j) \tag{3}$$

The probability of channel idle P_{idle} and transmission failure P_f also can be expressed as follows:

$$P_{idle} = \prod_{i=1}^{n} (1 - \tau_i) \tag{4}$$

$$P_f = 1 - \sum_{i=1}^{n} P_s^i - P_{idle} \tag{5}$$

Therefore, the throughput of STA i is as follows:

$$S_i = \frac{E(L)}{E(T_{slot})} = \frac{P_s^i s_i}{\sum_{j=1}^{n} (P_s^i T_s^i) + P_f T_f + P_{idle}\sigma} \tag{6}$$

Where, the $E(L)$ is the traffic load for STA, $E(T_{slot})$ is the average length of a unit time slot, the T_s^i is the time required for STA i to transmit data successfully and T_f is the maximum time cost incurred by a failure transmission. They can be expressed by the formulas (7), (8), (9) and (10) respectively.

$$E(L) = P_s^i s_i \tag{7}$$

$$E(T_{slot}) = \sum_{j=1}^{n} (P_s^i T_s^i) + P_f T_f + P_{idle}\sigma \tag{8}$$

$$T_s^i = T_{PCLP}^i + \frac{H + s_i}{R_i} + T_{SIFS} + T_{PCLP}^i + \frac{ACK}{R_i} + T_{DIFS} \tag{9}$$

$$T_f = T_{PCLP}^{1Mbps} + \frac{H + s}{R_{1Mbps}} + T_{DIFS} \tag{10}$$

Where, T_{PCLP}^i is the PLCP (Physical Layer Convergence Protocol) preamble transmission time of the STA, R_i is the rate of the STA, H is the MAC header size, T_{SIFS} is the time between SIFS (Short inter-frame space), ACK indicates the size of the ACK frame, T_{DIFS} is the time of the DIFS (Distributed Inter-frame Spacing). σ is the time of a time slot. T_f includes the time of 1Mbps STA transmitting a data frame to the destination STA and the time of the ACK frame responding to the sending STA.

Then, the total throughput of the system is as follows.

$$S = \sum_{i=1}^{n} S_i \tag{11}$$

After establishing the model, in order to study the relationship between the q (the STA's upper layer load is expressed by q and the aggregate throughput and collision rate, we made an experiment on q, network density, aggregate throughput and the change of collision rate. Got the law of system reaches to saturation. And we found that through q-value detection, it can be found that the aggregate throughput can reach a limit under different densities and the average conflict can reach a peak. It shows the network reaches saturation, the collision rate reaching a peak, so that the throughput will no longer increase, and we can use the STA's q value and the network average collision to detect whether the network is saturated.

3.3 Analysis of Network Model Under Different Density

In this subsection, we use the proposed network model to study the relationship of the aggregate throughput and collision rate affected by the q. We do some experiments based on the above analysis model and obtain the trend of aggregate throughput and collision rate by the variety of q under different network load and density. The results are shown in Fig. 2.

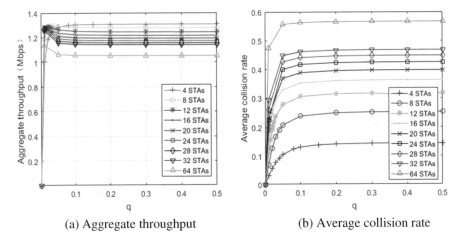

(a) Aggregate throughput   (b) Average collision rate

Fig. 2. The relationship of aggregated throughput and average collision affected by the q under different network density

As shown in Fig. 2(a) and (b), it can be found that the throughput will no longer change when the q value reaches to a certain critical value and the average collision rate of all stations will not increase evidently at this critical point too, which due to the network state changing from unsaturation to saturation. More specifically, all the critical state points almost are the constant value (q = 0.1) under different network density. Therefore, we carry out another experiment to find the relationship between the average collision rate and network state under different network density. The average q is fixed as 0.1, which is considered as the critical point of network state regardless of network density. We depict the average collision rates are occurred under different network density. Figure 3 shows the results. After the network reaches saturated state, the network collision rate hardly increases although the q always increases. Therefore, this phenomenon motivates us to construct a model which can detect the network state by collision rates under different network density.

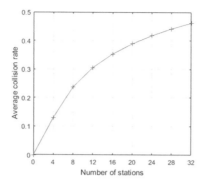

Fig. 3. The collision rate value for network saturated state under different network density

Figure 3 shows the relationship between the collision rate and the network density for network state detection, which indicates the saturated state can be detected by monitoring the average network collision rate under different network density.

4   Network State Detection and Channel Access Scheme

4.1   Network State Detection Mechanism

Through the previous description, we can summarize the saturated pattern detected by the collision rate for different network density. The critical collision rate is shown in Table 1.

Table 1. The critical collision rate under different density

n	Average collision rate
$0 < n \leq 4$	$p_{avg} < 0.032n$
$4 < n \leq 8$	$p_{avg} < (0.13 + 0.027(n - 4))$
$8 < n \leq 12$	$p_{avg} < (0.24 + 0.015(n - 8))$
$12 < n \leq 24$	$p_{avg} < (0.3 + 0.01(n - 12))$
$n > 24$	$p_{avg} < (0.42 + 0.005(n - 24))$

For implementation, each STA calculates the q, and then send it to AP with its date rate. AP can calculate the average collision rate of the network by formula (12). Therefore, the network state can be obtained according to Table 1.

$$p_{avg} = \frac{sum(p_i)}{n} \tag{12}$$

Moreover, due to network traffic variety, a weighted average algorithm shown in formula (13) can be used to get a stable q by the measuring q.

$$q = (1 - \alpha)q + \alpha q_{measure} \tag{13}$$

The pseudo-code of the network condition detection procedure is as follows.

Algorithm 1.   Network state Detection Mechanism

```
Input: Set of STA's q value {q₁,q₂,···,qₙ}
        Set of STA's rate {R₁,R₂,···,Rₙ}
          The number of STAs n
Output: Network status value NS
1:    AP calculates the collision rate pᵢ for each STA
based on the STA's Rᵢ and qᵢ values
2:    AP calculates p_avg = sum(pᵢ)/n based on all the collision

rates pᵢ
3:    if 0 < n ≤ 4
4:        if p_avg < 0.032n
5:            The network is not saturated   NS=0;
6:            else The network is saturated   NS=1;
7:            end if
8:     else if 4 < n ≤ 8
9:        if p_avg < (0.13+0.027(n-4))
10:            The network is not saturated   NS=0;
11:            else The network is saturated   NS=1;
12:            end if
13:     else if 8 < n ≤ 12
14:        if p_avg < (0.24+0.015(n-8))
15:            The network is not saturated   NS=0;
16:            else The network is saturated   NS=1;
17:            end if
18:     else if 12 < n ≤ 24
19:        if p_avg < (0.3+0.01(n-12))
20:            The network is not saturated   NS=0;
21:            else The network is saturated   NS=1;
22:            end if
23:     else if n > 24
24:        if p_avg < (0.42+0.005(n-24))
25:            The network is not saturated   NS=0;
26:            else The network is saturated   NS=1;
27:            end if
28:     end if
```

4.2   Optimal Channel Access Scheme

We propose a mechanism to distinguish the unsaturated condition and saturated condition of the network as a prerequisite for calling the airtime fairness algorithm.

This section describes an optimal channel access scheme. After the AP calculates the saturation ratio NS of the network, the AP will broadcast the NS value to all STAs which connected with this AP. And then each STA can get an appropriate opportunity to access the channel by adjusting the CW according to its own rate and the NS value. If the network is in saturated state, we make the channel access method to switch to airtime fairness algorithm, and the CW value is changed as $CW_i = CW_{default} \cdot 11/R_i$ [13] like DAFA, than the airtime of all STAs will be the same, which can solve the problem of performance anomaly, and get high aggregate throughputs and low collision rate; if the network is in unsaturated state, we make the channel access method to switch to IEEE 802.11 DCF, and the initial CW value is updated to the default value $CW_{default} = 32$, which can make each STA to get the same channel access opportunity and throughput, rather than reduce the delay of low-rate STAs when using the airtime fairness algorithm.

The pseudo-code of the optimal channel access scheme is as follows.

Algorithm 2.   Optimal Channel Access Scheme

```
Input: STA's rate R_i
       The value of CW_default
       The saturation ratio NS
Output: The value of CW_i
1:  Station receives the broadcast frame sent by the AP
and determines whether the network is in saturated
condition or not
2:  if NS=0
3:    set CW_i = CW_default  implementation of DCF algorithm
4:    else set CW_i = CW_default . 11/R_i implementation of airtime
fairness algorithm
5:    end if
```

---

5   Performance Evaluation

In this section, we verify the performance of our algorithm via Matlab, and the STA configuration is shown in Table 2. We make each group to contain 4 different rates of STAs, their data rates are 1 Mbps, 2 Mbps, 5.5 Mbps and 11 Mbps respectively. All STAs are associated with the same AP and they transmit the frame with the equal frame size for 1500 bytes. In addition, the parameters of m, W_0 and α are configured as 5, 32 and 0.125 respectively. We compare the proposed scheme with the legacy 802.11 DCF algorithm in terms of throughput, collision rate and fairness.

Table 2. IEEE 802.11PHY/MAC parameters used in simulations

Parameter	STA (1 Mbps)	STA (2 Mbps)	STA (5.5 Mbps)	STA (11 Mbps)
R	1 Mbps	2 Mbps	5.5 Mbps	11 Mbps
T_{PLCP}	192 µs	96 µs	96 µs	96 µs
H	34 bytes	34 bytes	34 bytes	34 bytes
ACK	14 bytes	14 bytes	14 bytes	14 bytes
σ	20 µs	20 µs	20 µs	20 µs
SIFS	10 µs	10 µs	10 µs	10 µs
DIFS	50 µs	50 µs	50 µs	50 µs

To verify the performance of the algorithm, we considered two scenarios. Initially, the q is configured as 0.01 and 0.5 to represent the low (referred to as scenario 1) and high load (referred to as scenario 2) respectively, and then we add the STAs to the scenario one group at a time.

5.1   Throughput

Figure 4 shows the trend of the aggregate throughput for two scenarios as the number of STA increases. For low load scenario, our proposed algorithm will obtains the better aggregate throughput compared to the legacy 802.11 DCF when the number of STA reaches to 24, which is due to the saturated state detected by our algorithm rapidly when the collision rate increases to a threshold for a certain network density. Therefore, our algorithm can switch to the channel access to the airtime fairness scheme to provide more transmission opportunity for high rate STAs and result in higher aggregate throughput. For high load scenario, the network enters early into the saturated state when the number of STA is 8, and our algorithm performs better that is the same as low load scenarios.

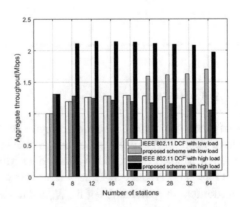

Fig. 4.   Comparison of aggregate throughput for several schemes

5.2   Collision Rate

Figure 5 shows the change of the average collision with the increase of the network density under low and high load scenarios. The network reaches to the saturated state after the number of STA increases to 8 and 24 for high and low load respectively. And then the airtime fairness algorithm for channel access is triggered by our algorithm, and thus resulting in lower collision rate compared to the 802.11 DCF mechanism.

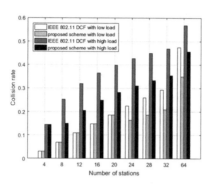

Fig. 5. Comparison of collision radio for several schemes

5.3   Fairness Index

We use the Jain's Fairness Index [14] to evaluate the channel access fairness of our algorithm by the simulations either. Figure 6 shows the change of the fairness index with the increasing density in the low and high load scenarios. Obviously, when the network saturated state is detected, our scheme can start up the airtime fairness algorithm and achieve a better access fairness.

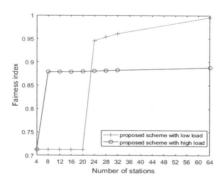

Fig. 6. Comparison of fairness index for several schemes

6   Conclusion

In this paper, we analyze the characteristics of the network and construct a network model to detect the network state, which can detect the network state under different density. We propose a channel access scheme that combines the legacy DCF and airtime fairness algorithm. This algorithm does not limit the throughput of low-rate STAs in unsaturated state and also obtains the better performance in terms of throughput, collision rate and fairness in saturated state. As a part of our future work, we plan to implement our algorithm in IEE802.11 hardware and evaluate its performance using real testbed.

Acknowledgments. This research is supported by the National Science Foundation of China (61602073) and Scientific and Technological Research Program of Chongqing Municipal Education Commission (KJ1400429).

References

1. Heusse, M., Rousseau, F., Bergersabbatel, G., Duda, A.: Performance anomaly of 802.11b. In: Joint Conference of the IEEE Computer and Communications, vol. 2(1), pp. 836–843(2015)
2. Tan, G., Guttag, J.V.: Time-based fairness improves performance in multi-rate WLANs. In: General Track: Usenix Technical Conference, vol. 12(2), pp. 269–282 (2004)
3. Dolińska, I., Jakubowski, M., Masiukiewicz, A., Szeszko, M.: Fairness calculation on the base of the station media access time in Wi-Fi networks. In: International Conference on Information and Digital Technologies. IEEE (2016)
4. Adnan, M., Park, E.C.: Hybrid control of contention window and frame aggregation for performance enhancement in multirate WLANs. Mob. Inf. Syst. **2015**, 16 (2015)
5. Krishnan, S., Chaporkar, P.: Stochastic approximation based on-line algorithm for fairness in multi-rate wireless LANs. Wirel. Netw. **23**(5), 1563–1574 (2017)
6. Le, Y., Ma, L., Cheng, W., Cheng, X., Chen, B.: A Time fairness based MAC algorithm for throughput maximization in 802.11 networks. IEEE Trans. Comput. **64**(1), 19–31 (2014)
7. Lei, J., Wu, Y., Zhang, X.: An improved active queue management algorithm for time fairness in multirate 802.11 WLAN. In: Computer Science On-line Conference (CSOC), pp. 161–171 (2017)
8. Ma, H., Roy, S.: Contention window and transmission opportunity adaptation for dense IEEE 802.11 WLAN based on loss differentiation. In: IEEE International Conference on Communications. IEEE (2008)
9. Hong, K., Lee, S.K., Kim, K., Kim, Y.H.: Channel condition based contention window adaptation in IEEE 802.11 WLANs. IEEE Trans. Commun. **60**(2), 469–478 (2012)
10. Huang, C., Shioda, S.: Detailed analysis for IEEE 802.11e EDCA in non-saturated conditions - Frame-transmission-cycle approach. In: International Symposium on Modeling & Optimization in Mobile, Ad Hoc & Wireless Networks (WiOpt), vol. 14(2), pp. 601–608 (2013)
11. Alkadeki, H., Wang, X., Odetayo, M.: Improving performance of IEEE 802.11 by a dynamic control backoff algorithm under unsaturated traffic loads. Int. J. Wirel. Mob. Networks (IJWMN) **7**(6) (2016)

12. Nithya, B., Gopinath, A.J., Kameswaran, V., Yogesh, P.: Optimized tuning of contention window for IEEE 802.11 WLAN. Int. J. Eng. Sci. Technol. **9**(2), 15–25 (2017)
13. Banchs, A., Serrano, P., Oliver, H.: Proportional fair throughput allocation in multirate IEEE 802.11e wireless LANs. Wirel. Netw. **13**(5), 649–662 (2006)
14. Jain, R., Chiu, D., Hawe, W.: A quantitative measure of fairness and discrimination for resource allocation in shared computer systems. Comput. Sci. (1998)

Reconstruction of 3D Permittivity Profile of a Dielectric Sample Using Artificial Neural Network Mathematical Model and FDTD Simulation

Mikhail Abrosimov, Alexander Brovko$^{(\boxtimes)}$ ⓘ, Ruslan Pakharev,
Anton Pudikov, and Konstantin Reznikov

Yuri Gagarin State Technical University of Saratov, Polytechnicheskaya st. 77, Saratov, Russia
brovkoav@gmail.com

Abstract. The paper presents a new method of determining 3D permittivity profile using electromagnetic measurements in the closed waveguide system. The method is based on the application of artificial neural network as a numerical inverter, and on the approximation of 3D profile with quadratic polynomial function. The neural network is trained with numerical data obtained with FDTD modeling of the electromagnetic system. Special criteria for choice of a number of hidden layer neurons are presented. The results of numerical modeling show possibility of determination of permittivity profile with a relative error less than 10%.

Keywords: Artificial neural network · FDTD method
Training algorithm for artificial neural network
Nondestructive evaluation and testing

1 Introduction

The problem of determining the spatial distribution (profile) of complex permittivity inside dielectric objects is of great interest due to its functional benefits for many applications, such as medical diagnostics (detection of inhomogeneity in distribution of tissues) [1], detection of defects and cracks in construction materials [2] and in composite panels [3], etc. Special interest in this technology is conditioned by the principal obstacle in the development of controllable microwave sintering of particulate and powder materials [4]. Because of substantial difficulties in measuring media parameters non-uniformly distributed inside the material heated up to 1,000 °C and higher, contactless microwave imaging may be an attractive approach in getting spatial profiles of effective complex permittivity of the sintered objects. A method for determination of 2D distribution of complex dielectric permittivity using artificial neural network (ANN) was developed in [5].

The purpose of this paper is extension of the approach presented in [5] on the case of determination of 3D distribution of dielectric permittivity inside the sample.

© Springer International Publishing AG, part of Springer Nature 2019
R. Silhavy (Ed.): CSOC 2018, AISC 765, pp. 272–279, 2019.
https://doi.org/10.1007/978-3-319-91192-2_27

2   Method

Determining of 3D dielectric permittivity profiles of the dielectric sample is much more complicated problem in comparison to 2D case due to presence of additional direction of possible dielectric permittivity variations, and, as a consequence, necessity to provide more measurable parameters on the input of ANN. The approach in this paper follows the ideas developed in [5], however it possess some significant differences.

As a measurement system we apply six-port turnstile junction (four ports in rectangular waveguides, and two ports corresponding to two orthogonal modes in circular waveguide), containing the dielectric sample, lying on the bottom wall of junction (Fig. 1). The measurement parameters are complex reflection and transmission coefficients (full S-matrix of the junction). This type of measurement system was chosen in order to provide sufficient amount of information to the input of ANN for the reconstruction of 3D permittivity profile in the sample. Furthermore, this type of measurement system permits to avoid rotation of the sample (performing measurements for several positions of the sample), as it was necessary in the system presented in [5].

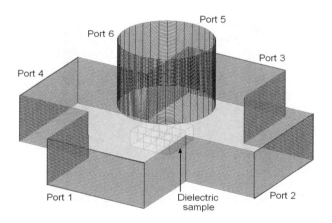

Fig. 1. Turnstile waveguide junction with sample inside.

The numerical treatment of the results obtained from measurements involves ANN technique. In order to provide more flexibility of the method in its ability to work with different spatial dielectric permittivity distributions, here we apply the combination of ANN with a regression mathematical model.

On the first step of the technique the 3D permittivity profile inside the sample is described mathematically with quadratic polynomial function which provides mapping between spatial coordinates inside the sample and complex permittivity values in the specified points. The function is defined with relatively small number of coefficients, and each set of the coefficients corresponds to a particular permittivity profile inside the sample.

On the next step of the technique the ANN with global cubic radial basis functions is applied as a numerical inverter. The numerical inverter sets mapping between measurable values (S-parameters of the electromagnetic system) and the coefficients of the function chosen on the first step. This ANN, after appropriate training, is used for reconstruction of the permittivity profiles corresponding to the measured S-matrix of the system.

The method proposed in this paper is intended to reconstruction of permittivity profiles in real experimental samples, which may have any arbitrary distribution of complex permittivity values in the points inside the sample. At the same time, we apply a model function, namely quadratic polynomial in this paragraph, to describe these profiles as precise as possible. In other words, we apply regression technique with three independent variables (space coordinates x, y, z), to fit the real experimental profile with quadratic polynomial function. According to optimal experimental design theory, to build the regression, we apply central composite orthogonal design [6], involving function values in fifteen points inside and outside the sample. We assume the sample of rectangular parallelepiped form. In this case, we apply eight points in the corners of the block, six star points outside the sample, and one point in the middle of the block (Fig. 2).

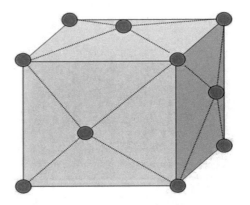

Fig. 2. Sample with base points.

The distance from the center of the block to star points is calculated according to the formula

$$d = \alpha \cdot ds/2 \tag{1}$$

where

$$\alpha = \sqrt{\sqrt{N \cdot 2^{k-2}} - 2^{k-1}} \tag{2}$$

ds is the size of the block side, k is the number of independent variables (k = 3 in our case), and N is the number of experiments of the plan, or base points (in our case N = 15).

The star points are located, therefore, outside the sample if we assume rectangular parallelepiped. In order to construct quadratic polynomial equation, describing the permittivity profile, we set the values of permittivity in star points so as it were the points in continuous media, but when we set material parameters for numerical simulation of the system, we set it only inside the sample, assuming air for the material surrounding the sample.

The values of permittivity in base points are used for reconstruction of quadratic polynomial surface. The reconstruction is performed with ANN with global cubic radial basis functions.

The proposed technique of the permittivity profile reconstruction is based on the application of the ANN as a numerical inverter. Here the idea of sophisticated function governing permittivity profiles is combined with the numerical model representing the experimental system. According to the proposed procedure, S-parameters come to the inverter from both FDTD solver, which is used on the ANN training stage, and network analyzer, which may be used for real experimental reconstruction of the permittivity profile in the sample with sufficiently trained ANN.

The ANN used for reconstruction of the permittivity profiles is a radial basis function (RBF) network with cubic basis function (Fig. 3).

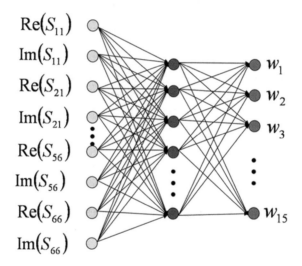

Fig. 3. ANN architecture.

The inputs of the ANN are complex S-parameters of the electromagnetic system, the outputs are the parameters governing the permittivity profile inside the sample (permittivity values in 15 base points). The mathematics behind the ANN training is described in details in [5], so it is not presented here.

However in [5] the technique required the number of neurons in the hidden layer to be equal to the number of training sets, and this lead to long computations during the ANN training. In this paper we present the technique which permits to reduce the number of hidden neurons without loss of prediction ability of ANN.

RBF network center vectors of hidden layer can be defined by cluster centroids that are found with application k-means algorithm [7]. The Euclidean distance for n center vectors and m variables is calculated with the following equations:

$$\begin{cases} \left(x_1 - w_{11}\right)^2 + \left(x_2 - w_{12}\right)^2 + \cdots + \left(x_m - w_{1m}\right)^2 = d_1^2 \\ \left(x_1 - w_{21}\right)^2 + \left(x_2 - w_{22}\right)^2 + \cdots + \left(x_m - w_{2m}\right)^2 = d_2^2 \\ \qquad\qquad\qquad \cdots \\ \left(x_1 - w_{n-11}\right)^2 + \left(x_2 - w_{n-12}\right)^2 + \cdots + \left(x_m - w_{n-1m}\right)^2 = d_{n-1}^2 \\ \left(x_1 - w_{n1}\right)^2 + \left(x_2 - w_{n2}\right)^2 + \cdots + \left(x_m - w_{nm}\right)^2 = d_n^2 \end{cases} \tag{3}$$

where X is input vector, W is a matrix of center vectors and D is distance vector. Our goal is making the number of centers as small as possible, but without data loss (i.e. the input data are restored with the same precision).

After opening brackets and substitution of equations the following linear system of $n-1$ equations can be derived:

$$\begin{cases} \left(w_{21} - w_{11}\right)x_1 + \left(w_{22} - w_{12}\right)x_2 + \cdots + \left(w_{2m} - w_{1m}\right)x_m = c_1 \\ \left(w_{31} - w_{21}\right)x_1 + \left(w_{32} - w_{12}\right)x_2 + \cdots + \left(w_{3m} - w_{2m}\right)x_m = c_2 \\ \qquad\qquad\qquad \cdots \\ \left(w_{n1} - w_{n-11}\right)x_1 + \left(w_{n2} - w_{n-12}\right)x_2 + \cdots + \left(w_{nm} - w_{n-1m}\right)x_m = c_{n-1} \end{cases} \tag{4}$$

where C is a new vector that contains the sums of variable-independent members.

With new equation system matrix designated as V and using Kronecker-Capelli theorem the following conditions for center vectors providing absence of data loss are met:

$$n \geq m + 1, \ n = \text{rank}(V). \tag{5}$$

The presented criteria for center vectors can be used with application of k-means algorithm and binary search algorithm to define the minimal set of center vectors that does not cause data loss in hidden layer.

In comparison to using all the data set points as center vectors the presented algorithm requires less time to process due to much less order of the system matrix and has lower error values in cases of learning set reduction.

As an illustration of the proposed approach, in this paper we apply ANN technique for reconstruction of real permittivity profiles assuming lossless media inside the sample.

3   Numerical Results

Numerical results presented in this section were obtained for the dielectric sample with a shape of rectangular parallelepiped with sizes $100 \times 100 \times 80$ mm. Reconstructions of the 3D dielectric permittivity profile are performed in the turnstile junction with rectangular waveguides cross section 248×124 mm and circular waveguide radius is 107.57 mm; working frequency is 915 MHz.

The sample is placed on the bottom wall of the turnstile junction, in the center of symmetry, as shown in Fig. 1. The medium of the sample is assumed lossless and dielectric permittivity varying smoothly, without abrupt changes, inside the sample in all three coordinates' axes in the ranges 2÷12.

The ANN is trained to recognize the coefficients of quadratic polynomial regression model. After training phase the S-parameters, computed for arbitrary test profiles, are provided to the input of ANN, which try to build a model permittivity distribution as close as possible to the test profile. The quality of reconstruction is estimated numerically with absolute and relative average errors calculated with the following formulas:

$$E_a = \frac{1}{N} \sum_{i=1}^{N} |\varepsilon_{ti} - \varepsilon_{ri}|, \tag{6}$$

$$E_r = \frac{1}{N} \sum_{i=1}^{N} \frac{|\varepsilon_{ti} - \varepsilon_{ri}|}{\varepsilon_{ti}} \cdot 100\%, \tag{7}$$

where ε_{ti} is permittivity value of test function in point i; ε_{ri} is permittivity value of reconstructed function in point i, N is the number of points in which the permittivity value is calculated.

Direct modeling of the measurement system by the FDTD method was performed with full-wave electromagnetic modeling software QuickWave-3D [8]. The FDTD model contains a non-uniform mesh with $3.3 \times 3.3 \times 4$ mm cells within the samples and $14 \times 14 \times 14$ mm cells outside them. The total numbers of cells in the model is nearly 181,000 (20 MB of RAM). In order to reach steady state, about 5,000 time steps for each of 6 ports are required – for an Intel Core 2 Duo 1.83 GHz PC, it takes about 6 min of CPU time.

The following results are obtained in reconstruction of permittivity profiles with ANN trained on quadratic polynomial functions. In Fig. 4 the test plane given as

$$\varepsilon' = \left(1/2 - x^2 + y^2\right) \exp\left(1 - x^2 - y^2/2 - z\right) + 5 \tag{8}$$

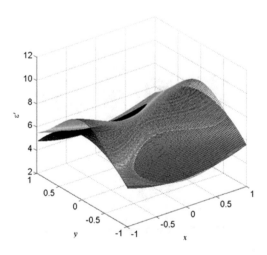

Fig. 4. Actual (transparent) and reconstructed (solid) profiles described by (8).

is shown along with reconstructed one. Average errors calculated according to (6) and (7) for the whole sample are $E_a = 0.5385$, $E_r = 9.17\%$.

In Fig. 5 corresponding pictures are presented for the test function given as

$$\varepsilon' = \ln\left(4(x - z/2)^2 + 4y^2 + 1\right) + 5. \qquad (9)$$

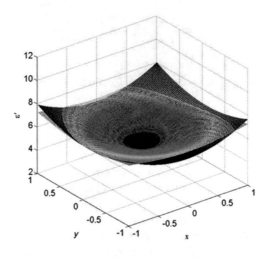

Fig. 5. Actual (transparent) and reconstructed (solid) profiles described by (9).

Dielectric permittivity profile was reconstructed with average errors $E_a = 0.3299$, $E_r = 5.32\%$.

Surfaces presented in Fig. 6 correspond to test function

$$\varepsilon' = 7 + 2\cos(3x)\sin((x - z)y). \qquad (10)$$

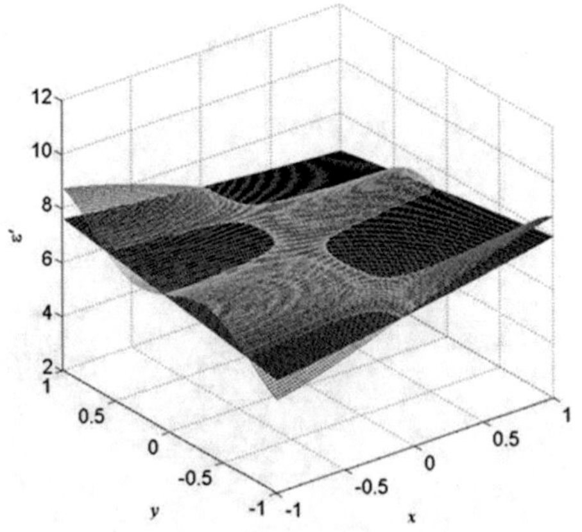

Fig. 6. Actual (transparent) and reconstructed (solid) profiles described by (10).

Average errors of the profile reconstruction are $E_a = 0.3976$, $E_r = 5.76\%$.

4 Conclusion

The proposed method for determination of 3D dielectric permittivity profiles using the results of electromagnetic measurements with application of turnstile waveguide junction as a measurement system and quadratic regression model as a function, defining 3D dielectric permittivity profile inside the sample, has demonstrated for quadratic regression model more precise results in comparison to a set of Gaussian functions as an approximating function. Application of multi-port junction permits to provide more information for determination of material parameters profile to the input of ANN, and also this permits to avoid rotation of the sample during the measurements. The relative error of real dielectric permittivity profile determination for presented tests does not exceed 10%, and for most of the tests is limited by 5%.

Application of turnstile waveguide junction as a measurement system requires development of commutation scheme for connection of waveguide ports to network analyzer for practical implementation of the method. Other multi-port waveguide junctions can be also used in the method; however, this will require numerical study of sensitivity of S-parameters to variations of the dielectric permittivity profile in three coordinate directions.

References

1. Li, X., Davis, S.K., Hagness, S.C., Van Der Weide, D.W., Van Veen, B.D.: Microwave imaging via space-time beamforming: experimental investigation of tumor detection in multilayer breast phantoms. IEEE Trans. Microw. Theor. Tech. **52**(8), 1856–1865 (2004)
2. Belkebir, K., Kleinman, R.E., Pichot, C.: Microwave imaging: location and shape reconstruction from multifrequency scattering data. IEEE Trans. Microw. Theory Tech. **45**(4), 469–476 (1997)
3. Qaddoumi, N., Carriveau, G., Ganchev, S., Zoughi, R.: Microwave imaging of thick composite panels with defects. Mater. Eval. **53**(8), 926–929 (1995)
4. Bykov, Y.V., Rybakov, K.I., Semenov, V.E.: High-temperature microwave processing of materials. J. Phys. D Appl. Phys. **34**(13), R55–R57 (2001)
5. Brovko, A.V., Murphy, E.K., Yakovlev, V.V.: Waveguide microwave imaging: neural network reconstruction of functional 2-D permittivity profiles. IEEE Trans. Microw. Theory Tech. **57**(2), 406–414 (2009)
6. Myers, R.H., Montgomery, D.C., Anderson-Cook, C.M.: Response Surface Methodology Process and Product Optimization Using Designed Experiments, 3rd edn. Wiley, New York (2009)
7. Schwenker, F., Kestler, H.A., Palm, G.: Three learning phases for radial-basis-function networks. Neural Netw. **14**, 439–458 (2001)
8. QuickWave-3 DTM, QWED Sp. z o.o., ul. Nowowiejska 28, lok. 32, 02-010 Warsaw, Poland. http://www.qwed.com.pl/. Accessed 16 Dec 2017

Novelty Detection System Based on Multi-criteria Evaluation in Respect of Industrial Control System

Jan Vávra[✉] and Martin Hromada

Faculty of Applied Informatics, Tomas Bata University in Zlin, Zlin, Czech Republic
jvavra@fai.utb.cz

Abstract. The industrial processes and systems have become more sophisticated and also adopted in diverse areas of human activities. The Industrial Control System (ICS) or Internet of Things (IoT) have become essential for our daily life, and therefore vital for contemporary society. These systems are often included in Critical Information Infrastructure (CII) which is crucial for each state. Consequently, the cyber defense is and will be one of the most important security field for our society. Therefore, we use the novelty detection approach in order to identify anomalies which can be a symptom of the cyber-attack in ICS environment. To achieve the main goal of the article One-Class Support Vector Machine (OCSVM) algorithm was used. Moreover, the anomaly detection algorithm is adjusted via multi-criteria evaluation and classifier fusion.

Keywords: Cyber security · Novelty detection · Anomaly detection
Industrial control systems · Multi-criteria evaluation

1 Introduction

Emerging development in information and communication technology (ICT) caused critical changes in understanding of the ICT nature. Therefore, increasing interconnection, interdependencies, and complexity of the ICT resulted in increasing of effectiveness in a considerable number of human activities. On the other hand, this development is accompanied by new cyber threats which can result in global crisis. The newly formed "global cyber organism" has become much more vulnerable to sophisticated malware which analogic to a global human population in case of biologic viruses. The rapid development in ICT has an eminent influence on recently isolated industrial control systems (ICS) which are vital for our society. Therefore, ICS cyber security has been subject to fundamental changes which resulted in reconfiguration of "status quo". Furthermore, the malware Stuxnet was the main milestone in ICS cyber security which the led to necessary changes in cyber security.

ICS is developed in order to control of industrial processes. Moreover, according to "Guide to Industrial Control Systems (ICS) Security" [1] we can divide ICS into two main subgroups. The first is geographically independent Supervisory Control and Data Acquisition (SCADA) system, and the second is a geographically dependent system

© Springer International Publishing AG, part of Springer Nature 2019
R. Silhavy (Ed.): CSOC 2018, AISC 765, pp. 280–289, 2019.
https://doi.org/10.1007/978-3-319-91192-2_28

known as Distributed Control System (DCS) [2]. The boundary between these systems is often relatively insufficiently defined, which leads to the mutual misinterpretation of the groups. However, a considerable number of experts use the terminology SCADA instead of DCS. This misinterpretation occurs frequently and therefore is mostly acceptable by the experts.

The detection of cyber-attacks is one of the crucial factors of cyber security or cyber defense. Moreover, there is a considerable number of cyber security solution which can be adapted in case of ICS. However, one of the most progressive method how to defense ICS is anomaly based detection. Therefore, we are focusing on cyber defense system based on anomaly detection algorithms which can be easily adopted for intrusion detection systems (IDS). The anomaly detection involves the problem of finding patterns in a dataset that do not match the expected behavior. Moreover, every anomaly can be a symptom of the cyber-attacks [3]. Thus, the there are three main subgroups: Supervised anomaly detection, Semi-supervised anomaly detection, and Unsupervised anomaly detection which are based on differently structured datasets. This distribution is supported by a considerable number of authors [3–8]. Taking into account the importance of various input data is crucial for every anomaly detection system. However, the anomaly detection systems have been deployed in various fields of human activities. Akoglu et al. [9] investigated the areas in which are anomaly detection system often used. We can highlight some of them: medical problems, image processing, insurance fraud, data center monitoring, image/video surveillance, etc. [9].

Stouffer et al. [1] pointed to historical developments in ICS where systems and devices are often used more than 20 years. In addition, a considerable number of ICS systems had been developed before private networks and the Internet deployment that we know today. However, these commonly used technologies are now interconnected with ICS which led to the creation of new vulnerabilities. Moreover, it is evidenced by an increasing number of vulnerabilities which are reported to ICS-CERT (753% in recent years). Pollet [10] predicted increasing interdependencies between ICS and ICT, and therefore the percentage of industrial companies providing the IDS for ICS will continually grow. Horkan [11] concluded that the IDS going to be an essential part of the ICS systems in following years. The application of IDS in ICS environment was examined by a considerable number of researchers: Verba and Milvich [12], Zhu and Sastry [13], Yang et al. [14], Maglaras and Jiang [15]. Moreover, Maglaras and Jiang [15] investigated the possibility of the OCSVM deployment in ICS environment. Unfortunately, the authors did not cover how they set Gamma parameter for OCSVM in deep. Furthermore, the computational cost of anomaly detection system was not considered.

On this basis, we established Semi-supervised anomaly detection system also known as Novelty detection. We carried out a multistep procedure in order to achieve the objectives of the research, and therefore obtain reliable as well as low computational cost of anomaly detection system. Moreover, presented predicted model is modified according to multi-criteria evaluation where we take into account computational cost.

The rest of the article is organized as follows. Section 2 is focused on a description of anomalies. Classification algorithm used in the research is analyzed in Sect. 3. Section 4 gives a necessary insight into methods which were used in the research. The Sect. 5 includes results. Finally, Sect. 6 provides the conclusion of the article.

2 Anomaly as a Symptom of Cyber-Attack

Anomaly detection is a progressive method to find and separate patterns that deviate from normal behavior. Computer intrusion includes hacking, viruses, computer worms etc. However, the intrusion represents only a small percentage of total network and computer capacity [5]. Anomalies are relatively rare events in computer systems or networks, which can be divided into two main groups. The first group is anomalies caused by intentional human activities that involve cyber-attacks. The second group included anomalies that were caused by unintentional human activity (poor handling of the cybernetic system) or natural disasters and mistakes caused by technical error, lack of technical equipment or unintended human action.

According to Knapp [16] we can distinguish ICS anomalies into four main groups. The first group includes the monitoring of network traffic which includes source and destination Internet Protocol (IP) address, TCP/UDP ports, traffic volume etc. The second groups can be characterized as a user activity which includes logins and logoffs of the users and other user activities. The third main group of monitored system behavior is Process and Control behavior which is also subject to this article. Moreover, this specific group is focused on system behavior which involves configuration of the system. Finally, the last group is focused on event and incident activity and handling, monitoring criticality, total number and severity of the incidents etc. [16].

3 Support Vector Machine

Support Vector Machine (SVM) is one best-suited classification algorithm for wide range applications. It is also an exceptional choice for high dimensional data and nonlinear separation. Moreover, SVM is considered as a straightforward solution for anomaly detection system based on unbalanced dataset. All the advantages of the SVM are needed to build reliable detection system in multidimensional space for a nonlinear dataset. The predictive model is built on SVM. It classify the data into one of the predefined class. Moreover, the OCSVM is usually used for binary classification cases which are classified as +1 or −1. The SVM creates the widest margin near the boundary between two sets of data.

The Fig. 1 illustrates how SVM algorithm operates with boundary. The circles and asterisks represent two classes in two-dimensional space. Each data point is represented by $(\bar{x}, y)$ where $\bar{x}$ are feature values and y is a label (asterisk, circle or −1, 1). Moreover, the boundary is calculated in order to maximize the margin space [17]. The boundary is calculated according to Eq. (1).

$$f(\bar{x}) = \bar{w}\bar{x} + b \tag{1}$$

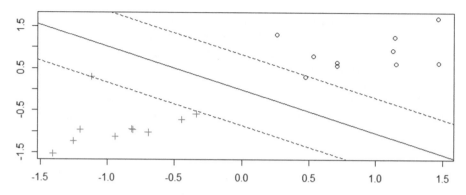

Fig. 1. The SVM boundary with margins

The main boundary is also known as hyperplane which is defined as $\bar{w}\bar{x} + b = 0$ and the margin width is defined as $\max \dfrac{2}{\|w\|}$. According to gutter constraint, we can set the margins on $+1$ and -1. The relationship is represented by equation as $\bar{w}\bar{x}_i + b = y_i$, where $\bar{x}_i \in \{-1, +1\}$. The OCSVM algorithm solves dual optimization problem in order to optimize constrained system. The final function can be seen in (2) [17].

$$L_{(\alpha)} = min\frac{1}{2} \sum_i^n \sum_j^n \alpha_i\alpha_jK(x_ix_j) \qquad (2)$$

Where $0 \leq \alpha_i \leq \dfrac{1}{vm}$ and $\sum_i^n \alpha_i = 1$. Moreover, α_i is a Lagrange multiplier, v is a trade-off parameter, m represents the total number of datapoints in a training dataset and $K(x_i, x_j)$ is a kernel function which is dot product in higher dimensional space [17]. There is a necessity to separate the dataset. However, the separation of the datasets are computational demanding process in most cases. The solution for this problem is the transformation of data into higher dimensional space. Thus, the kernel function K is described by the Eq. (3).

$$K\left(x_i, x_j\right) = \left(\varnothing\left(x_i\right), \varnothing\left(x_j\right)\right) \qquad (3)$$

There are four commonly used kernels (Linear kernel, Polynomial kernel, Radial Basis Function (RBF) and Sigmoid kernel). However, we decided to use RBF which is suitable for the purpose of the research. Moreover, the kernel nonlinearly maps samples into a higher dimensional space [18]. Where γ represents Gamma parameter.

$$K\left(x_i, x_j\right) = exp\left(-\gamma\left\|x_i - x_j\right\|^2\right), \ \gamma > 0 \qquad (4)$$

3.1 Gamma Parameter

Gamma (γ) is the main parameters for nonlinear RBF also for SVM. The predictive model is set up for the best suited boundary in order to maximize space between margins.

However, the shortage of the approach is the misclassification which can lead to poorly assembled predictive model. Therefore, Cortes and Vapnik [17] developed soft margins which allow to change or excluded data points for the purpose of minimize the number of errors. Gamma is the parameter of the nonlinear classification due to RBF kernel. Moreover, this parameter is a trade-off between error due to bias and variance of the predictive model. Therefore, there are two main problems, a problem of overfitting of the model and the boundary does not correspond with the complexity of data.

4   Methods

The purpose of the article is to create time efficient and accurate detection system in ICS environment. The OCSVM with RBF kernel is used in order to fulfill the main goal of the article and therefore develop a confidential predictive model. However, a considerable number of ICS devices which have limited computational power due to their long life cycle. Therefore, every anomaly detection system has to take into account requirements for computational power. Additionally, we can conclude that computational power is increasing due to growing Gamma value. Hence, there must be the specific equilibrium between the detection capabilities and computational complexity. The multi-criteria evaluation is one of the possible ways how to establish accurate and low computational cost detection system. The multi-criteria evaluation is based on the reference point of the multiple criteria (Accuracy, Sensitivity, Specificity, Precision, False Positive Rate (FPR) and Time).

- Accuracy - It represents the correct classification of the model. Moreover, accuracy is calculated as correct classification divided by correct and incorrect classification.
- Sensitivity - Sensitivity is also known as recall or true positive rate. Moreover, it is based on true positive condition and predicted positive condition. The criterion expresses how much relevant results are retrieved by the predictive model.
- Specificity - Specificity is also known as True negative rate. This criterion represents the measure of how correctly the negatives examples are classified.
- Precision - The criterion is also known as positive predictive value, takes into account true positive value and false positive value. The precision gives us information about how many relevant and irrelevant results give us the predictive model.
- FPR - This criterion is commonly known as false alarm rate. The predictive model improperly identifies normal harmless behavior as an anomaly which may lead to disruption of ICS. Therefore, FPR is highly important for critical infrastructure because the availability of the services is the most important criterion for ICS.
- Time - Time represents necessary time period for creation and evaluation of the predictive model.

The predictive model is based the Mississippi State University and Oak Ridge National Laboratory SCADA dataset [19]. The dataset consisting of 37 power system event scenarios. The dataset is structured as follow natural events (8), no events (1) and attack events (28). Normal operation of the system is represented by "no events". The "natural events" can be characterized as a natural fault of the system. The "attack events" can be described as the system under the cyber-attack. Furthermore, four Intelligent Electronic Devices (IED) were monitored. We investigated cyber-attack type: Data injection.

5 Results

Preprocessed dataset is divided into four subsets which representing data for each IED. We created seven hundred and fifty predictive models for each subset and different value of gamma parameter in order to evaluate the detection system. Moreover, the criteria for each predictive model are calculated (Accuracy, Sensitivity, Specificity, Precision, FPR and Time). The best fitting value of gamma parameter is determined by multi-criteria evaluation (reference point). Moreover, the weight for each criterion is selected according to its priority for ICS system. Therefore, we established three groups. The first and least important group include Accuracy and Sensitivity due to their focus only on positive classification. The second group includes Specificity, Precision and Time. The first two criteria which partially involving false positive identification, and time to build the predictive model which is very important for ICS. The last group involving false positive rate as the most important criterion due to the possible availability disruption of the ICS.

The Fig. 2 shows the results for Accuracy, Sensitivity, Specificity, Precision, False Positive Rate (FPR) and Time for the first IED. The results decompose in the interval: Accuracy from 0.832 to 0.967, Sensitivity from 0.819 to 1, Specificity from 0.382 to 0,941, Precision from 0.931 to 0.992, FPR from 0.008 to 0.069 and Time from 6.279 to 19.997 ms. Moreover, the best outcomes for each criterion according to gamma parameter is calculated as follow: Accuracy - 0.232 gamma, Sensitivity - 0.232 gamma, Specificity - 0.002 gamma, Precision - 0.008 gamma, FPR - 0.008 gamma and Time - 0.008 gamma.

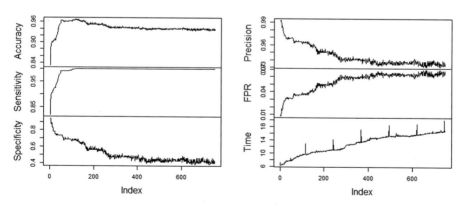

Fig. 2. The results for the first IED

The graphs in Fig. 3 represents the results for the second IED. The results decompose in the interval: Accuracy from 0.859 to 0.975, Sensitivity from 0.849 to 1, Specificity from 0.338 to 0.941, Precision from 0.927 to 0.992, FPR from 0.008 to 0.073 and Time from 5.092 to 19.236 ms. The gamma parameter for the best outputs is as follow: Accuracy - 0.124 gamma, Sensitivity - 0.124 gamma, Specificity - 0.002 gamma, Precision - 0.006 gamma, FPR - 0.006 gamma and Time - 0.026 gamma.

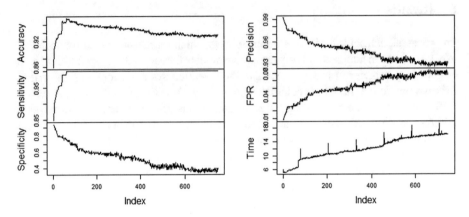

Fig. 3. The results for the second IED

The Fig. 4 shows the results for the third IED which are decomposed in the interval: Accuracy from 0.863 to 0.947, Sensitivity from 0.854 to 0.979, Specificity from 0.515 to 0,941, Precision from 0.944 to 0.992, FPR from 0.056 to 0.008 and Time from 6.374 to 16.881 ms. The gamma parameter for the best outputs of the criteria is as follow: Accuracy - 0.116 gamma, Sensitivity - 0.814 gamma, Specificity - 0.002 gamma, Precision - 0.008 gamma, FPR - 0.008 gamma and Time - 0.008 gamma.

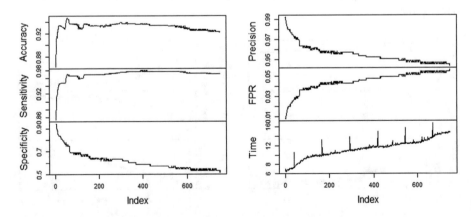

Fig. 4. The results for the third IED

The last results are shown in Fig. 5. The results for each criterion is spread as follow: Accuracy from 0.884 to 0.964, Sensitivity from 0.877 to 1, Specificity from 0.485 to 0,941, Precision from 0.942 to 0.992, FPR from 0.008 to 0.058 and Time from 4.954 to 12.758 ms. The gamma parameter for the best outputs of each criterion is as follow:

Accuracy - 0.25 gamma, Sensitivity - 0.266 gamma, Specificity - 0.002 gamma, Precision - 0.004 gamma, FPR - 0.004 gamma and Time - 0.01 gamma.

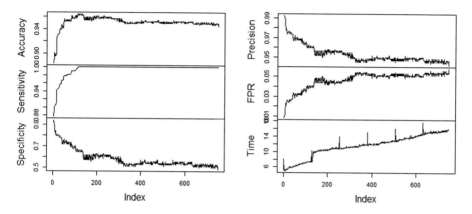

Fig. 5. The results for the fourth IED

In Table 1 can be seen all values for selected criteria according to chosen Gamma parameter. The parameter Gamma was computed for each IED according to the Reference point. Moreover, it calculates the best choice for each criterion and compares it to the actual state of the criteria according to their weights. At the end of the multi-step procedure, the results are fused into one via Majority vote technique. The final results affected by fusion are as follows: Accuracy - 0.898, Sensitivity - 0.888, Specificity - 0.985, Precision - 0.998, FPR - 0.002.

Table 1. The overall results for the computed gamma parameters

	Accuracy	Sensitivity	Specificity	Precision	FPR	Time (ms)	Gamma
IED 1	0.892	0.891	0.897	0.986	0.014	6.491	0.01
IED 2	0.906	0.905	0.912	0.989	0.012	5.235	0.014
IED 3	0.898	0.893	0.941	0.992	0.008	6.374	0.008
IED 4	0.901	0.9	0.911	0.988	0.012	5.179	0.012

6 Discussions

The presented paper is focused on improvement of detection capabilities of predictive models via choosing an appropriate value of the Gamma parameter. The Gamma parameter is one of the determining parameters for Radial kernel of the SVM. We established novelty detection system based on one-class SVM. Moreover, four IED under cyber-attack were used in order to create and evaluate the proposed solution. Furthermore, seven hundred and fifty predictive models with a different value of Gamma parameter were used.

The results presented in Figs. 2, 3, 4, 5 are assigned to four IED. The overall results indicate relatively high values for Accuracy, Sensitivity, Specificity, Precision and low values for FPR and Time especially for the low value of Gamma parameter. Moreover, the progress of graphs for is similar within a group of IED. The most significant results are situated in the first quarter of each graph (Figs. 2, 3, 4, 5) as result of high FPR and Time parameter in the rest of the data. Therefore, it is important to note that every miscalculation of Gamma parameter could have the serious impact on ICS. All relevant criteria achieve relatively high values in case of Accuracy, Sensitivity, Specificity, Precision and contrary FPR, Time criteria achieve considerably low values. The results for all predictive models show the best results for the relatively low value of Gamma parameter. Thus, proposed system based on multi-criteria evaluation calculated low values of Gamma parameter (0.01, 0.014, 0.008, 0.012). Moreover, the classifier fusion of the subsets resulted in improvement of detection capabilities of the detection system, especially for FPR parameter.

Acknowledgments. This work was funded by the Internal Grant Agency (IGA/FAI/2018/003) and supported by the project ev. no. VI20152019049 "RESILIENCE 2015: Dynamic Resilience Evaluation of Interrelated Critical Infrastructure Subsystems", supported by the Ministry of the Interior of the Czech Republic in the years 2015–2019 and also supported by the research project VI20172019054 "An analytical software module for the real-time resilience evaluation from point of the converged security", supported by the Ministry of the Interior of the Czech Republic in the years 2017-2019. Moreover, this work was supported by the Ministry of Education, Youth and Sports of the Czech Republic within the National Sustainability Programme project No. LO1303 (MSMT-7778/2014) and also by the European Regional Development Fund under the project CEBIA-Tech No. CZ.1.05/2.1.00/03.0089. Finally, we thank our colleagues from Mississippi State University and Oak Ridge National Laboratory which provides SCADA datasets.

References

1. Stouffer, K., Lightman, S., Pillitteri, V., Abrams, M., Hahn, A.: Guide to Industrial Control Systems (ICS) Security. NIST special publication, 800(82) R2, 16-16 (2015)
2. Macaulay, T., Singer, B.: Cybersecurity for Industrial Control Systems: SCADA, DCS, PLC, HMI, and SIS, 193 p. CRC Press, Boca Raton (2012). ISBN 14-398-0196-7
3. Chandola, V., Banerjee, A., Kumar, V.: Anomaly detection: a survey. ACM Comput. Surv. (CSUR) **41**(3), 15 (2009)
4. Dewa, Z., Maglaras, L.A.: Data mining and intrusion detection systems. Int. J. Adv. Comput. Sci. Appl. **7**(1), 62–71 (2016)
5. Pathan, A.S.K.: The State of the Art in Intrusion Prevention and Detection. Auerbach Publications, Boca Raton (2014)
6. Goldstein, M., Uchida, S.: A comparative evaluation of unsupervised anomaly detection algorithms for multivariate data. PLoS One **11**(4), e0152173 (2016)
7. Ebrahimi, M., Suen, C.Y., Ormandjieva, O., Krzyzak, A.: Recognizing predatory chat documents using semi-supervised anomaly detection. Electron. Imaging **2016**(17), 1–9 (2016)
8. Sharma, V., Suryawanshi, V.: Network anomaly detection through hybrid algorithm. Int. J. Comput. Sci. Trends Technol. (IJCST) **5**, 74–78 (2017)
9. Akoglu, L., Tong, H., Koutra, D.: Graph based anomaly detection and description: a survey. Data Min. Knowl. Disc. **29**(3), 626–688 (2015)

10. Pollet, J.: SCADA 2017: The Future of SCADA Security. 8th Annual ICS & SCADA Security Summit, SANS, 12–13 February 2013. https://files.sans.org/summit/euscada12/PDFs/RedTigerSecurity_SCADA_2017.pdf

11. Horkan, M.: Challenges for IDS/IPS Deployment in Industrial Control Systems. SANS Institute (2015). https://www.sans.org/reading-room/whitepapers/ICS/challenges-ids-ips-deployment-industrial-control-systems-36127

12. Verba, J., Milvich, M.: Idaho national laboratory supervisory control and data acquisition intrusion detection system (SCADA IDS). In: IEEE Conference on Technologies for Homeland Security, pp. 469–473. IEEE (2008)

13. Zhu, B., Sastry, S.: Intrusion detection and resilient control for SCADA systems. In: Securing Critical Infrastructures and Critical Control Systems: Approaches for Threat Protection: Approaches for Threat Protection, vol. 352 (2012)

14. Yang, Y., McLaughlin, K., Littler, T., Sezer, S., Wang, H.F.: Rule-based intrusion detection system for SCADA networks. In: 2nd IET Renewable power generation conference (RPG 2013), pp. 1–4. IET (2013)

15. Maglaras, L.A., Jiang, J.: Intrusion detection in SCADA systems using machine learning techniques. In: Science and Information Conference (SAI 2014), pp. 626–631. IEEE (2014)

16. Knapp, E.: Industrial Network Security: Securing Critical Infrastructure Networks for Smart Grid, SCADA, and Other Industrial Control Systems, vol. xvii, 341 p. Syngress, Waltham (2011). ISBN 15–974-9645-6

17. Cortes, C., Vapnik, V.: Support-vector networks. Mach. Learn. **20**(3), 273–297 (1995)

18. Hsu, C.W., Chang, C.C., Lin, C.J.: A practical guide to support vector classification. BJU Int. **101**, 1396–400 (2008)

19. Hink, R.C.B., Beaver, J.M., Buckner, M.A., Morris, T., Adhikari, U., Pan, S.: Machine learning for power system disturbance and cyber-attack discrimination. In: 2014 7th International Symposium on Resilient Control Systems (ISRCS), pp. 1–8. IEEE, August 2014

TRMA: An Efficient Approach for Mutual Authentication of RFID Wireless Systems

R. Anusha[✉] and V. Veena Devi Shastrimath

Department of Electronics and Communication Engineering,
NMAM Institute of Technology NITTE, Udupi, India
anu4research@gmail.com

Abstract. In today's world, transmission and reception of data secrecy are the foremost concerns in wireless communication. Recently Radio Frequency Identification (RFID) has perceived the massive amount of attention as it inhibits privacy invasions and data disclosure. An entire generation of new security and privacy attacks might arise as the RFID technologies have a widespread deployment all over. One of the significant drawbacks in RFIDs are the unsteady authentication, either the password itself is leaked else data among the reader and a tag. Also, researchers have observed that there are threatening security drawbacks in Electronic Product Code-Class I Generation 2 (EPC-C1G2). To overcome this issue, a function called as the Pad Generation is used that assists in refining the security parameter in the mutual authentication scheme. This paper stresses on Tag-Reader Mutual Authentication (TRMA) - system of RFID tag-reader with the modified version of PadGen. The modified Pad-Gen design uses XOR operation for the RIFD-TRMA protocol. For the cost-effective hardware implementation, the proposed framework is simulated on FPGA Artix7-XC7A100T-3CSG324 device and physically verified on Chip scope pro tool.

Keywords: Security · Authentication · RFID · Mutual authentication scheme
Tag identification

1 Introduction

RFID has witnessed the rapid amount of deployment in the recent years. Various applications such as transportation management, product tracking information, inventory maintenance, and logistics need an RFID to function. Conventionally, an RFID design has three prime components namely RFID reader, RFID tags, and a server. The reader queries the Tag Identity (TID) and sends it to the back-end server. The tag in RFID is an integrated circuit, and an antenna which stores the information and further forwards it to the tag reader. Individually the tags are accommodated with a unique identification to recognize the object attached to it [1]. A primary system of RFID makes it possible for the reader to intercept the data transmitted between the air interfaces quickly. The Generation 2 (G2) is a tag that comprises of Pseudo-Random Number Generator (PRNG) and secures the information integrity via the Cyclic Redundancy Check. The memory space is sub-classified into four categories, reserved memory, user memory, tag identify

© Springer International Publishing AG, part of Springer Nature 2019
R. Silhavy (Ed.): CSOC 2018, AISC 765, pp. 290–299, 2019.
https://doi.org/10.1007/978-3-319-91192-2_29

memory and the EPC. In July 2006, EPCglobal Class-1 generation-2 was approved to be ISO18000-6C. It is firmly believed that tags respective to generation-2 would be prevailing the development in RFID implementations as the reading range is comparatively broad [2]. EPCglobal is a combined venture of EAN and UCC, the organization responsible for regulation of barcode in the USA and the other countries as well [3]. The EPCglobal Class-1 generation-2 (C1G2) standard for RFID Ultra High Frequency (UHF) describes the passive technology of RFID that happens to be a global protocol with an option of open access. The EPC global C1G2 scheme scrutinizes the protocol of RFID in the range of UHF scale. The range is defined as the frequency between the values of 860 until 960 MHz. The specifications of EPC C1G2 do not entirely work in for protecting security invasion and privacy concerns of the data. The standard stresses that the RFID tag should have two passwords, i.e., 32-bit access and kill password that will operate disabling RFID tag permanently. To attain security of the data being developed, the notion of PadGen was presented. It is used to create a pad to cover the code for the masking purpose of tag's enabling password before the data transmission [4]. In spite of many prospective mechanisms, the security of RFID is an exponentially growing and extensive matter of concern. Significant of the designed protocols are not carried out for their verification in hardware. Hardware prototyping of Field Programmable Gate Arrays (FPGA) is flexible to initiate the check of protocol in real-time scenarios. The hardware verification using FPGA can be extended for fields like robotics, radio communication, digital signal processing and robotics [5].

The paper describes the efficient approach of mutual authentication of wireless RFID systems, called the TRMA protocol. The function of PadGen is altered, and the manipulated version is made to perform XOR operation. The hardware prototyping is simulated over FPGA Artix7-XC7A100T-3CSG324 device, and the physical verification is built on a tool named Chip scope pro. Following is the manner in which the paper is organized.

Section 2 foregrounds on the background work carried for the EPC-C1G2 standard. Section 3 highlights the related work that recognizes the research work associated with the various security attacks on RFID. The problem statement is highlighted in Sect. 4. Section 5 discusses the proposed methodology to extend the Pad-Gen function in FPGA Artix7-XC7A100T-3CSG324 and Chip scope pro tool. The outcomes of the proposed method are discussed in Sect. 6 whereas Sect. 7 is the conclusion of the paper.

2 Background

2.1 Standard of EPCglobal-C1G2

EPC-C1G2 was initially introduced in the year 2004, to establish a path for the tag-reader communication globally. The prior objective of the designers was to address the requirements for achieving economic and high throughput enabling applications in a supply chain. The standard concentrates over RFID interaction protocol in the UHF spectrum that lies in the range starting from 860–960 MHz.

EPC-C1G2 identifies the necessity of an acquiescent RFID tag must comprise of an individual 32-bit kill and access password. However, the standard does not entirely work in favor of guaranteeing that privacy isn't compromised else data corruption should not take place. The basic instructions of kill and access present on the specifications of the EPC-C1G2 are not sufficient to yield secure authentication task and data or privacy leakage prevention. The following mentioned are minimum non-volatile memory characteristics in an RFID tag,

- A password of 32-bit each to perform the operation of enabling and disabling the tag forever are the password for accessing and killing, are the contents of reserved memory.
- The memory allocated to the EPC section has, Cyclic Redundancy Code (CRC) parameters (16-bits), bits pertaining in Protocol Control (PC) (16-bits) and an EPC that would recognize the device/object to which the tag is attached or will be attached.
- The memory allocated for the TID has adequate data to identify the reader characteristics which the tag maintains and a vendor determined data.
- The user-peculiar information is stored in the user memory.

Complex functions namely, hash operation, symmetric encryption and public encryption are incapable to be managed by EPC-C1G2, due to its drawback of resources and cost. The essential functions that can be accomplished are:

- PRNG of 16-bit capacity.
- CRC of 16-bit capacity to generate checksum code for the verification activity of transmitted information and control the integrity.
- Exclusive OR operation.

RFID tags are more often seen from the storage media point of view and very rarely as smart tags. Therefore, the access towards the computational resources is constrained. RFID tags consist of nearly 1000–5000 logic gates, out of which 400-200 can be deployed in the security-relevant operations. Also, has function and encryption mechanism are not suitable for the tags that exploit the EPCglobal-C1 standard. None of the protocols confine to the EPCglobal-C1 scheme. Less attention is given to privacy and security concerns in the EPC-C1G2 specification. Further, when specific schemes were formulated carrying out the analysis of EPC-C1G2 specification, severe issues in the aspect of security was demonstrated.

2.2 Cost-Effective RFID Tag Vulnerabilities

Sound business promotion can viably avail with the technology of low-cost RFID due to their potential of enhancing processes by saving costs. RFID systems provide significance in securing risks, majorly because of cost restricted implementations and insecure communication paths on which the users interact. EPC-C1G2 assures less security, and hence this tends to make RFID tag highly susceptible towards cloning attacks, information leakage, and password disclosure. The EPC-C1G2 weakness can be overruled by implementing an alternative scheme where the principal component is a PadGen function. The below-discussed security issues occur very often in RFID networks,

(a) **Data Integrity:** Integrity function ensures that the resourcefulness of the message being transmitted and guarantees that no manipulation takes place while transitioning.

(b) **Forward Data Secrecy:** In the case wherein an intruder discovers an internal state of the reader at a specific instant of time, it should further be able to know the reader's interactions which happened at the same instant of time. The protocol should enable forward secrecy to avoid the compromise of secret data.

(c) **Denial of Service (DoS) Attack:** An intruder can alter or create a hindrance to the data being transmitted to the reader. Similarly, the intruder could be a main cause for disturbing the synchronization between the RFID reader and the tag. This is the reason for DoS attack on the server.

(d) **Data Privacy:** One of the prime matters of concerns in RFID systems is data privacy, precisely when the technology creates interconnections between smart and personal devices like mobile phones, tablets, e-passports, credit cards, etc. Illegitimate tracking must be stopped entirely; this must be confirmed by an authentication protocol. Also, it should prevent the intruder form acquiring the tag message.

(e) **Replay Attack:** Here, the intruder will be able to play back the eavesdropped data from the reader side to the tag without any notification, therefore conducting a strong authentication either at reader or tag section [5, 6].

3   Related Work

Lai et al. [7] proposed an Optimal Query Tracking Tree protocol (OQTT) that initiated towards the separation of the RFID tags into minor sets to minimize the rate of the collision at the starting of identification. The proposed work utilized three approaches namely, query tracking tree, bit estimation, optimal partition. Lui et al. [8] introduced a protocol of anti-collision of wireless RFID network. Since the basis of the algorithm was derived from the protocols of Collision Tree (CT) and Improved Collision Tree, the proposed framework was named as Adaptive Collision Tree (ACT).

Maguire and Pappu [9] studied the analysis of an algorithm, i.e., medium access controlling of the standard ISO/IEC 180000-6C RFID, with a procedural way to attain optimized read rates. Ning et al. [10] proposed protocol, distributed KAAP that gave the opportunity to categorize the security protection. The synthetical analysis of KAAP is carried out considering aspects such as performance, security, and logic.

Preradovic et al. [11] implemented an utterly passive RFID wireless system which was printable and chipless. To secure the documents the tag is integrated with multi-resonator spiral microstrip and a polarized microstrip organized in a manner of cross configuration, both for transmitting and reception sections monopole antennas. These antennas were disc loaded and consisted of ultra-wideband. Kazuya et al. [12] reframed the design of RFID architecture by splitting a radio frequency reader into two different modules, i.e., Trusted Shield Device (TSD) and an RF activator. After which a novel scheme for coding was proposed, the Random Flipping Random Jamming (RFRJ) that secured the contents of RFID tag.

Sun et al. [13] found two methodologies to desynchronize the attacks for the ultra lightweight RFID protocol to break. To assure that the information in the RFID system is authenticated and reserved along the minimum transmission computational resources, the de-synchronization scheme is necessary. Tan et al. [14] proposed an authentication protocol that yielded protection without the requirement of a central server. What a protocol for that dealt with the secure seeks of RFID tag was also implemented.

Yin et al. [15] executed a passive System on Chip (SoC) UHF RFID tag embedded with a sensor which measures temperature. Yu et al. [16] introduced a protocol to enable tag detection via Bloom filter-based Missing Tag Detection (BMTD).

Zhang et al. [17] implemented two protocols of anti-collision in the tag, functioning on the base of Manchester encoding Assigned Tree Slotted Aloha (ATSA) with Improvised ATSA (ImATSA) were studied. Zhu and Yum [18] proposed a model to read the Markov Chain process and induce the optimized reading scheme via analyzing the first-time process. The optimal model scheme can be easily integrated with the standard of EPCglobal to provide the significant improvement in performance.

The work of Zheng and Li [19] included designing of algorithms to give appropriate solution for stringent delay necessary in the generation of protocols that implement faster search process. Following which the process formulates issue of tag search in the in large-scale RFID networks. The framework of a small approximator that effectively unites the enormous value of RFID tag data and interchanges the data among the protocol of two phase approximator is proposed. Oren and Shamir [20] implemented a hugely significant category, i.e., power analysis in the channel attacks as it tried to gain passwords and cryptographic keys by investigating the device's power consumed.

Myung et al. [21] presented an anti-collision scheme in a flexible tag of two: protocol of Adaptive Query Splitting (AQS) and protocol of Adaptive Binary Splitting (ABS) which is an enhanced version of query tree and derived from binary tree algorithm for RFID protocol respectively. Luo et al. [22] modified the existing protocol along with the time required for it. An implementation was proposed with a fraction of the time to minimize the tag search process and also to analyze a new aspect of energy preserving in the issue of missing tag detection.

Huang et al. [23] addressed an organized perspective towards enabling authentication via three parameters, biometrics, smart cards, and password. Han et al. [24] proposed an efficient model for evaluating the tag population, which was anonymous. The model leveraged the location of the initial response from a cluster of tags present in a frame.

The study of Doss et al. [25] presented two ownership authenticated transfer schemes. The open loop and the closed system were adopted. The transfer by the aid of ownership was assured. Fabian et al. [26] proposed SHARDIS, a service that provides enhanced privacy for the RFID data based on the paradigm of peer-to-peer communication.

4   Problem Statement

Theft Identities, hackers, viruses, unauthorized access to credit and debit cards have raised the issue of data confidentiality. Ensuring the data security is the vital matter of concern for government organizations, corporate world, and individuals. Adoption of RFID at a very high rate gives scope to the latest frontier for information threats and data measures in security. Widely speaking RFID possesses the capability of complete spectrum in wireless devices pertained various energy capabilities, also involving EPC tags, vehicle immobilizers, etc. The technology of EPC Gen2, tags don't exist for batteries on-board. They happen to be passively supplied power via radio frequency signals from interrogators. Like other various communication systems, the RFID stage of EPC Gen2 protocol is prone to threats bothering the data secrecy and user privacy held by tagged objects. Due to this, there is essentiality to compute the nature of such risks and recognition of possible attackers, so that appropriate counter-measures can be performed to limit the use of message which is unauthorized.

5   Modification PAD-GEN Design

The TRMA protocol based on the XOR operation requires a password for accessing the information before the process of interchange between the RFID tag and the corresponding reader. The password used for accessing will be a 32-Bit value present in the memory of reserved space in tag's memory. Setting up this password would require the reader to indicate the confirmation of valid password at the beginning of data interchange between the tag and the reader. RFID manufacturer, reader, tag are involved in the schematic design of TRMA protocol. Here, two passwords ensure the data secrecy in the system, i.e., the Access Password (AP) and the Kill Password (KP). Each password has a bit containing the capacity of 32-bits. Following is the representation of MSB and LSB corresponding to the AP and KP for 16 bit each as illustrated in the Fig. 1,

$$AP = a_{p0}a_{p1} \ldots \ldots \ldots a_{p31} \tag{1}$$

$$AP_L = a_{p0}a_{p1} \ldots \ldots \ldots a_{p15} \tag{2}$$

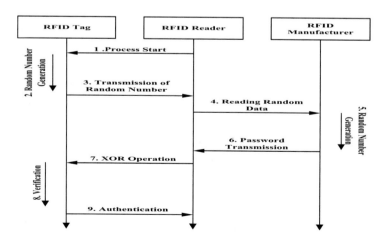

Fig. 1. Design flow of RFID-TRMA

$$AP_M = a_{p16}a_{p17} \cdots \cdots a_{p31} \tag{3}$$

$$KP = k_{p0}k_{p1} \cdots \cdots \cdots k_{p31} \tag{4}$$

$$KP_L = k_{p0}k_{p1} \cdots \cdots \cdots k_{p15} \tag{5}$$

$$KP_M = k_{p16}k_{p17} \cdots \cdots k_{p31} \tag{6}$$

The value of the Access Password (AP) and Kill Password (KP) are obtained by performing concatenation operation on the Most Significant Bit (MSB) and Least Significant Bit (LSB) value accordingly. Brief descriptions of the algorithm steps utilized in the TRMA protocol are:

1. Start the process ← Reader requests (R) the tag.
2. Generation of the random number by Linear-Feedback Shift Register (LFSR) ← a 16-bit number by the tag is generated.
3. Transmission of random data ← at the reader from tag via EPC
4. The process of random data reading ← Reader reads the data and forwards to the manufacturer.
5. Generation of the random number by LFSR ← a 16-bit number by the manufacturer is generated.
   - Perform the XOR operation on the tag and manufacturer generated random numbers.
   - Perform the AP operation over the pad generated a random number of tag and manufacturer respectively and the output attained is considered to be Rn.
   - Execution of the operation AP carried on random tags generated bits and random bits obtained as an outcome from tag and manufacturer. The output is the variable Rp.
   - KP operation is made to implement over Rn and Rp. The resultant is PAD1.
   - KP performs in the favor Rn, and Rn operated with Rp through xor operation. Resultant acquired is PAD2.
   - The Code-Covering passwords in for mutual authentication scheme protocol an Access Password (AP) are implemented.
6. The passwords of computed code-covering are transmitted to RFID-reader via EPC.
7. At the RFID tag, the passwords which are code covered are stored and are operated with PAD1 and PAD2 through XOR implementation to develop the MSB, LSB of AP.
8. Verification ← the tag checks for the validation of the correct password.
9. Authentication ← If the tag is authenticated appropriately, the reader with the EPC is authenticated.

6   Results

Top Module of TRMA using XOR operation is illustrated in Fig. 2. It includes the global clock (CLK), asynchronous reset (RST), control signals like enable (EN) and CRC_EN. The 32 bit AP and KP are the input ports whereas the modified pad generated outputs are PAD1 Out and PAD2 Out having 16 bits each. Similarly, the cover code passwords are the CCPL and CCPM (32 bit each) that are concatenated to form the Authenticated outputs (XC1 and XC2).

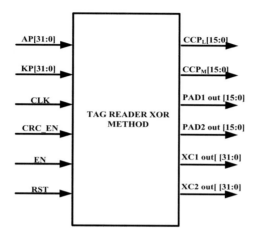

Fig. 2. Top Module of TRMA using XOR operation

The design is modeled in Xilinx 14.7 ISE utilizing Verilog code. The simulation is implemented using Modelsim 6.5f and executed on low-cost Artix7 FPGA. As the clock is activated, the reset is set to active low, and the en signal is made high that is used to generate random signals in RM and RT modules. The crc_en is used to create the crc_out using the LFSR's. The value of Apsd and Kpsd are 32'h values individually that are used for the modified pad generation function to obtain the results. The final output is attained as XC_Out1 and XC_Out2 which is the concatenated outcome of CCPM with crc_out1 and CCPL with crc_out2 respectively is shown in below Fig. 3.

Signal	Value	Waveform values
/Tag_Reader_XOR_Method_tb/clk	0	
/Tag_Reader_XOR_Method_tb/rst	0	
/Tag_Reader_XOR_Method_tb/en	1	
/Tag_Reader_XOR_Method_tb/crc_en	0	
/Tag_Reader_XOR_Method_tb/Apsd	abcdefab	00000000 · abcdefab
/Tag_Reader_XOR_Method_tb/Kpsd	efababcd	00000000 · efababcd
/Tag_Reader_XOR_Method_tb/PAD1_out	a1a1	0000 · a1a1 · 81a1 · 0020 · a1a1
/Tag_Reader_XOR_Method_tb/PAD2_out	a1f1	0000 · a1f1 · 81d1 · 00d1 · a1f1
/Tag_Reader_XOR_Method_tb/CCPsdM	0a6c	0000 · 0a6c · 2a6c · abed · 0a6c
/Tag_Reader_XOR_Method_tb/CCPsdL	4e5a	0000 · 4e5a · 6e7a · ef7a · 4e5a
/Tag_Reader_XOR_Method_tb/XC_Out1	0a6ca001	00000000 · 0a6c0000 · 0a6c3d68 · 0a6cb218 · 2a6c9132 · abed1bcd · 0a6ca0ca · 0a6cfdde
/Tag_Reader_XOR_Method_tb/XC_Out2	4e7abe0f	00000000 · 4e5a0000 · 4e5a25d9 · 4e5af90f · 6e7ab3f4 · ef7a4d2b · 4e5acdec · 4e5a09be
/Tag_Reader_XOR_Method_tb/uut/RM	f800	0000 · 0002 · 0006 · 000e · 000e · 001e · 003e · 003e
/Tag_Reader_XOR_Method_tb/uut/RM1	f000	0000 · 0002 · 0006 · 000e · 001e · 003e · 007e
/Tag_Reader_XOR_Method_tb/uut/RT	f800	0000 · 0003 · 0006 · 000e · 000e · 001e · 003e · 003e
/Tag_Reader_XOR_Method_tb/uut/RT1	f000	0000 · 0003 · 0006 · 000e · 001e · 003e · 007e
/Tag_Reader_XOR_Method_tb/uut/Rv	f1f1	0000 · f1f1 · f1e1 · e1e1 · f1f1
/Tag_Reader_XOR_Method_tb/uut/Rw	f1f1	0000 · f1f1 · e1f1 · f1f1
/Tag_Reader_XOR_Method_tb/uut/RMt	0000	0000 · 0001 · 0000
/Tag_Reader_XOR_Method_tb/uut/Rvw	0000	0000 · 0010 · 0000
/Tag_Reader_XOR_Method_tb/uut/crc_out1	a001	0000 · 3d68 · b218 · 9132 · 1bcd · a0ca · ffde · bda3
/Tag_Reader_XOR_Method_tb/uut/crc_out2	be0f	0000 · 25d9 · f90f · b3f4 · 4d2b · cdec · 09be · 105d

Fig. 3. Simulation results of TRMA using XOR operation

The TRMA design is combined with IP cores like ILA, ICON to form the standard prototyping of architecture over FPGA platform. The Implemented design is physically debugged on My ILA to observe the verified results on chipscope pro tool with reset pulled to 1 and 0 has shown in the Figs. 4 and 5.

Bus/Signal	X	O	1008	1009	1010	1011	1012	1013	1014	1015
PAD1_out	A1A	A1A1							A1A1	
PAD2_out	A1F	A1F1							A1F1	
CCPsdM	0A6	0A6C							0A6C	
CCPsdL	4E5	4E5A							4E5A	
XC_Out1	000	0000001							00000000	
XC_Out2	000	0000001							00000000	

Fig. 4. Modified Pad-Gen function implemented on Chip scope pro tool using XOR operation, when RST = 1, EN = 0, and CRC_EN = 0

Bus/Signal	X	O	22	23	24	25	26	27	28	29	30	31	32	33
PAD1_out	A1A	A1A1	A1A1		81A1	0020							A1A1	
PAD2_out	A1F	A1F1	A1F1		81D1	00D1							A1F1	
CCPsdM	0A6	0A6C	0A6C		2A6C	ABED							0A6C	
CCPsdL	4E5	4E5A	4E5A		6E7A	EF7A							4E5A	
XC_Out1	0A6	0A6C69DE	0A6...	0A6C...	0A6C...	2A6C...	ABE..	0A6C...	0A6C...	0A6C...	0A6...	0A6C...	0A6C...	0A6C...
XC_Out2	4E5	4E5AD857	4E5...	4E5A...	4E5A...	6E7A..	EF7...	4E5A..	4E5A..	4E5A..	4E5..	4E5A..	4E5A..	4E5A..

Waveform - DEV:0 MyDevice0 (XC7A100T) UNIT:0 MyILA0 (ILA)

Fig. 5. Modified Pad-Gen function implemented on Chipscope pro tool using XOR operation, when RST = 0, EN = 1, and CRC_EN = 1

7   Conclusion

The hardware realization of TRMA mutual authentication protocol is verified using the modified pad generation function and implemented on an FPGA prototype. The proposed modification is feasible in improvising the weakness of EPC-C1G2 authentication standard. For the physical verification of the design, the Chipscope pro tool is utilized. The tool works for serving the purpose of hardware debugging the design. The proposed TRMA system's simulation results are analyzed. The function of the CRC generator in the design is to improvise the security of message transmission at the next level between the tag and the reader. To conclude, the main characteristic of the new approach is addressed in acquiring security management to the further stages of the authentication protocol.

References

1. Chen, M., et al.: An efficient tag search protocol in large-scale RFID systems with noisy channel. IEEE/ACM Trans. Network. (TON) **24**(2), 703–716 (2016)
2. Gao, L., et al.: An ultra lightweight RFID authentication protocol with CRC and permutation. J. Netw. Comput. Appl. **41**, 37–46 (2014)
3. Juels, A.: RFID security and privacy: a research survey. IEEE J. Sel. Areas Commun. **24**(2), 381–394 (2006)

4. Huang, Y.-J., et al.: Hardware implementation of RFID mutual authentication protocol. IEEE Trans. Ind. Electron. **57**(5), 1573–1582 (2010)
5. Feldhofer, M.: An authentication protocol in a security layer for RFID smart tags. In: Proceedings of the 12th IEEE Mediterranean Electro-technical Conference, MELECON 2004, vol. 2. IEEE (2004)
6. Maarof, A., et al.: Authentication protocol conforming to EPC class-1 Gen-2 standard. In: International Conference on Advanced Communication Systems and Information Security (ACOSIS). IEEE (2016)
7. Lai, Y.-C., et al.: A novel query tree protocol with bit tracking in RFID tag identification. IEEE Trans. Mob. Comput. **12**(10), 2063–2075 (2013)
8. Liu, X., et al.: An adaptive tag anti-collision protocol in RFID wireless systems. China Commun. **11**(7), 117–127 (2014)
9. Maguire, Y., Pappu, R.: An optimal Q-algorithm for the ISO 18000-6C RFID protocol. IEEE Trans. Autom. Sci. Eng. **6**(1), 16–24 (2009)
10. Ning, H., et al.: Scalable and distributed key array authentication protocol in radio frequency identification-based sensor systems. IET Commun. **5**(12), 1755–1768 (2011)
11. Preradovic, S., et al.: Multiresonator-based chipless RFID system for low-cost item tracking. IEEE Trans. Microw. Theor. Tech. **57**(5), 1411–1419 (2009)
12. Sakai, K., et al.: A novel coding scheme for secure communications in distributed RFID systems. IEEE Trans. Comput. **65**(2), 409–421 (2016)
13. Sun, H.-M., Ting, W.-C., Wang, K.-H.: On the security of Chien's ultralightweight RFID authentication protocol. IEEE Trans. Dependable Secure Comput. **8**(2), 315–317 (2011)
14. Tan, C.C., Sheng, B., Li, Q.: Secure and serverless RFID authentication and search protocols. IEEE Trans. Wirel. Commun. **7**(4), 1400–1407 (2008)
15. Yin, J., et al.: A system-on-chip EPC Gen-2 passive UHF RFID tag with embedded temperature sensor. IEEE J. Solid-State Circ. **45**(11), 2404–2420 (2010)
16. Yu, J., et al.: Finding needles in a haystack: missing tag detection in large RFID systems. IEEE Trans. Commun. (2017)
17. Zhang, L., Zhang, J., Tang, X.: Assigned tree slotted aloha RFID tag anti-collision protocols. IEEE Trans. Wirel. Commun. **12**(11), 5493–5505 (2013)
18. Zhu, L., Yum, T.-S.P.: The optimal reading strategy for EPC Gen-2 RFID anti-collision systems. IEEE Trans. Commun. **58**(9), 2725–2733 (2010)
19. Zheng, Y., Li, M.: Fast tag searching protocol for large-scale RFID systems. IEEE/ACM Trans. Network. (TON) **21**(3), 924–934 (2013)
20. Oren, Y., Shamir, A.: Remote password extraction from RFID tags. IEEE Trans. Comput. **56**(9), 1292–1296 (2007)
21. Myung, J., et al.: Tag-splitting: adaptive collision arbitration protocols for RFID tag identification. IEEE Trans. Parallel Distrib. Syst. **18**(6), 763–775 (2007)
22. Luo, W., et al.: Efficient missing tag detection in RFID systems. In: IEEE INFOCOM 2011 Proceedings (2011)
23. Huang, X., et al.: A generic framework for three-factor authentication: preserving security and privacy in distributed systems. IEEE Trans. Parallel Distrib. Syst. **22**(8), 1390–1397 (2011)
24. Han, H., et al.: Counting RFID tags efficiently and anonymously. In: IEEE INFOCOM 2010 Proceedings (2010)
25. Doss, R., Zhou, W., Shui, Yu.: Secure RFID tag ownership transfer based on quadratic residues. IEEE Trans. Inf. Forensics Secur. **8**(2), 390–401 (2013)
26. Fabian, B., Ermakova, T., Muller, C.: SHARDIS: a privacy-enhanced discovery service for RFID-based product information. IEEE Trans. Industr. Inf. **8**(3), 707–718 (2012)

SC-MANET: Threats, Risk and Solution Strategies for Security Concerns in Mobile Ad-Hoc Network

C. K. Vanamala[1(✉)] and G. Raghvendra Rao[2]

[1] Department of Information Science and Engineering, NIE, Mysuru, India
ckvanamala@nie.ac.in
[2] Department of Computer Science and Engineering, NIE, Mysuru, India

Abstract. A MANET is a form of the wireless network among the mobile, wireless nodes. The presence of various significant attributes in MANET like end to end communications, dynamic topology and simple setup, leads to difficulties like routing, security, and clustering. Network security is an important aspect for both wired and wireless communication. This paper, provide a detail review on different security attacks over the MANETs. Further discussion is carried out by providing of prior solutions to overcome the security attacks and gives a bench mark for future study. An extensive survey of existing researches towards security in MANET is addressed. Later on, a research gap in current state of art in MANET security is discussed. Especially, this survey study mainly focus on security problems, challenges and solution strategies for security concerns in mobile ad-hoc networks.

Keywords: MANET · OLSR · Security attack

1 Introduction

With the growing advancements in wireless technology and expanding fame of wireless gadgets, it is difficult to handle and protect the network infrastructure from the attackers. Therefore, security and privacy is major issue for both wired and wireless communication systems. MANET is independent network which exhibits the amazing qualities like dynamic topology, open network boundary, end to end communication, etc. These attributes of MANET made it famous, particularly in disaster management and military applications. Because of exceptional features, the MANET faces different issues. The integration with web [1], peer to peer applications [2], security [3], and keeping up network topology [4] and vitality [5, 6] are the absolute most essential challenges in MANET. In MANET all nodes are allowed to connect and disconnect the network, likewise called open network boundary [7]. Every single intermediate hub between a source and destination participate in routing, additionally called "hop-by-hop" communications. In wireless communication, each node will receive the packets within the range. Because of these qualities, every node can undoubtedly gain access to different nodes packets or infuse fault packets to the network. Hence, the protection of MANET against malicious attacks most essential challenge [8].

In light of MANET's unique attributes, there are some essential metrics meant for MANET security which are imperative among all the security approaches; i.e., "Security Parameters," the unaware of these parameters can lead to the failure of security approach in MANET. The parameter "Network Overhead" means the number of control packets generated due to security approaches. Due to existence of shared wireless communication, additional control packets can cause the congestion or collision in MANET resulting "Packet lost". Hence the existence of high packet overhead may lead to high packet loss and re-transmission of packets which efficiently drains the energy of nodes and networks resources. Thus, every security mechanism needs processing time to identify bad conduct and kill malicious nodes. The existence of dynamic topology in MANET communication among the two nodes can break as portability results. Therefore, security mechanisms need to have low processing time to have enhanced flexibility in MANET. MANET nodes have constrained energy supply thus upgrading energy consumption is highly challenge-able concern. Hence the higher consumption of energy may lead to a reduction of network's lifetime. Every security protocol must know about these three essential parameters. In some cases a trade-off among these parameters is presented to a satisfaction level in every one of them.

In this paper, we investigate the different security related problems and challenges in MANETs. Furthermore, the paper also contains a discussion on solution strategies and defeating approaches. Therefore, the paper work is organized as: the Sect. 2, briefly describes about the concepts of MANET, followed by existing methods for MANET are discussed in Sect. 3. Later, the existing research works in MANET security are highlighted in Sect. 4. Section 5 represents the research gap which illustrates the solution methods for MANET security. Finally, Sect. 7 concludes survey paper and presents the future scope of work towards MANET security particularly with OLSR.

2 Background

This section discusses different concepts of the MANET; types of MANET, issues related to MANET, Challenges associated with MANET security, Different types of attack on MANET and existing security solution for MANET.

2.1 Mobile Ad-Hoc Networks (MANETS)

This is a special system which can vary the regions and system design itself. Since MANETS are versatile, they utilize remote associations with an interface with different systems. The MANET can be of the type Wi-Fi collaboration, or an alternative mechanism, for instance, satellite or a cell transmission system [1, 2, 9] (Fig. 1).

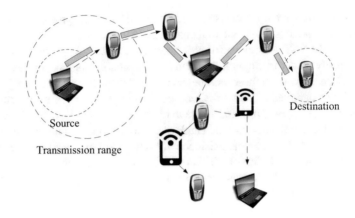

Fig. 1. Structure of MANET

A few numbers of MANETs are confined to the operation of neighborhood remote gadgets, (for example, a gathering of smart phones); others may be associated with the Internet [1–9]. Given the element nature of MANETs, they are normally not exceptionally secure, so it is vital to be cautious of what kind of data is sent on a MANETs [9].

2.2 Security Attacks in MANETs

There are various security attacks [1–13] exists and are explained below:

- **Passive attack:** In this sort of attack, the intruder just plays out some observing on specific connections to get information about the traffic without infusing any fake information. This sort of attack serves the attacker to gain information and makes the footprint of the attacked arrange to apply the attack efficiently. The sorts of passive attacks are found in, traffic analysis and snooping.
- **Denial of service attack:** These are aimed at complete routing information disruption and therefore the whole ad-hoc network operation.
- **Traffic Analysis:** In MANETs the information bundles and traffic design both are imperative for adversaries. For instance, secret information about network topology can be determined by investigating traffic designs. Traffic analysis can likewise be directed as a dynamic attack by devastating hubs, which animates self-organization in the network, and important information about the topology can be assembled. Traffic analysis in ad hoc networks may uncover the following sort of information.
- **Snooping:** This is not restricted to gain access to data at the instant of transmission which consists of casual e-mail-observance appearing over another computer as what another user is typing.
- **Active attack:** In this attack, the intruder performs an effective violation on either the network resources or the data transmitted; this is done by International Journal on New Computer Architectures and Their Applications causing routing disruption, network resource depletion, and node breaking. In the following are the types of active attacks over MANET and how the attacker's threat can be performed.

- **Flooding attack**: In this, the attacker exhausts the network resources like bandwidth and to consume a node's resources like computational and battery power or for routing operation disruption to cause network performance degradation.
- **Black hole Attack:** The process "Route discovery" in AODV is vulnerable to the "black hole attack." The mechanism where any intermediate node can respond to the route request (RREQ) message when it exists fresh enough routes, devised to minimize the delay in routing which is utilized by a node which is malicious to cooperate the system.
- **Jamming:** This is a disk operating system (DOS) attack type initiated by the malicious node after determination of communication frequency. Here, the jammer will transfer the signals along with security threats. The attacks prevent the legitimate packets reception.
- **Malicious code attacks:** This kind of attack includes the Viruses, Spywares, Trojan horses, and Worms, can attack both OS and user application.

These Malicious hackers regularly use the snooping mechanism to monitor key strokes, hack the passwords and login information and for interception of e-mail and other data transmissions.

The corporations snoop on employees logically to have eye business computers usage and to have vigilance over internet usage. Even though governments also snoop on every citizen to collect information to control crime and terrorism. Although the snooping has a negative aspect in computer technology snooping is a program/utility which performs a monitoring functions.

3 Existing Security Solution Approaches for MANETs

This section discusses the existing security solutions which are implemented for MANETs and are represented in Fig. 2.

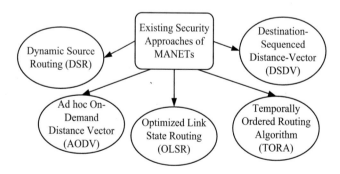

Fig. 2. Security approaches for MANET

3.1 Dynamic Source Routing (DSR)

It is a reactive protocol which doesn't utilize infrequent advancements. It figures the routes when vital and after that cares for them. In Source routing method the group of sender chooses the entire arrangement of node via which the bundle needs to pass; the sender explicitly records this course in the bundle's header, perceiving each sending "hop" by the location of the accompanying node to which to transmit the packet to destination host. The DSR exhibit two working stages like Route Discovery and Route Maintenance. A host beginning a course disclosure telecasts a course request distribute may be gotten by those hosts within wireless transmission scope of it. The course requests bundle perceives the host, suggested as the objective of the course disclosure, for which the course is inquired. If the course disclosure is productive the beginning host gets a course answer distribute an arrangement of framework jumps through which it may accomplish the objective [14, 15].

3.2 Routing Algorithm of a DSDV- Destination-Sequenced Distance-Vector

Routing Algorithm relies upon the customary Bellman-Ford Routing Algorithm with specific changes. Each flexible station keeps up a steering table those rundowns every single accessible destination, the amount of jumps to go to the destination and the arrangement number doled out by the destination center point. The arrangement number is utilized to perceive unattained job from new ones and in this way avoid the advancement of circles. The stations irregularly transmit their guiding tables to their incite neighbors. A station moreover transmits its coordinating table if a massive variation has been observed in the tabular form from the final update forwarded. This tends to overhaul in both cases of time-driven and event-driven scenarios. The steering table that is redesigned can be transferred in two manners: a "full dump (sends the full guiding table to the neighbors and could navigate various packets)" or an "incremental update (upgrade only those segments from the coordinating table are sent that has a metric change)" ensuing to the last overhaul and it must fit in a packet [16].

3.3 Ad Hoc On-Demand Distance Vector (AODV)

This offers low framework utilizes and utilizes destination arrangement number to ensure loop opportunity. It is a reactive protocol proposing that it requests a course when required and it doesn't maintain routes for those nodes that don't viably share in a correspondence. An AODV component is the one which utilizes a destination grouping number, which compares to a destination node requested by a coordinating sender node. The destination itself furnishes the number close by the route it needs to take to reach from the solicitation sender node up to the destination. If there are different jobs from a solicitation sender to a destination, the sender carries the route with a higher grouping number. This ensures the impromptu system protocol remains loop-free [15, 17].

3.4   Optimized Link State Routing (OLSR)

Protocol is a proactive routing protocol where the routes are persistently quickly accessible when required. OLSR is a headway type of an unadulterated connection state protocol in which the topological changes make the flooding of the topological information each and every accessible host in the framework. OLSR may streamline the reactivity to topological changes by decreasing the best time between times for intermittent control message transmission. Furthermore, as OLSR incessantly takes care of routes to all destinations in the framework, the protocol is useful for movement outlines where a broad subset of nodes are comparing with another considerable subset of nodes, and in which [source, destination] sets will be changing later. OLSR protocol is proper for the application which does not allow high delay in data transmission. The best working environment for OLSR tradition protocol is a thick framework, where the most correspondence is focused between considerable amounts of nodes.

OLSR diminish the control overhead driving the MPR to spread the updates of the association state. Furthermore, the viability is grabbed diverged from customary association state tradition when they picked Multi Point Relay (MPR) set is as smaller as possible [15, 18].

3.5   Temporally Ordered Routing Algorithm (TORA)

It is an extremely flexible, competent and adaptable passed on routing algorithm considering the possibility of association reversal. A vital component of TORA is that control messages are restricted to a little arrangement of nodes near the occasion of a topological change. The tradition has three key limits: Route creation, Route upkeep, and Route cancellation. Route creation in TORA is made using query packet (QRY) and User Datagram Protocol (UDP) parcels. The route creation algorithm starts by setting the stature of destination to 0 and for every other node to NULL. The source telecasts a QRY distribute the destination node's id in it. A center with a non-NULL stature responds with a UDP packet that has its height in it. A node getting a UDP packet sets its height is seen as upstream and a node with bringing down stature downstream. Along these lines, a coordinated acrylic diagram is developed from source to the destination. The subsequent improvement of course on TORA is done by trading demand from source and getting an answer from destination [19].

4   Review of Literature

This section discusses the various recent researches meant for security aspects in MANETs. Research focusing on MANETs with Cognitive Radios is found in Yu et al. [20]. This research was developed based on recent advancement in the Consensus algorithm which is inspired by self –organizing behavior of bees, ants, fish, etc. The effectiveness of the method was evaluated by simulation results and found effective against Spectrum Sensing Data Falsification (SSDF) attack.

To overcome the dynamic shortest path routing issues in MANETs, author Yang et al. [21] has presented the system which implements the genetic algorithms with immigrants and memory schemes. Li and Liu [22] addressed the security method for MANET by using cluster-based key management. The dynamic nature of network causing security issues in communication is considered which uses threshold cryptography and ID-based secret key management scheme which eliminates the secure authentication. It provides the scalable, economic, autonomous key management mechanism. Li and Wu [23] have proposed both proactive and reactive mechanisms for trust convergence and minimization of uncertainty in the context of MANET for handling tradeoff between delay, uncertainty and cost. Bu et al. [24] presented the security authentication in MANETs by using prevention based mechanism and intrusion detection (ID) mechanisms to overcome the issues associated with large networks. Defrawy et al. [25] discussed the routing protocol against both outsider and insider adversaries of privacy attack in MANET. Liu et al. [26] expressed ID and authentication mechanism for continuous authentication in MANETs. Dhurandher et al. [27] presented a friend-based ad-hoc routing system which gives a list of trusted nodes to the source node through which data transmission takes place. Cho et al. [28] presented a survey research on trust management for MANET which can be implemented for cognitive and social networks. Djahel et al. [29] addressed the packet dropping issues in MANETs and suggested for in-depth depth solution to tackle the issue. Xu et al. [30] discussed the policy enforcing mechanism to tackle the issues of enforcing policies and offered trusted platform module. Chan [31] has discussed the distributed private key generation for Identity Based Cryptography (IBC) in MANET and given a protocol which distributes the identities among all the users without using any centralized server. Venkataraman et al. [32] considered the critical factors which affect the performance of the network in the presence of the malicious nodes and introduced a regression-based trust model over "AODV" and "OLSR" protocol in MANET. Ahn and Wu [33] considered risk aware mitigation problem and presented risk-aware response mechanism that handles attacks of routing and also extended research with Dempster-Shafer mathematical theory for performance evaluation. Wei et al. [34] presented a unified trust management for security enhancement in MANETs which found effective with throughput and packet delivery ratio (PDR). Combined work of Vanamala and Raghavendra [35] expressed the strategies for secure routing mechanism in OLSR protocol for MANET and provided effective tradeoff among delay, processing time and throughput.

5   Research Gap

From the research survey performed over recent MANET security mechanism by different author found followings:

- Lack of research towards security solution: Most of the researches are focused towards survey work. Also found that very rare works in the design works meant for security assurance.
- Lack of effectiveness: The researches expressed in the recent past are presented to provide the effective outcomes towards security.
- No real-time researches: The presented researches in recent past are not considered real time scenario. All the researches are performed on theoretical assumptions.

- No benchmarking: There is no such work which can be considered as benchmarking. All the research techniques have implemented and presented different security approaches with different results.

Thus the above-discussed points are needed to be resolved to get the effectiveness in the current MANET security mechanisms.

6   Comparative Analysis Among Different Routing Protocols

This section briefly summarizes about five different routing protocols such as DSR, DSDV, AODV, OLSR, and TORA. The following Table 1 shows the performance comparison of routing protocols. From the case study, can conclude the an 'AODV' has better performance by comparing with other routing protocols in terms of data rates without attacks as well as packet delivery ratio. Nevertheless, it has been notices that 'AODV' protocol is more vulnerable to defeat the attacks by the comparison of other protocols. While, 'DSR' also perform better but only this is it has less data rate and low mobility. From the attack scenario, 'OLSR' is the most prominent protocol which greatly performs compared with other routing protocols. Moreover, 'DSDV' performs quit less comparatively TORA and other protocols. Also it performs very low under the high node-mobility and high data-rate.

Table 1. Performance comparison of routing protocols

Routing protocols	Security attacks	Packet delivery ratio	Average of end to end delay	Normalized routing load
DSR	Without	High	High	High
	With	Low	High	Average
DSDV	Without	Average	Average	Average
	With	Low	Average	Average
AODV	Without	High	High	Average
	With	Low	Average	Average
OLSR	Without	High	Average	High
	With	Average	Average	High
TORA	Without	Average	Average	Average
	With	Low	Average	Low

7   Conclusion

This paper discussed about survey study on security aspects of MANETs. In this study, have studied on different security attacks faced by MANET also discussed the solution strategies for security concerns. This paper provides an extensive case study over the existing researches in MANET security problems by considering current researches. Additionally, the work overviewed on the research gap in current MANET security concerns. Finally, have presented the comparative analysis report of different routing protocols which shows the performance rate of the routing protocol. The survey study will be of help full for further research work which is addressed in following section.

8   Future Line of Research

This research survey can be used for future research to tackle the problems of detour attack (lethal attack) in MANETs mainly considering for OLSR protocol. The future research can be obtained by following steps:

- To propose a security model that can potentially understand the misbehaved node behavior in OLSR protocol.
- To design a framework to study the misbehaved nodes behavior in MANET.
- To achieve benchmarking of the proposed schema by comparing the accomplished results with other techniques of mitigating attacks in OLSR.

References

1. Gantes, A., Stucky, J.: A platform on a mobile ad hoc network challenging collaborative gaming. In: International Symposium on Collaborative Technologies and Systems (2008)
2. Khan, K.U.R., Zaman, R.U., Reddy, A.V.G.: Integrating mobile ad hoc networks and the internet challenges and a review of strategics. In: Presented at the 3rd International Conference on Communication Systems Software and Middleware and Workshops, COMSWARE (2008)
3. Suguna, M., Subathra, P.: Establishment of stable certificate chains for authentication in mobile ad hoc networks. In: Presented at the International Conference on Recent Trends in Information Technology (ICRTIT) (2011)
4. Nishiyama, H., Ngo, T., Ansari, N., Kato, N.: On minimizing the impact of mobility on topology control in mobile ad hoc networks. IEEE Trans. Wirel. Commun. $11(3)$, 1158–1166 (2012)
5. Rango, F.D., Fotino, M., Marano, S.: EE-OLSR: energy efficient OLSR routing protocol for mobile ad-hoc networks. In: Presented at the Military Communications Conference, MILCOM (2008)
6. Du, K., Yang, Y.: Policy-based time slot assignment algorithm in a MANET (PBTSA). In: Presented at the 3rd International Conference on Anti-counterfeiting, Security, and Identification in Communication, ASID (2009)
7. Sheikh, R., Chande, M.S., Mishra, D.K.: Security issues in MANET: a review. In: Presented at the Seventh International Conference on Wireless and Optical Communications Networks (WOCN) (2010)
8. Deng, H., Li, W., Agrawal, D.P.: Routing security in wireless ad hoc networks. IEEE Commun. Mag. 10, 70–75 (2002)
9. Papadimitrates, P., Hass, Z.J.: Secure routing for mobile ad hoc networks. In: Proceeding of SCS Communication Networks and Distributed System Modeling and Simulation Conference (CNDS), San Antonio, TX, (2002)
10. Lou, W., Fang, Y.: A survey of wireless security in mobile ad hoc networks: challenges and available solutions. In: Chen, X., Huang, X., Du, D. (eds.) Ad Hoc Wireless Networks, pp. 319–364. Kluwer Academic Publishers, Dordrecht (2003)

11. Yi, S., Kravets, R.: Composite key management for ad hoc networks. In: Proceedings of the 1st Annual International Conference on Mobile and Ubiquitous Systems: Networking and Services (MobiQuitous 2004), pp. 52–61 (2004)
12. Zhou, L., Haas, Z.: Securing ad hoc networks. IEEE Netw. Mag. **13**(6), 24–30 (1999)
13. Yi, S., Naldurg, P., Kravets, R.: Security aware ad hoc routing for wireless networks. In: Report No. UIUCDCS-R-2290, UIUC (2002)
14. Johnson, D.B., Maltz, A., Broch, J.: DSR: The Dynamic Source Routing Protocol for Multi-Hop Wireless Ad Hoc Networks, vol. 5, pp. 139–172. Addison-Wesley, Boston (2001)
15. Gupta, A.K., Sadawarti, H., Verma, A.K.: Performance analysis of AODV, DSR & TORA routing protocols. Int. J. Eng. Technol. **2**(2), 226 (2010)
16. He, G.: Destination-sequenced distance vector (DSDV) protocol, pp. 1–9. Networking Laboratory, University of Technology, Helsinki (2002)
17. Perkins, C., B-Royer, E., Das, S.: Ad hoc on-demand distance vector (AODV) routing. No. RFC 3561 (2003)
18. Weiss, E., Hiertz, G.R., Xu, B.: Performance analysis of temporally ordered routing algorithm based on IEEE 802.11 a. In: IEEE 61st Vehicular Technology Conference, VTC-Spring, vol. 4. IEEE (2005)
19. Park, V.D., Macker, J.P., Corson, M.S.: Applicability of the temporally-ordered routing algorithm for use in mobile tactical networks. In: 1998 Proceedings of Military Communications Conference, MILCOM 98, vol. 2. IEEE (1998)
20. Yu, F., Huang, M., Tang, H.: Biologically inspired consensus-based spectrum sensing in mobile ad hoc networks with cognitive radios. IEEE Netw. **24**(3), 26–30 (2010)
21. Yang, S., Cheng, H., Wang, F.: Genetic algorithms with immigrants and memory schemes for dynamic shortest path routing problems in mobile ad hoc networks. IEEE Trans. Syst. Man Cybern. Part C (Appl. Rev.) **40**(1), 52–63 (2010)
22. Li, L.-C., Liu, R.-S.: Securing cluster-based ad hoc networks with distributed authorities. IEEE Trans. Wirel. Commun. **9**(10), 3072–3081 (2010)
23. Li, F., Wu, J.: Uncertainty modeling and reduction in MANETs. IEEE Trans. Mob. Comput. **9**(7), 1035–1048 (2010)
24. Bu, S.: Structural results for combined continuous user authentication and intrusion detection in high security mobile ad-hoc networks. IEEE Trans. Wirel. Commun. **10**(9), 3064–3073 (2011)
25. E-Defrawy, K., Tsudik, G.: Privacy-preserving location-based on-demand routing in MANETs. IEEE J. Select. Areas Commun. **29**(10), 1926–1934 (2011)
26. Bu, S.: Distributed combined authentication and intrusion detection with data fusion in high-security mobile ad hoc networks. IEEE Trans. Veh. Technol. **60**(3), 1025–1036 (2011)
27. Dhurandher, S.K.: FACES: friend-based ad-hoc routing using challenges to establish security in magnets systems. IEEE Syst. J. **5**(2), 176–188 (2011)
28. Cho, J.-H., Swami, A., Chen, R.: A survey on trust management for mobile ad hoc networks. IEEE Commun. Surv. Tutor. **13**(4), 562–583 (2011)
29. Djahel, S., N-Abdesselam, F., Zhang, Z.: Mitigating packet dropping problem in mobile ad hoc networks: proposals and challenges. IEEE Commun. Surv. Tutor. **13**(4), 658–672 (2011)
30. Xu, G., Borcea, C., Iftode, L.: A policy enforcing mechanism for trusted ad hoc networks. IEEE Trans. Dependable Secure Comput. **8**(3), 321–336 (2011)
31. Chan, A.C.F.: Distributed private key generation for identity-based cryptosystems in ad hoc networks. IEEE Wirel. Commun. Lett. **1**(1), 46–48 (2012)
32. Venkataraman, R., Pushpalatha, M., Rao, T.R.: Regression-based trust model for mobile ad hoc networks. IET Inf. Secur. **6**(3), 131–140 (2012)

33. Zhao, Z.: Risk-aware mitigation for MANET routing attacks. IEEE Trans. Dependable Secure Comput. **9**(2), 250–260 (2012)
34. Wei, Z.: Security enhancements for mobile ad hoc networks with trust management using uncertain reasoning. IEEE Trans. Veh. Technol. **63**(9), 4647–4658 (2014)
35. Vanamala, C.K., Rao, G.R.: Strategic modeling of secure routing decision in OLSR protocol in mobile ad-hoc network. In: Software Engineering Trends and Techniques in Intelligent Systems, pp 201–208 (2017)

DSMANET: Defensive Strategy of Routing Using Game Theory Approach for Mobile Adhoc Network

K. Pradeep Kumar[1]([⊠]) and B. R. Prasad Babu[2]

[1] JNTUK, Kakinada, India
pradeepkumarkrishnappa@gmail.com
[2] Department of CSE, RRCE, Bengaluru, India

Abstract. The Ad hoc technology enables a unique network namely mobile Ad hoc Network (MANET), where the establishment of the routing is a challenging task because of various characteristics such as mobility of the node and resource constraints of individual nodes. The self-configurability of the nodes makes the network vulnerable to initiate different forms of attacks in a very easy way. The malicious activities by the nodes are identified by its behavior of deceiving the other nodes data using participating into routing then dropping the packets, at the same time similar activities of packet drop could happen by non-malicious nodes due to some fault in the nodes. It is a challenging task to distinguish these two categories of nodes. Most of the current approaches focus on the trust building by the previous behavior of the nodes and black listing it, this approach leads to minimize the size of the network and sometimes permanent clustered network partition which leads to network failure. This paper introduces a novel strategy of defensive approach by the regular nodes as opposed to the malicious node and the method is named as "Defensive Strategy in MANET (DSMA-NET)." The performance parameters like throughput and routing overhead is compared with one of state of the art secure routing method DSMANET and found exhibiting better performance.

Keywords: MANET · Malicious nodes · Packet drop · Routing

1 Introduction

The wireless network established with ad hoc technology, where the deployed nodes exhibit unique characteristics of self-configurability and data packet communication takes place nodes to nodes without having any fixed infrastructure is named as mobile ad hoc network (MANET) where the nodes had mobility too [1]. The application of MANET exists from simple rescue application in both civil and military to Internet of Things (IoT) [2]. The mission-critical applications require a better Quality of service (QoS) in their routing protocol. To standardize IP routing for MANET, internet engineering task force (IETF) has setup a MANET working group. Many requests for comment (RFCs) has been released for a proactive and reactive protocol like DSR,

© Springer International Publishing AG, part of Springer Nature 2019
R. Silhavy (Ed.): CSOC 2018, AISC 765, pp. 311–320, 2019.
https://doi.org/10.1007/978-3-319-91192-2_31

DSDV, AODV, ZRP, OLSR, and TORA, etc.; All these routing protocols focus on finding the path between source and destination with consumption minimum resources like energy and bandwidth to get optimal QoS. It is observed that the MANET working group lacks the consideration of security aspects in their routing protocol. The security vulnerability arises in MANET because of its characteristics of self-configurability which leads the nodes with mischievous intension to be part of the network. The conventional study on the security aspect in MANET provides specific nomenclature of the nodes such as the node behaves correctly: Regular node and the node which behaves maliciously as a malicious node. The tag of malicious comes once there is an evidence of dropping the packets while routing the data, whereas it is very much possible that some regular nodes may drop the packets due to system failure. Thus distinguishing the regular node and malicious node becomes a challenging task. Therefore, the proposed study formulated a novel routing mechanism which applies defensive strategy to handle the problems associated with detection of malicious nodes as mentioned above. The proposed strategy has been conceptualized by a game theory approach which helps the entire system to track down the suspicious activities being exercised by an intruder for packet dropping within a specified route. The performance efficiency of the proposed algorithm thereby approved and validated in a numerical computing tool from different aspects. The outcomes of the study further compared with the conventional secure routing mechanisms for MANET and found superior regarding different performance parameters such as throughput, routing overhead, routing latency, etc. The overall pattern of this research manuscript is planned as follow, Sect. 2 deals with exhibiting the current research trends adopted into the secure routing of MANET. On the other hand Sect. 3 will discuss the conceptual design methodology of the proposed DSMANET routing strategy from an analytical view point followed by an illustration of experimental outcomes obtained by simulating proposed algorithm in Sect. 4, where a comparative performance analysis exhibited the effectiveness of the proposed defense routing strategy in MANET. Finally, Sect. 5 concludes the overall study.

2 Review of Literature of Related Work

The very basic effect of malicious nodes is packet dropping while data packet routing, the details of the challenges of reducing the data packet drop is discussed by Djahel et al. [3]. One of the work by Casado et al. [4] have proposed a method for detecting malicious nodes by cross-layer approach for a specific application in MANET and further, a generic framework is proposed by Song et al. [5]. These approaches may be incorporated for brining agility into the routing but cannot facilitate as a pure security paradigm. There exist much literature of security models for specific attacks such as selective forwarding attacks [6], BlackHole attack [7, 8, 9], Jelly fish attack [10], etc. The study of Arepalli et al. [11] mostly emphasized on the operational issues associated with secure multicast communication and collaboration among nodes deployed in a MANET. Although the design constraints make the multicast routing protocol PUMA very challenging still, the authors claimed to find a scope of enhancement to integrate it to a certain extent with an ability to defend Man in the Middle Attack (MIMA). Therefore, study further presents an analytical design namely Elliptic Curve Group Diffie-Hellman (ECGDH) mechanism to defend the attack aforementioned. The experimental outcomes exhibited that the ECGDH attains better simulation outcomes even in the presence of MITM attack. A study in the similar direction has been performed by Krishnappa and Prasad [12], the same author of this manuscript where they have extensively studied the potential features of Particle Swarm Optimization (PSO) towards mitigating the conventional security loopholes present in MANET. Further, the study extensively elaborated the discussion on different trust management, authentication and attack models adopted in some of the recent studies done in the dame area of research. Finally, the study found extracting open research problems followed by an identification of review.

3   Conceptual Design Methodology

The proposed system model consists of (1) Node Positioning (2) Node Strategy mechanism (3) secure routing algorithm. This segment of the study introduces three different scenarios which are created by incorporating a numerical simulation, and it includes deployment of a set of mobile nodes within a specific region. The detail numerical interpretation associated with node localization or positioning is as follows:

3.1   Node Positioning

The study performed numerical computation considering three different scenarios where the deployment of nodes considered being some 25, 50 and 100. The proposed system further also initializes and defines the number of the source node (SN) and the destination node (DN) followed by the active participation of malicious nodes (AN). The incorporation of node deployment is subjected to the establishment of connectivity among themselves (Table 1).

Table 1. Description of symbols used

Sl. No	Symbol	Description
1	A	Area Component
2	N	Total Number of Nodes
3	ρ	% of malicious nodes $0 \leq \rho \leq 50$
4	D	Area offset
5	N_{mal}	Number of Malicious nodes

The modeling of node positing depends upon three independent variables such as Area Component (A), Total Number of Nodes (N), and % of malicious nodes (ρ). The offset for the visibility is set to 'D.' The initial computation of malicious node (N_{mal}) is done by Eq. (1)

$$N_{mal} = f_{round}[(N \times \rho)/100] \tag{1}$$

And the regular node is computed by the Eq. (2)

$$N_{reg} = N - N_{mal} \tag{2}$$

The algorithm intended to perform node localization represented as follows (Table 2):

Table 2. Description of symbol used in following mathematical algorithm

ρ_A	Simulation area components
n	Number of nodes
n_M	Number of malicious nodes
A	Simulation area
$B(A)$	Boundary of area
G_n	Number of genuine nodes
α_R	Regular node
α_m	Malicious node

Algorithm 1. Node Positioning

Input: ρ_A, n, n_M
Output: Node positioning
START
1. **Init** $\rho_A, n, n_M, B(A)$
2. Compute the number of attacker nodes

$$n_M = f_{round}(\frac{n \times n_M}{100}) \dots eq.\ 5.1$$

3. Compute the number of genuine nodes
$$G_n = (n - n_M)$$
4. Assign $0 \rightarrow \alpha_R \wedge 1 \rightarrow \alpha_m$
5. Generate random x and y coordinate under the boundary value
$$L_x = B(A) + (A - 2 \times B(A)) \times f_{round}(n,1)$$
$$L_y = B(A) + (A - 2 \times B(A)) \times f_{round}(n,1)$$
6. Perform localization of
$$\rho_A, n, n_M$$
END

The above algorithm integrated with the proposed system to perform node positioning before the node strategy mechanism. A case study has been evaluated using exhibiting different node localization patterns concerning different simulation parameters as follows (Table 3):-

Table 3. A case study for node localization

The above-mentioned case study implies that there exist 5 attacker nodes deployed in a specific region by an intruder.

3.2 Node Strategy Mechanism

The further segments discuss and give insight into different aspects of node strategy mechanism followed by both genuine and attacker nodes. The study also considered different scenarios with a combination of different node deployment strategic mechanisms. Which are thereby *(1) Protective, (2) Planned, (3) Steady*. The study also performed another 2 different case studies for static or dynamic node movements concerning different aspects. The description of a node includes different attributes and maintains a record of source node ID, Destination node ID along with Transmission Range T_R of that respective node. The proposed study preformed evaluation of three different routing algorithms where each algorithm defined with their respective strategy to perform defensive routing against any vulnerable attacks. The study evaluated each algorithm along with our proposed defensive routing based strategy to effectively recognize the attacker node which intentionally drops the forwarded data packets during the communication scenario. The simulation outcomes also exhibit the routing table maintained by each algorithm during the route discovery and connectivity establishment process. The following Fig. 1 shows the planned system model of the proposed system where three different MANET routing algorithms are simulated, and their performance efficiency has been tracked down by underlying node movements. The node strategy mechanisms adopted by both normal and malicious nodes are thereby (1) protective (2) planned and (3) steady. Along with this the system also configures each node by specifying its communication range. The proposed algorithm is well capable of detecting the malicious node which can be a part of the established route for communication between a source node and destination node. While detecting suspicious node ID within a specific route, the system notifies its admin about the malicious activity and immediately terminates the communication at the node where packets are dropped. The routing strategy also adopted Dijkstra's algorithm to formulate optimal route for commuting between the source and the destination node. During the route establishment and communication scenario every time the routing table updated with the new route and its updated values concerning the nodes participating in the communication.

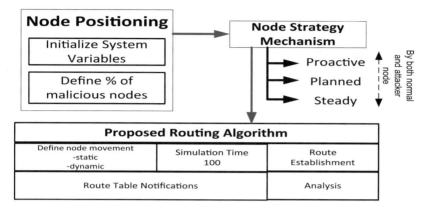

Fig. 1. Proposed DSMANET system

The system evaluated the proposed algorithm along with two other different algorithms with respect to the different round of simulation. The proposed algorithm evaluated for 50 iterations and found to achieve considerable throughput and very minimum routing overhead. The system also computed the probability malicious nodes by simulating the following Eq. 3.

$$\theta = \frac{\beta}{\alpha + \beta} \tag{3}$$

In the above algorithm α, β denotes the networking co-efficient assigned with some initial values. The probability of a node being malicious has been initially considered as 0.5 during the experimental prototyping. The proposed algorithm also computes the degree of uncertainty factor concerning the following Eq. 4

$$D_{u} = \frac{12 \times \alpha \times \beta}{(\alpha \times \beta)^{2 \times (\alpha + \beta + 1)}} \tag{4}$$

The following algorithm exhibits the design methodology of the proposed routing schema from an analytical view point.

Algorithm of Proposed DSMANET

Input: $\rho_{A, n}, n_M$

Output: Detection of malicious node (K_{node})

START

1. Init ρ_A, n, n_M
2. Init θ , Gain of attack (G_a) , cost of attack (C_a) , gain of cooperation (G_c), cost of cooperation (C_c)
3. Init threshold for regular node, variance for false alarm
4. Init α , β values with 1 for all the regular and other nodes

      FOR i$\leftarrow$1: n

        i. **Compute** α , β

        ii. **Compute probability of a node being malicious**

$$\theta = \frac{\beta}{\alpha + \beta}$$

        iii. **Compute Uncertainty**

$$D_u = \frac{12 \times \alpha \times \beta}{(\alpha \times \beta)^{2 \times (\alpha + \beta + 1)}} \quad \ldots \text{ Eq. 4}$$

        iv. Belief and disbelief

      END

5. Deploy node strategy mechanism for both regular and malicious node
6. Define communication range of each node C_r
7. Process inputs for route establishment
8. Show current position of each node
9. Calculate the possible communication links
10. Obtain the connection matrix
11. Create connection matrix
12. Find the optimal path considering Dijkstra's algorithm
13. Plot network topology
14. Perform communication between the source and destination
15. Obtain the status value of transmitter (Tx) node
16. Compute probability of cooperation
17. Obtain the statues value of receiver node
18. Update routing table with node ID, action, and type
19. Detect the malicious node
20. Perform analysis

END

The next segment of the study will discuss the extensive simulation outcomes obtained after simulating the proposed algorithm in a numerical computing platform.

4 Experimental Outcomes

This section exclusively discusses the numerical outcomes obtained after simulating the proposed DSMANET system in a computing environment. The experimental proto-typing in this regards exclusively carried out to evaluate the comparative analysis which ensures the effectiveness of the proposed system as compared to the conventional baselines. The study outcomes thereby obtained that the proposed algorithm achieves higher network throughput by transmitting as much as secure data packets to the destination node by its defensive strategy. The dynamic simulation outcomes evaluated for three different routing mechanism exhibits the performance efficiency of the proposed DSMANET regarding high throughput which is represented by the following Fig. 2.

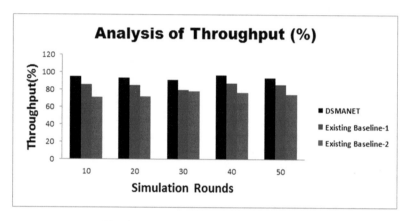

Fig. 2. Comparative performance analyses

The above figure depicted that the proposed algorithm DSMANET achieved superior outcomes regarding throughput as compared to the conventional baselines. It also achieved a higher percentage of normal node gain by defending the attacker nodes in every possible scenario with alternative route establishment. The outcomes obtained also show that the proposed system exhibits negligible computational complexity in detecting the attacker node becoming a part of the communication link which exhibits its effectiveness.

5 Conclusion

The study has mostly emphasized on addressing the routing problems associated with MANET which involves both static and dynamic based topology formation. To overcome the challenges associated with secure data packet transmission, the study conceptualized an analytical methodology which employs a defensive routing mechanism to establish secure communication in a MANET topology. The experimental outcomes further conveyed the effectiveness of the proposed study regarding throughput and computational efficiency.

References

1. Loo, J., Lloret, J., Ortiz, J.H.: Mobile Ad Hoc Networks: Current Status and Future Trends. CRC Press, New york (2011)
2. Bellavista, P., Cardone, G., Corradi, A., Foschini, L.: Convergence of MANET and WSN in IoT Urban Scenarios. IEEE Sens. J. **13**(10), 3558–3567 (2013)
3. Djahel, S., Nait-Abdesselam, F., Zhang, Z.: Mitigating Packet Dropping Problem in Mobile Ad Hoc Networks: Proposals and Challenges. IEEE Commun. Surv. Tutor. **13**(4), 658–672 (2011)
4. S'nchez-Casado, L., Maci'-Fern'ndez, G., García-Teodoro, P.: An efficient cross-layer approach for malicious packet dropping detection in MANETs. In: IEEE 11th International Conference on Trust, Security and Privacy in Computing and Communications, Liverpool, pp. 231–238 (2012)
5. Song, R., Tang, H., Mason, P.C., Wei, Z.: Cross-layer security management framework for mobile tactical networks. In: MILCOM 2013 - 2013 IEEE Military Communications Conference, San Diego, CA, pp. 220–225 (2013)
6. Ren, J., Zhang, Y., Zhang, K., Shen, X.: Adaptive and channel-aware detection of selective forwarding attacks in wireless sensor networks. IEEE Trans. Wirel. Commun. **15**(5), 3718–3731 (2016)
7. Sherif, A., Elsabrouty, M., Shoukry, A.: A novel taxonomy of black-hole attack detection techniques in mobile Ad-hoc network (MANET). In: IEEE 16th International Conference on Computational Science and Engineering, Sydney, NSW, pp. 346–352 (2013)
8. Singh, H.P., Singh, R.: A mechanism for discovery and prevention of cooperative black hole attack in mobile ad hoc network using AODV protocol. In: International Conference on Electronics and Communication Systems (ICECS), Coimbatore, pp. 1–8 (2014)
9. Araghi, T.K., Zamani, M., Manaf, A.B.A., Abdullah, S.M., Bojnord, H.S., Araghi, S.K.: A secure model for prevention of black hole attack in wireless mobile ad hoc networks. In: Proceedings of the 12th WSEAS International Conference on Applied Computer and Applied Computational Science (2013)
10. Laxmi, V., Lal, C., Gaur, M.S., Mehta, D.: JellyFish attack: analysis, detection and countermeasure in TCP-based MANET. J. Inf. Secur. Appl. **22**, 99–112 (2015). ISSN 2214-2126
11. Arepalli, G., Erukula, S.B.: Secure multicast routing protocol in MANETs using efficient ECGDH algorithm. Int. J. Electr. Comput. Eng. (IJECE) **6**(4), 1857 (2015)
12. Pradeep Kumar, K., Prasad Babu, B.R.: Investigating open issues in swarm intelligence for mitigating security threats in MANET. Int. J. Electr. Comput. Eng. (IJECE), 5(5) (2017)

OCSLM: Optimized Clustering with Statistical Based Local Model to Leverage Distributed Mining in Grid Architecture

M. Shahina Parveen[1(✉)] and G. Narsimha[2]

[1] Department of Computer Science and Engineering, JNTU, Hyderabad, India
shahina1phd.jntu@gmail.com
[2] Professor of CSE and HOD, JNTUH-CES, Sulthanpur, Sanga Reddy (D) 502 273,
Telangana, India

Abstract. Grid computing offers significant platform of technologies where complete computational potential of resources could be harnessed in order to solve a complex problem. However, applying mining approach over distributed grid is still an open-end problem. After reviewing the existing system, it is found that existing approaches doesn't emphasized on data diversity, data ambiguity, data dynamicity, etc. which leads to inapplicability of mining techniques on distributed data in grid. Hence, the proposed system introduces Optimized Clustering with Statistical Based local Model (OCSLM) in order to address this problem. A simple and yet cost effective machine-learning based optimization principle is presented which offers the capability to minimize the errors in mined data and finally leads to accumulation of superior quality of mined data. The study outcome was found to offer better sustainability with optimal computational performance when compared to existing clustering algorithms on distributed networking system.

Keywords: Grid computing · Grid network · Clustering · Data mining
Machine learning

1 Introduction

Grid computing is one of the techniques in networking technology that could offer proper utilization of the resources involved in the network [1]. The prime logic about using grid computing is to save significant amount of resource as well as harness the computational potential of every latent component. Some of the demanding characteristics about grid computing is that it offers secure environment where multiple forms of resources can be shared among the dynamic user with higher degree of flexibility in order to solve any complex problems [2]. Another potential reason of increasing adoption of grid computing is its capability to offer virtualized environment for channel capacity, storage, processing, etc. [3]. All the nodes connected in grid network are featured with unique operating platform, discrete architecture, version etc. [4]. However, with changing times, there are certain complexities and challenges that are exponentially increases over grid computing e.g. (i) lack of any effective standard (ii) incorporating distributive networking capabilities over grid, (iii) absence of any dedicated software framework to control grid, (iv) resource sharing, etc. [5]. At present, there has been various studies carried out in addressing the problems associated with grid computing e.g. [6, 7, 8]; however, there are certain problems that are quite unspoken and unaddressed. We discuss the problems associated with applying conventional mining algorithms over grid network. The first unaddressed problem is *complexity of data*. Owing to different users on different side of grids, there will be massive generation of data diversity. In presence of higher diversities of data, it is near to impossible to store the data in proper storage system. While if the data are not stored efficiently, it is imperative that mining algorithms cannot be applied on the top of it. Moreover, if a grid architecture is repositing a real-time data than it is not feasible to ascertain the quality of data. It is because massive number of incoming data with unpredictable level of data diversity will not allow the decision making component to forward correctly to the grid clusters. This will not only result in unacceptable data and resource allocation but also will lead to degraded quality of services that will ultimately minimize user experience. The second unaddressed problem is *adoption of stereotypes cases*. A closer look into majority of existing system will show that they are inclined towards either smart grid or power management and deals with applying mining algorithms in it. Moreover, framework to address such case studies is not reusable for solving other problems and hence there is a big narrowness in implementing data mining algorithms over other case studies of grid computing. The third unaddressed problem is *less availability of optimization scheme*. Inspite of presence of various optimization algorithms in distributed network [9], the area of grid computing has been less explored using optimization. These entire problems sum up to offer inference that existing grid structure is not ready to service the distributed computing of data and hence offers an increasing research scope. Although, there are existence of various other cloud-based techniques, but adoption of grid based technique always offers full exploitation of resources. Therefore, an effective distributed mining mechanism is required to be constructed in order to develop a framework or application which runs over upcoming applications over grid computing architecture. This paper has presented one such solution where the core idea is to address all the major problems that are found to be unaddressed in existing literatures in order to fill the research gap. The idea is also to apply a simple and yet cost effective optimization principle for mining algorithm on grid network. The organization of the paper is- Related approaches are briefly outlined in Sect. 2 while its associated problems to be addressed are discussed in Sect. 3. The brief outline of proposed solution is discussed in Sect. 4 accompanied by all the four algorithms to address four discrete problems in data warehousing in terms of data quality issue. Section 5 discusses about the result obtained from algorithm implementation followed by conclusion in Sect. 6.

1.1 Background

Our prior review work has investigated various forms of the grid architectures [10] while this section further enriches with more updates on some recent implementation of mining approaches found over grid environment. The recent work discussed by Zhang et al. [11] has discussed about trends of applying analytics for energy-based application in order to investigate its consumption. Asad and Chaudhry [12] have discussed about the emergence of big data approach on grid environment aiming for energy efficiency. Ergin et al. [13] have addressed the problem associated with retrieving of positional information from grid cells. Study of grid computing and analytics considering healthcare sector has been carried out by Goyal et al. [14] with similar aim of energy assessment. Hochbaum and Baumann [15] have presented a machine learning approach for analytica operation on data over grid environment. Jindal et al. [16] has also presented similar machine learning based approach for identifying specific events over smart grid. Lee et al. [17] have used developed a platform for analyzing grid using large scale data with similar aim of energy-based analysis over data centers. Aouad et al. [18] have discussed about grid-based techniques and its possible applicability on knowledge discovery process. Similar approach toward energy management study over grid has also been carried out by Giri [19]. Kar et al. [20] have constructed a mining framework over distributed generation of energy with a capability of identifying faults in the network. Kumar and Jumnal [21] have investigated about usage of sharing scheme provided by grid in educational institutions. Loia et al. [22] has addressed the problems associated with data uncertainty over smart grids while Luan et al. [23] have addressed the problems associated with navigational system by developing an analytical tool for smart metering system. Xu et al. [24] have used similar mining approach for predicting weather-based information using neural network. Galan et al. [25] have used fuzzy logic integrated with bio-inspired algorithm for performing knowledge discovery from grid environment. The work carried out by Chung et al. [26] have performed analysis of effective resource required for exploring better services over grid network. Dan et al. [27] have studied different forms of libraries in grid environment that utilizes semantics. Adoption of mining was also seen in work of Pan et al. [28] exclusively meant of leveraging the security system over power management. The work of Tlili and Slimani [29] have implemented conventional association rule over grid network for studding the level of heterogeneity in the network. Usage of semantics was also seen in work of Guanet al. [30] for enhancing service delivery. Hence, it can be seen that majority of the existing work is directed towards energy management considering the case study of smart grids with very less number of standard work on implementing mining approaches over grid environment. Hence, there is an open scope of research for developing a distributing mining algorithm exclusively for grid architecture in order to address this gap. The next section outlines the problems associated with existing system followed by proposed solution and algorithms constructed to achieve the goal.

1.2   Problem in Existing System

The significant problems associated with existing system are as follows:

- Presence of diversified forms of data from multiple domains cannot be stored effectively over the grid and applying of any existing mining can never result in knowledge discovery.
- Existing techniques are not capable of identifying an ambiguous data especially of the incoming traffic is massive in size that potential degrades the knowledge results.

- None of the existing mining approaches in distributed grid have ever considered the problems associated with data overloading resulting from massive size of data.
- Involvement of computational resources and complexity increases with size of incoming traffic and existing distributed mining approaches are not found to solve this problem.

Therefore, the identified problem of the study is *"Constructing a novel distributed mining framework that offers cost effective optimization performance while in the process of extracting knowledge from complex data from distributed network as well as computing environment."*

1.3 Proposed Solution

With an extension of our prior implement [31–34], we present a novel mechanism of distributed mining that is exclusively meant for grid architecture. The complete design principle of proposed system uses analytical research methodology. Figure 1 exhibits the schematic diagram of proposed system.

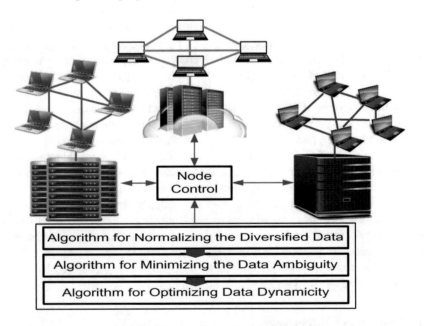

Fig. 1. Proposed system of OCSLM

The complete algorithm discussed in proposed system is meant to be executed in the controller component of grid architecture with an assumption that all the communication links are well established and secured. Considering different communication domains on grid, diversified data is surfaced that is primarily addressed in proposed study. Further, it also reduces the problems associated with data ambiguity over the grid architecture followed by data dynamicity. The optimization of the proposed algorithm is carried out considering a novel machine learning approach to control the computational performance of distributed mining. The next section illustrates about all these algorithm implementation.

2   Algorithm Implementation

The proposed algorithm mainly emphasizes on the optimization process for improving the clustering performance for distributive data over grid architecture. The complete construction of algorithm considers addressing the problems associated with existing approaches as contribution by introducing three discrete and sequential set of algorithms. For a given set of data, the proposed system chooses to initially address the problem associated with (i) *diversity of data*, (ii) *ambiguity in data*, and (iii) *data dynamicity*. In order to formulate this concept, we consider a case study of distributed enterprise system where dynamic data is generated across the grid and statistical-based approach is considered for evolving up with a distributed and perceive clustering model. This section overviews three sequential algorithm as follows:

2.1   Algorithm for Normalizing the Diversified Data

The prime role of this algorithm is to construct a distributed software framework that can effectively maintain the diversified data of various forms in terms of distributed archiving over grid architecture. The operational flow of proposed algorithm is as shown in Fig. 2.

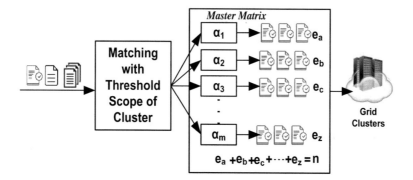

Fig. 2. Operational flow of algorithm for normalizing the diversified data

As majority of the incoming data are originated from different sources, it posses various form of data diversity. Even a data stream generated from same source may also be quite diverse in nature. Therefore, it is required to store the diversified data very effectively and precisely so that applying mining approaches in later stages becomes less challenging. This algorithm considers the input of d (raw data), α (cluster), T

(threshold) that after processing leads to generated of d_1 (normalized data). The algorithmic steps are as follows:

Algorithm for Normalizing the Diversified Data
Input: d (raw data), α (cluster), T (threshold)
Output: d_1 (normalized data)
Start
1. $d \leftarrow data_{stream}$
2. $\alpha \mid \alpha_m = \{e_n, \ \forall \ m \neq n \ \& \ m=n=integer\}$
4. **If** $d \subseteq \alpha_T$
5.   **If** $e_a = 0$
6.   Allocate d to α_o, where $o \subseteq m$
7. $d_1 \rightarrow$ Extract pos(d) and d_{ind} i.e.
8. **Else**
9. $d_1 \rightarrow$ Add d to α_{m+1}
End

Referring to Fig. 2, various streams of incoming traffic are initially tested if their categories belong to any present cluster of the enterprise application. For this purpose, we consider that there are threshold of T number of categories that are supported by the present data within the records and hence the first task is to ensure if the incoming data corresponds to T. If it is could matching than the next task to check the corresponding data with specific cluster α considering that proposed system could support maximum of m number of cluster. Within the cluster lies sub-matrix denoted by e_a, e_b, e_c ...ez, where z is integer and a, b, c are dimensions of sub-matrix with specific position of the cell along with cell index. Hence, maximum of n dimensional are permissible. This will also mean that size of the master matrix is defined by m and n. After the data stream is allocated to a ingress port (Line-1) than it matches with the existing threshold scope. In case the data lies within the threshold scope of existing cluster i.e. α_T (Line-4) than it further checks for empty location of its sub-matrix i.e. e (Line-5). In case of vacant position, the incoming data is stored followed by generation of position and index of storage within the master matrix i.e. pos (d) and d_{ind} (Line-7). However, if the incoming data stream is found to posses new category than the proposed algorithm adds up the new category to the new position of cluster i.e. α_{m+1}. (Line-9). Therefore, a highly structured formed of distributed storage is obtained by implementing the proposed normalization algorithm.

2.2 Algorithm for Minimizing the Data Ambiguity

The prior algorithm assists in normalizing the diversity scale of the incoming traffic in distributed grid architecture. While doing so, it may also lead to storage of null data or vague data that finally results in higher degree of ambiguity. The proposed study uses a threshold-based context that offers precise substitution of any form of ambiguous data prior to perform any form of mining operation. A data structure over grid could be rendered less ambiguous if all the data in the sub-matrix are very much well defined and there is a dedicated location to keep a track of it. The operational flow of proposed algorithm is highlighted in Fig. 3.

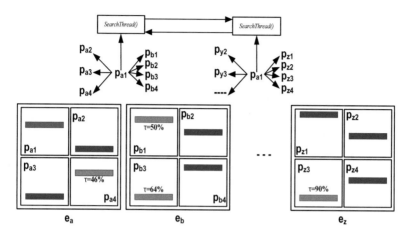

Fig. 3. Operational flow of algorithm for minimizing data ambiguity

This algorithm is takes the input of d_1(normalized data) and d_{1ind}(data index) that after processing yields finally allocated data d_2 on the cell with either missing value or error-prone value. The proposed system applies statistical approach in order to substitute the ambiguous data. The steps of algorithm are as follows:

Algorithm for Minimizing the Data Ambiguity
Input: d_1(normalized data), d_{1ind}(data index)
Output: d_2 (Allocated Data)
Start:
1. Obtain pos (d_1) and d_{1ind}
2. init Search Thread();
3. **For** e_i=1: n
4.   **If** pos (d_1) =0   //empty cell
5.   Allocate w
6.   **else**
7.   Calculate T_F of p_{nh}
8. $\tau \rightarrow \arg_{max}(T_F)$ //context
9.   generate searchThread(1)
10.   **If** τ<thresh
11. $d_2 \rightarrow$unselect(τ)
12.   **else**
13. $d_2 \rightarrow$select(τ)
14.**End**
End

Referring to Fig. 3, this algorithm initiates by identifying a cell with ambiguous data within a sub matrix e_a. Prior to start, it will extract the positional information of this ambiguous data with respect to address $pos(d_1)$ and index of data d_{1ind} (Line-1). The algorithm then checks for term frequency associated with other part of the data because we consider that only certain part of the data is missing or ambiguous. In case the complete cell is allocated with null data than we substitute with a weight that is easy to track the missing value (Line-4). Otherwise, the algorithm computes the term frequency (T_F) and extracts context τ from maximum value of it (Line-8). The first *search Thread*() searches the ambiguous cell content p_{a1} with all the other cell contents and reposits in a memory of master matrix via *search Thread*() itself. While performing this search operation, any context value found lesser than a thresholded value (Line-10) is rejected showing that they bear less correlation and any context found more than thresholded value (Line-13) is considered. The selected context is the actual replacement of the ambiguous value in p_{a1}. The *search Thread*() is continued until it reaches the last position of sub-matrix e within master matrix. This procedure is not required to be iterated as all the *search Thread*() are stored in master matrix and any new search may finally get a lower response time as majority of the information pertaining to the context is stored within the *search Thread*(). Hence, the proposed algorithm offers less computational burden.

2.3 Algorithm for Optimizing Data Dynamicity

The output obtained from the previous algorithm could have possibilities of estimation errors as well as channel-based errors. Presence of such errors is typically impossible to model with potential impact on outcome and hence we apply machine learning approach to solve this problem. We perform convergence test by considering hypothetical error applied to machine learning in order to obtain the best possible knowledge out of it. The operational flow of this algorithm is highlighted in Fig. 4.

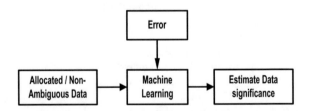

Fig. 4. Operational flow of algorithm for optimizing data dynamicity

This algorithm takes the input of e(error) and d_2 (allocated/non-ambiguous data) that after processing leads to generation of d_{sig} (data significance). A thresholded error e_{th} is considered against whom the experimental values are observed in order to obtain a data with least errors. The steps of the algorithm are as follows:

Algorithm for Optimizing Data Dynamicity
Input: e(error), d_2 (allocated/non-ambiguous data)
Output: d_{sig} (data significance)
Start
1. init e
2. **For** $e \leq e_{th}$
3.   Apply $f_{train}(x) \rightarrow d_2$
4.   Apply $f_{min}(d_2)$
5.   Sort new(d_2)
6.   **If** $e(new(d_2)) < e_{th}$
7. $q = arg_{min}(new(d_2))$
8. $d_{sig} \rightarrow \mu(q)$
9. **End**
End

According to the algorithmic steps, first the hypothetical errors e are compared with respect to thresholded error (Line-2) followed by applying Multi-Layered Perception based machine learning function $f_{train}(x)$ in order to obtained hypothetical values of d_2 (Line-3). Machine learning with error minimization function f_{min} is applied (Line-4) followed by sorting newly obtained data d_2 (Line-5). The system performs re-check of error values of the newly obtained data and compares with the updated thresholded value i.e. e_{th} (line-6). Only the newly formed data with reduced errors are considered. Finally, the system associates weight μ with the newly obtained data in order to extract empirical value of data significance d_{sig} (Line-8). Hence, the algorithm results in highly optimized mined data without involving any extra computational burden.

3   Result Analysis

The discussion presented in previous section for algorithm implementation clearly highlights that proposed system OCSLM offers a non-conventional distributed mining approach with novel optimization technique on grid environment. Proposed system also addresses some of the challenging problems which existing mining approaches are reported to be quite incapable. The scripting of the proposed logic was carried out in MATLAB and mainly emphasizes on the outcomes obtained from convergence testing. The study outcome of the proposed system was assessed using error in y-axis and incoming traffic load in x-axis considering 500-1000 nodes. As the performance parameters are static for proposed system, we consider certain assessment environment in order to evaluate its performance. We believe that an effective mining algorithm will need to be (*i*) scalable, (*ii*) have the potential of identifying the data diversity and assists in structured storing mechanism over distributed system, (*iii*) should be able to overcome any form of errors or challenges related to data while performing mining, and (*iv*) should have higher accuracy of the mined data. For this purpose we enhance our machine learning scheme by adding up non-linear processing units in order to extract significant information about the mined outcome i.e. knowledge while performing clustering. It also has the capability to perform clustering using supervised and unsupervised learning algorithms using gradient descent approach meant for error minimization. Apart from this, the study outcome was compared with the most frequently used mining approach i.e. Linear Regression based approach (LR) and Multi-Layered Perception (MLP). The proposed study computes error as the degree of difference between the experimental outcome of data significance and hypothetical outcome of it using machine learning. This section outlines the error score and assesses the trend of it with increasing traffic load considering multiple assessment environments.

3.1   Analysis-1: Impact of Data Diversity

In order to assess the impact of data diversity, we apply random modeling of diversity on incoming traffic which will mean that rate of data diversity will keep on increasing with every incoming traffic where there is no predefined rule or any form of apriority information about the diversity score of the data. This poses one of the challenging environments in order where it is challenging to ensure an effective structurization of the incoming data without knowing its diversity. Figure 5 exhibits the trend of error score with respect to increasing data diversity.

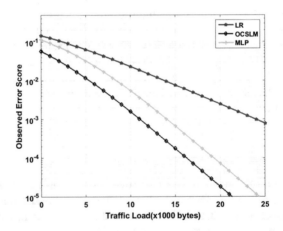

Fig. 5. Analysis of error with respect to data diversity

The proposed system has exhibited a smoother trend of minimization of error which is quite predictive in nature as compared to existing system. The prime reason behind this is- The LR based technique offers an impressive statistical modeling with better supportability until the data are homogeneous in nature. However, real-time data are less homogeneous and more diversified causing the LR technique to fail in establishing any significant relationship between its variables. This also results in significant consumption of computational resources and processing time. The situation becomes more challenging when the incoming traffic on grid reaches its peak condition. In other words, the LR technique fails to identify the appropriate positions of incoming data over the existing grids causing either failure in effective data forwarding to clusters or depends on third party to take the decision. Hence, LR technique is found only suitable for centralized mining approach as well as capable of processing homogeneous data too irrespective of increasing size of it. The MLP technique, on the other hand, offers improved performance as compared to LR technique when there is data diversity. It is because it significantly supports processing multiple and different task at same time (by assigning suitable weights to each inputs). With an aid of activation function and iterative learning process, the MLP technique also offers significant reduction of error. After initializing the statistical error value more than 0.7, the MLP technique is found to reduce down the error to 0.5 successfully. However, in order to do so, MLP technique has higher inclusion of iterative steps as compared to LR technique which offers slight increase in trials in order to reduce the error. The proposed system OCSLM offers a simple and non-iterative data structuring principle that uses threshold-based approach not only to identify the type of incoming data but also allocates the incoming data in a novel matrix form. Structurization of the data in the matrix form not only offers better data management but also allows any mining approach to be applied on distributed data. Hence, even with an increase of data diversity, the proposed system maintains a master matrix for all its grid lets that can further identify the data diversity factor and therefore, it contributes to minimization of error score.

3.2   Analysis-2: Impact of Ambiguous Data

This assessment environment was reconstructed by randomly manipulating the segments of the data in different parts in order to map the ambiguous data. The level of data ambiguity is increased on every increment in traffic load in order to evaluate the performance of error. With higher incorporation of error, it is anticipated of bigger computational challenge to minimize error. Figure 6 highlights the error score with respect to data ambiguity.

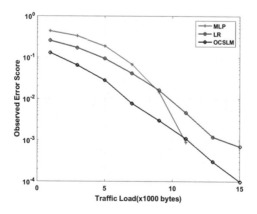

Fig. 6.   Analysis of errors with respect to ambiguous data

The outcome shows that proposed OCSLM offers further better outcome for solving the ambiguous data even compared to its capability to sustain data diversity in contrast to existing mining approaches. The prime reason behind this is proposed system uses context factor using term frequencies that explicitly understand the content of the cell in sub-matrix as well as it also understand the appropriate value to fill up. The process is further optimized by applying machine learning approach using similar number of layers in order to balance the minimization of errors with reduced dependencies of computational resources. Hence, proposed system is more capable of performing mining in presence of various challenges associated with distributed form of data over grid. The existing system of LR shows its constraint of limiting the error score with increase of incoming traffic and hence it shows that it cannot optimize further. On the other hand, inclusion of higher degree of iterative steps in MLP also consumes more resources (e.g. buffer) and yet there is more scope of minimizing the error found in it.

3.3   Analysis-3: Impact of Data Overload

We state that when all the sub-matrix within the distributed grid is populated by the data than it is an obvious situation that more number of incoming data finds itself difficult for being stored in distributed matrix. The initial problem is to explore the empty location of grid followed by problem associated with consistent searching for an empty cell within any sub-matrix. An effective design of distributed mining should have better data management in order to deal with such situation that poses intensive challenges while applying analytics on real-time data. Figure 7 shows the trends of data overload of different approaches.

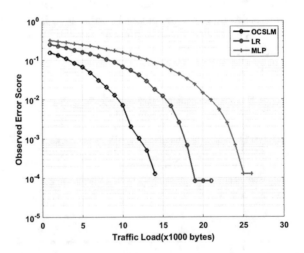

Fig. 7.   Analysis of error with respect to data overload

Proposed system uses a master matrix where all the information related to position and index of the data is being retained. It also gets updated upon any new searching thread on any specific sub-matrix (and not all of them). This unique quality of updating results in taking an effective decision of structuring the data in highly distributive manner with further update of indexes resulting in faster query processing. Hence, increase of traffic has actually no potential impact on proposed system.

3.4   Analysis-4: Impact of Data Dynamicity

With increasing number of dynamic users and source of data originations, data dynamicity will definitely increase and poses a challenge in mining algorithms. Hence, we construct such environment by altering the rate of transmission in random manner and adopted multiple communication channels in order to transmit the distributed data to the grid. The idea was to check the impact of data dynamicity trend as shown in Fig. 8.

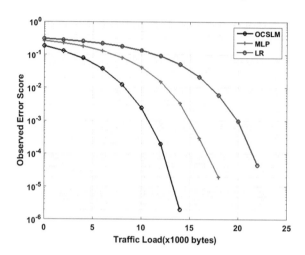

Fig. 8. Analysis of error with respect to data dynamicity

A closer look into the trend will show that the existing approaches of MLP are found to offer good error minimization performance as compared to LR technique. However, its success rate is very less as the data is highly heterogeneous in nature. So, until and unless that data bears good correlation, LR technique offers good error minimization performance. Although, this problem is solved by MLP technique, but their convergence performance is limited to specific number of steps. Unlike these existing systems, proposed system applies predictive mining approach, where the unseen errors are not only computed by also compensated in the final stage and hence proposed OCSLM offers better capability to address data dynamicity problems in grid.

3.5   Analysis-5: Algorithm Processing Time

Algorithm response time is one of the critical parameter when it comes to distributed mining algorithms. It is quite normal to anticipate that algorithm processing time will positively increase with increase in incoming traffic and the increment may be considerably high if the data is distributed in the grid environment. Therefore, we compute the algorithm processing time as total time required by the proposed algorithm in order to compute data significance value. It is because data significance value is the actual empirical representation of quality of the distributed mined data over grid.

The outcome shown in Fig. 9 exhibits that proposed OCSLM offers significantly lower algorithm processing time in comparison to existing LR and MLP scheme. The prime reason behind this is – unlike existing approaches of MLP, proposed system performs less iterative. Moreover, the approach uses dual step of minimization of errors for which reason convergence and accuracy of proposed system is much better as compared to the existing approaches. This outcome also proves that it is suitable for performing mining on any complex forms of distributed data over grid.

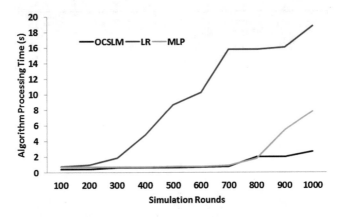

Fig. 9. Analysis of algorithm processing time

4   Conclusion

Developing a non-conventional form of mining algorithm in order to balance knowledge exploration process and dynamic demands of user is quite a challenging task. The proposed system introduces a novel form of machine learning mechanism that is improved form of conventional machine learning in order to achieve less iteratative steps and progresses more towards error minimization. This characteristic offers proposed system to achieve approximately 85% of improvement in algorithm processing time and more than 45% of improvement in obtaining error-free mined data. The proposed algorithm is highly suitable for the applications e.g. healthcare analytics, mining of sensory data, knowledge extraction of traffic-related data with good temporal performance. The algorithm also offers significantly less computational complexity which is essential for it to be working under uncertain networking conditions. Therefore, the supportability of the grid architecture is quite high for this algorithm with true optimization steps of obtained better management and structurization of mined data without involving any form of expensive computational resources or other dependencies.

References

1. Mahmood, Z.: Software Project Management for Distributed Computing: Life-Cycle Methods for Developing Scalable and Reliable Tools. Springer, Cham (2017)
2. Kacsuk, P., Kranzlmüller, D., Németh, Z., Volkert, J.: Distributed and Parallel Systems: Cluster and Grid Computing. Springer, New York (2017)
3. Magoules, F., Pan, J., Tan, K.A., Kumar, A.: Introduction to Grid Computing. CRC Press, New York (2009)
4. Kosar, T.: Data Intensive Distributed Computing: Challenges and Solutions for Large-scale Information Management: Challenges and Solutions for Large-scale Information Management. IGI Global, Hershey, PA (2012)
5. Wang, L., Jie, W., Chen, J.: Grid Computing: Infrastructure, Service, and Applications. CRC Press, New York (2009)
6. Suwan, A., Siewe, F., Abwnawar, N.: Towards monitoring security policies in grid computing: a survey. In: SAI Computing Conference (SAI), London, pp. 573–578 (2016)
7. Setia, H., Jain, A.: Literature survey on various scheduling approaches in grid computing environment. In: IEEE 1st International Conference on Power Electronics, Intelligent Control and Energy Systems (ICPEICES), Delhi, pp. 1–4 (2016)
8. Liu, Y., Rong, Z., Jun, C., Ping, C.Y.: Survey of grid and grid computing. In: International Conference on Internet Technology and Applications, Wuhan, pp. 1–4 (2011)
9. Aggarwal, S., Agarwal, N., Jain, M.: Uncertain data mining: a review of optimization methods for UK-means. In: 3rd International Conference on Computing for Sustainable Global Development (INDIACom), New Delhi, pp. 3672–3677 (2016)
10. Shahina Parveen, M., Narsimha, G.: Scaling effectivity of research contributions in distributed data mining over grid infrastructures. Commun. Appl. Electron. 3(8), 17–27 (2015). Published by Foundation of Computer Science (FCS), NY, USA
11. Zhang, Y., Yang, R., Zhang, K., Jiang, H., Zhang, J.J.: Consumption Behavior Analytics-Aided Energy Forecasting and Dispatch. IEEE Intell. Syst. 32(4), 59–63 (2017)
12. Asad, Z., Chaudhry, M.A.R.: A two-way street: green big data processing for a greener smart grid. IEEE Syst. J. 11(2), 784–795 (2017)
13. Ergin, M.O., Handziski, V., Wolisz, A.: Grid-based position discovery. In: International Conference on Localization and GNSS (ICL-GNSS), Barcelona, pp. 1–8 (2016)
14. Goyal, A., et al.: Asset health management using predictive and prescriptive analytics for the electric power grid. IBM J. Res. Dev. 60(1), 4:1–4:14 (2016)
15. Hochbaum, D.S., Baumann, P.: Sparse computation for large-scale data mining. IEEE Trans. Big Data 2(2), 151–174 (2016)
16. Jindal, A., Dua, A., Kaur, K., Singh, M., Kumar, N., Mishra, S.: Decision tree and SVM-based data analytics for theft detection in smart grid. IEEE Trans. Industr. Inf. 12(3), 1005–1016 (2016)
17. Lee, S., et al.: CloudSocket: Smart grid platform for datacenters. In: IEEE 34th International Conference on Computer Design (ICCD), Scottsdale, AZ, pp. 436–439 (2016)
18. Aouad, L.M., An-LeKhac, N., Kechadi, T.: Grid-based approaches for distributed data mining applications. J. Algorithms Comput. Technol. 3(4), 517–534 (2009)
19. Giri, J.: Proactive Management of the Future Grid. IEEE Power Energy Technol. Syst. J. 2(2), 43–52 (2015)
20. Kar, S., Samantaray, S.R., Zadeh, M.D.: Data-mining model based intelligent differential microgrid protection scheme. IEEE Syst. J. 11(2), 1161–1169 (2017)

21. Kumar, S.M.D., Jumnal, A.: A real-time grid enabled test bed for sharing and searching documents among universities. In: Second International Conference on Advances in Computing and Communication Engineering, Dehradun, pp. 604–609 (2015)
22. Loia, V., Terzija, V., Vaccaro, A., Wall, P.: An affine-arithmetic-based consensus protocol for smart-grid computing in the presence of data uncertainties. IEEE Trans. Industr. Electron. **62**(5), 2973–2982 (2015)
23. Luan, W., Peng, J., Maras, M., Lo, J., Harapnuk, B.: Smart meter data analytics for distribution network connectivity verification. IEEE Trans. Smart Grid **6**(4), 1964–1971 (2015)
24. Xu, Q., et al.: A short-term wind power forecasting approach with adjustment of numerical weather prediction input by data mining. IEEE Trans. Sustain. Energy **6**(4), 1283–1291 (2015)
25. García-Galán, S., Prado, R.P., Expósito, J.E.M.: Swarm fuzzy systems: knowledge acquisition in fuzzy systems and its applications in grid computing. IEEE Trans. Knowl. Data Eng. **26**(7), 1791–1804 (2014)
26. Chung, W.C., Hsu, C.J., Lai, K.C., Li, K.C., Chung, Y.C.: Direction-aware resource discovery service in large-scale grid and cloud computing. In: IEEE International Conference on Service-Oriented Computing and Applications (SOCA), Irvine, CA, pp. 1–8 (2011)
27. Dan, Z.: Study on the construction of digital library model under semantic Grid environment. In: International Conference on Computer Application and System Modeling (ICCASM 2010), Taiyuan, pp. V1-494–V1-498 (2010)
28. Pan, S., Morris, T., Adhikari, U.: Developing a hybrid intrusion detection system using data mining for power systems. IEEE Trans. Smart Grid **6**(6), 3104–3113 (2015)
29. Tlili, R., Slimani, Y.: Executing association rule mining algorithms under a grid computing environment. ACM J., 55–61 (2011)
30. Guan, T. et al.: Enhancing Grid service discovery with a semantic wiki and the concept matching approach. In: Fifth International Conference on Semantics, Knowledge and Grid, Zhuhai, pp. 208–215 (2009)
31. Cesario, E., Talia, D.: A failure handling framework for distributed data mining services on the grid. In: 19th International Euromicro Conference on Parallel, Distributed and Network-Based Processing, Ayia Napa, pp. 70–79 (2011)
32. Shahina Parveen, M., Narsimha, G.: Distributed data mining approaches as services on the grid infrastructure. In: National Conference on Soft Computing and Knowledge Discovery (2012)
33. Shahina Parveen, M., Narsimha, G.: Optimized clustering with statistical-based local model for replica management in DDM over grid. Software Engineering Perspectives and Application in Intelligent Systems, vol. 2. Springer, Cham (2016)
34. Shahina Praveen, M., Narsimha, G.: SADM: Sophisticated architecture of distributed mining over grid infrastructure. International Journal of Computer Science and Electronics Engineering (IJCSEE) **4**(3), 129–134 (2016)

New Numerical Investigation Using Meshless Methods Applied to the Linear Free Surface Water Waves

Mohamed Loukili[(✉)] and Soumia Mordane

Polymer Physics and Critical Phenomena Laboratory, Faculty of Sciences Ben M'sik,
University Hassan II, P.O.BOX 7955, Casablanca, Morocco
md.loukili@gmail.com

Abstract. A new investigation using meshfree methods to resolve the linear free surface water waves problem. Two methods are used in this current investigation, the method of fundamental solutions (MFS) and multiquadric (MQ) radial basis function. The problem is solved by collocation of boundary points since the governing equations are satisfied automatically in order to provide information on fast and accurate method for an efficient hydrodynamic prediction.

Keywords: Meshless method · Free surface · Gravity waves
Radial basis functions · MQ · MFS · Hydrodynamic prediction

1 Introduction

Radial basis function (RBF) methods have become the primary tool for solving partial differential equations (PDEs) in complexly shaped domains. Classical methods for the numerical solution of PDEs (finite difference, finite element, finite volume, and pseudospectral methods) are usually limited by their algebraic convergence rates. The previous works such as the comparison of the MQ collocation method with the finite element method in the reference [1] have been done that illustrate the superior accuracy of the MQ method when compared to local polynomial methods. The large number of recent books which include [2–6], literature review shows various studies have been published on numerical gravity waves propagation problem using meshless methods, the uses of the fundamental solution of Laplace's equation has been proposed for simulating the nonlinear free surface water waves and wave–structure interactions [7, 8], The method is further employed by Xiao et al. [9] to simulate nonlinear irregular waves in shallow water by introducing a new form of the numerical wave tank (NWT). Senturk [10] proposed the localized meshless RBF method in the simulation of free surface waves by breaking down the computational domain into a number of subdomains, leading to a sparse global system matrix. This approach is particularly advantageous in mitigating the time-consuming nature of the simulation process.

The rapid development in offshore activities, along with increasingly extreme environments, demands more accurate and fast prediction of the hydrodynamic performance of offshore structures. In this work, we focus in the numerical investigation of meshfree

© Springer International Publishing AG, part of Springer Nature 2019
R. Silhavy (Ed.): CSOC 2018, AISC 765, pp. 337–345, 2019.
https://doi.org/10.1007/978-3-319-91192-2_33

methods applied to the linear gravity waves, two methods are used in this investigation MQ RBF function and the method of fundamental solutions (MFS), in order to illustrate the fast and accurate method for a rapid hydrodynamic prediction.

2   Position of the Problem

We consider a monochromatic incident wave of small amplitude propagating in the presence of a flat bottom of the numerical wave tank (NWT), which is derived from the physical ocean engineering basin, as shown in Fig. 1.

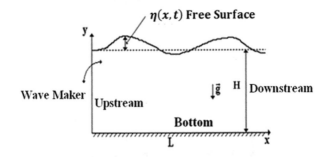

Fig. 1. Sketch of numerical wave tank (NWT).

As a part of the linear wave theory, the movement is supposed to be plane and periodic in time, irrotational, incompressible and the fluid is assumed to be inviscid. The formulation can be made in terms of the velocity potential φ and the elevation of the free surface η, these two variables will be assumed complex, with harmonic dependence on time. The velocity potential φ(x, y, t) can be written as:

$$\varphi(x, y, t) = \phi(x, y)e^{i\omega t}$$

Under these assumptions, the problem of propagation of linear waves is reduced to resolve the system of equations:

$$\Delta \phi(x, y) = 0 \quad \text{in the fluid domain} \tag{1}$$

$$\vec{U} = \vec{\nabla}\phi(x, y) \quad \text{in the fluid domain} \tag{2}$$

$$\vec{\nabla}\phi(x, y) \cdot \vec{n} = 0 \quad \text{at the bottom} \tag{3}$$

$$-\omega^2\phi(x, y) + g\frac{\partial\phi(x, y)}{\partial y} = 0 \quad \text{at the mean free surface} \tag{4}$$

Where (x, y) Cartesian coordinates, H the water depth; g the acceleration of gravity and $\vec{\nabla}$ the operator gradient. The elevation of the free surface is then:

$$\eta = \frac{-i\omega}{g}\varphi \tag{5}$$

Downstream, it is assumed that the transmitted waves propagate without reflection; this condition numerically translates into the radiation condition [11]:

$$\frac{\partial\phi(x, y)}{\partial n} = -ik\phi(x, y) \tag{6}$$

Where $k = 2\pi/\lambda$, λ being the wavelength.
The condition on the upstream boundary is expressed as:

$$\frac{\partial\phi(x, y)}{\partial n} = 2ikf(y) - ik\phi(x, y) \tag{7}$$

$$\text{Where} \quad f(y) = \frac{ag}{\omega}\frac{\cosh(ky)}{\cosh(kH)} \tag{8}$$

3 Numerical Formulations of the Meshfree Methods

Consider a domain Ω defined in $\mathbb{R}^2$, limited by a boundary $\partial\Omega$. The problem to solve is written in the form:

$$\Delta\,\phi(x, y) = 0 \text{ In } \Omega \tag{9}$$

$$\frac{\partial\phi(x, y)}{\partial n} = g(x \cdot y) \text{ On } \partial\Omega \tag{10}$$

Ω: represents the computational domain and $\partial\Omega$ the Neumann boundaries.
The meshfree numerical approximation used in this present work defined as follows:

- **The Method of Fundamental Solutions (MFS):**

The fundamental solution for the Laplace 2D equation is:

$$\phi(x, y) = \sum_{j=1}^{N} \alpha_j G(r_{ij}) \quad \text{(MFS)} \tag{11}$$

Where $G(r_{ij}) = \frac{-1}{2\pi}\ln(r_{ij})$ is the Green's function.

$r_{ij} = \sqrt{\left((x_i - \beta_j)^2 + (y_i - \delta_j)^2 + c^2\right)}$ is the distance between a field points and a boundary points.

- **The Method of Multiquadric Radial Basis Function Approximation (MQ RBF):**

The multiquadric radial basis function used in this word written as:

$$\phi(x, y, \varepsilon) = \sum_{j=1}^{N} \alpha_j \varepsilon \sqrt{1 + \varepsilon^2 r_{ij}^2} \quad \text{(MQ)} \tag{12}$$

$r_{ij} = \sqrt{\left(x_i - \beta_j\right)^2 + \left(y_i - \delta_j\right)^2}$ is the distance between a field points and a boundary points;

Where N is the number of points on the boundaries and α_j the undetermined coefficients that represent the strengths of singularities, respectively; $\vec{x_i} = (x_i, y_i)$ is the position of the field points; $\vec{s_j} = (\beta_j, \delta_j)$ is the location of the boundary points and, c the spatial parameter introduced to be free from the ill-conditioning effect. To determine the unknowns α_j of the problem, we will use a collocation method to the boundary conditions. Once these coefficients are determined, the potential $\varphi(x, y)$ and the elevation of the free surface $\eta(x, t)$ can be obtained by linear combination of the fundamental solutions. In this work, we typically select the location of boundary field points, and centre points as illustrated in Fig. 2.

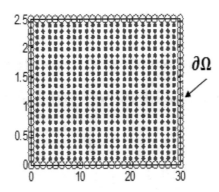

Fig. 2. Distribution of field points (blue symbols) and boundary points (red symbols)

4   Results and Discussion

In this section we consider a numerical wave tank (NWT) of constant depth H = 2.5 m and length L = 30 m, the present numerical schemes proposed in this work are evaluated by comparing the analytical solutions (Stokes at the order 1) [14] and the numerical solutions.

The numerical accuracy of the MFS and the MQ are studied and the results are compared as shown in Figs. 3, 4 and 5. The ill-conditioning of the matrix is generally observed with the MFS method. Our numerical experiments indicates that the system of resulting equations is not badly conditioned, when the value of the parameter c is selected between 0.03 m and 0.1 m. outside this range the matrix happens to be ill conditioned. We have subsequently chosen the parameter c = 0.03 m.

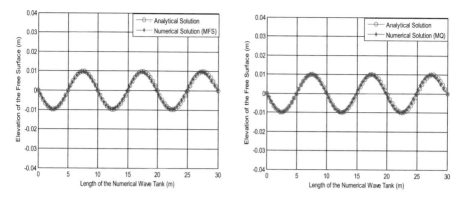

Fig. 3. Elevation of the free surface for MFS and MQ along the numerical wave tank (NWT)

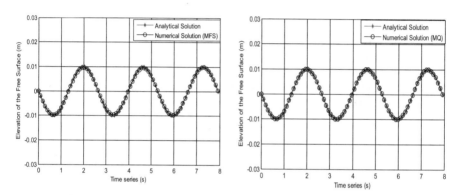

Fig. 4. Elevation of the free surface for MFS and MQ along the time series

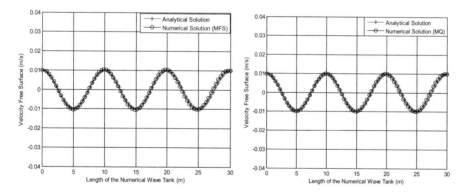

Fig. 5. Velocity free surface for MFS and MQ along the length of the numerical wave tank (NWT).

For the MQ, the Shape parameter ε is suggested to be proportional to the camber of the free surface and interpolation node number $as\ \varepsilon = \left(\frac{a}{\lambda}\right) * \frac{10}{\sqrt{N_t}}.$

In Figs. 3, 4 and 5 a comparison of the linear numerical solutions and the analytical solutions of the elevation and the velocity of the free surface, presented for $\lambda = 10$ m and $a = 0.01$ m.

The results shown in this section are noted that there is a good agreement between the analytical and numerical approaches, which makes it possible to affirm that the numerical approaches used are valid for progressive waves in a water domain of homogeneous depth.

The treatment of these problems, are of great importance to offshore engineering but are difficult to deal with. They do not only have the features of general fluid dynamic problems but also have an unknown moving free surface. Any new numerical method, even successfully applied to other fluid problems, is usually required to be carefully refined when it is applied to water waves problem.

4.1   Shape Parameter for Multiquadratic

Many types of RBFs used in these problems contain a shape parameter, and there is much experimental evidence showing that accuracy strongly depends on the value of this shape parameter. The authors [12] used arbitrary precision computations to experimentally derive a formula for the error dependence on the shape parameter c and the nodal spacing h. Hardy [15] suggested the use of $c = 0.815D$, Franke [13] recommended $c = 1.25 \dfrac{D}{\sqrt{N_t}}$, where D is the diameter of the smallest circle containing all data points ($D = h$ for equispaced nodes). In this paper we recommended an efficient shape parameter to compute the optimal value of the elevation of the free surface that minimizes the approximation error such as:

$$\varepsilon = \left(\frac{a}{\lambda} \right) * \frac{10}{\sqrt{N_t}}$$

Where a: Amplitude of the wave. λ: Wavelength. N_t: Total interpolation node numbers.

The optimal value of ε was strongly dependent on the interpolation number node and proportional to the camber of the free surface, The value of the optimal shape parameter was studied and explained in the next section.

4.2   Root Mean Square Error (RMSE)

To give a more quantitative understanding of accuracy, the root-mean-square errors (RMSE) of the elevation of the free surface water wave are depicted in Tables 1 in order to show that **Shape parameter** ε suggested is adequate to solve the linear free surface water waves problem.

Regarding a good agreement of the numerical approaches with the analytical solution presented in the previous section we illustrate the RMSE of the elevation of the free surface using MFS and MQ approach for different interpolation node numbers in order to investigate on accurate method for resolving the linear free surface water waves problem. The RMSE is regarded as an index for the accuracy measure and defined as follows:

$$RMSE \equiv \sqrt{\frac{\sum_{i=1}^{N}(\text{numerical result} - \text{exact solution}))_i^2}{N}}$$

Table 1. RMSE of the elevation of the free surface using MQ approach for different Shape parameter ε.

$\varepsilon \times 10^{-5}$	7.7	8.3	8.9	9.6	1.4	1.13	1.25	1.38	1.56
ERROR $\times 10^{-9}$	3.6	3.8	4.1	4.3	4.6	5	5.6	6.4	7.7

Table 2. RMSE of the elevation of the free surface using MFS and MQ approach for different interpolation node numbers.

Interpolation node numbers	1600	2500	3600	4900	6400	8100	10000
RMSE_MFS	$1.5 \cdot 10^{-3}$	$1.4 \cdot 10^{-3}$	$1.5 \cdot 10^{-3}$	$1.6 \cdot 10^{-3}$	$1.6 \cdot 10^{-3}$	$1.8 \cdot 10^{-3}$	$1.7 \cdot 10^{-3}$
RMSE_MQ	$4.8 \cdot 10^{-8}$	$1.9 \cdot 10^{-8}$	$8.8 \cdot 10^{-9}$	$4.7 \cdot 10^{-9}$	$2.7 \cdot 10^{-9}$	$8.6 \cdot 10^{-7}$	$5.6 \cdot 10^{-7}$

The results illustrated in the Table 1, show a good capacity and reliability of the shape parameter ε recommended to the MQ approach applied to the linear free surface water waves. In the Table 2 the RMSE of the computational results are obtained by using different interpolation node numbers to establish a convergent method. Comparing two numerical approaches with analytical solutions, the RMSE of the numerical results for different interpolation node numbers shows that the MQ approach has a good accuracy compared to the MFS approach.

4.3   The Run Time

In this part to illustrate the fast method for a rapid hydrodynamic prediction a CPU time is calculated as function of different interpolation node numbers, for two approaches MQ and MFS as shown in Fig. 6. The simulations are made on a laptop with 2.40 GHz Intel CPU and 3 Go RAM.

The present numerical simulation are performed by using different interpolation node numbers, the results of the Fig. 6, show that the MQ approach is powerful for analyzing the linear free surface water waves and fast for a rapid hydrodynamic prediction.

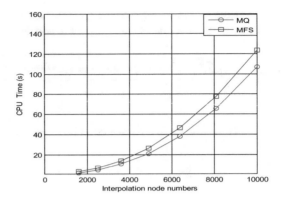

Fig. 6. Time consuming

5   Conclusion

In this work a numerical schemes was proposed to resolve the linear water waves propagation equations using two methods: the method of fundamental solutions (MFS) and Multiquadric (MQ) radial basis function in order to illustrate the fast and accurate method for an efficient hydrodynamic prediction. Two axes 2D situations are discussed:

(1) propagation of progressive monochromatic waves in a NWT with a flat bottom where the MFS approach coincides with the analytical solution when the values of the parameter c is considered to be in the range 0.03 m and 0.1 m. Outside this range the matrices happen to be ill-conditioned. For the MQ approach we suggested a Shape parameter ε to be proportional to the camber of the free surface and interpolation node number. The parameter recommended shown to be good and adequate for analyzing the free surface water waves problem.
(2) The Multiquadric (MQ) collocation method shown in this current investigation that is highly efficient and fast for a rapid hydrodynamic prediction compared to the MFS approach.

References

1. Li, J., Cheng, H.-D., Ching, C.-S.: A comparison of efficiency and error convergence of multiquadric collocation method and finite element method. Eng. Anal. Bound. Elem. **27**, 251–257 (2003)
2. Atluri, S.N.: The Meshless Method (MLPG) for Domain & BIE Discretizations. Tech Science Press, Forsyth (2004)
3. Atluri, S.N., Shen, S.: The Meshless Local Petrov-Galerkin (MLPG) Method. Tech Science Press, Encino (2002)
4. Chen, Y., Lee, J., Eskandarian, A.: Meshless Methods in Solid Mechanics. Springer, New York (2006)
5. Ferreira, A., Kansa, E.J., Fasshauer, G.E., Leito, V.: Progress on Meshless Methods. Springer, Dordrecht (2008)

6. Griebel, M., Schweitzer, M.: Meshfree Methods for Partial Differential Equations. Springer, Heidelberg (2002)
7. Xiao, L.F., Yang, J.M., Peng, T., Li, J.: A meshless numerical wave tank for simulation of nonlinear irregular waves in shallow water. Int. J. Numer. Methods Fluids **61**, 165–184 (2009)
8. Loukili, M., Mordane, S.: Modélisation de l'interaction houle-marche rectangulaire par la méthode des solutions fondamentales. In: 13ème Congrès de Mécanique Meknès, Maroc (2017)
9. Xiao, L.F., Yang, J.M., Peng, T., Tao, T.: A free surface interpolation approach for rapid simulation of short waves in meshless numerical wave tank based on the radial basis function. J. Comput. Phys. **307**, 203–224 (2016)
10. Senturk, U.: Modeling nonlinear waves in a numerical wave tank with localized meshless RBF method. Comput. Fluids **44**, 221–228 (2011)
11. Orlansky, I.: A simple boundary condition for unbonded hyperbolic flows. J. Comput. Phys. **21**(3), 251–269 (1976). https://doi.org/10.1016/0021-9991(76)90023-1
12. Huang, C.S., Lee, C.-F., Cheng, A.H.D.: Error estimate optimal shape factor, and high precision computation of multiquadric collocation method. Eng. Anal. Bound. Elem. **31**, 614–623 (2007)
13. Franke, R.: Scattered data interpolation: tests of some method. Math. Comput. **38**, 181–200 (1982)
14. Mordane, S.: Contribution numérique à la résolution du problème d'interaction houle-obstacles. Thèse de Doctorat d'Etat, Université Hassan II - Mohammedia, Casablanca, Maroc (2001)
15. Hardy, R.L.: Multiquadric equations of topography and other irregular surfaces. J. Geophys. Res. **76**(8), 1905–1915 (1971)

Framework for Capturing the Intruders in Wireless Adhoc Network Using Zombie Node

Jyoti Neeli[1]([✉]) and N. K. Cauvery[2]

[1] Department of Information Science and Engineering,
Global Academy of Technology, Bengaluru, India
jyotirneeli@gmail.com
[2] Department of Information Science and Engineering,
RV College of Engineering, Bengaluru, India

Abstract. Wireless Adhoc Network has a bigger range of application where some are already in commercial market and majority of them are still under roof of investigation. The degree of vulnerability of wireless adhoc network is quite high irrespective of presence of various existing techniques, which are too much specific to attack or orchestrated scenario of attack. Therefore, the proposed study introduces a framework that introduces a zombie node to capture the attention of attacker by advertising possession of many hops. The study also introduces broker node that is meant for balancing transmission demands along with robust security requirements. The message of zombie node is registered in hop table which makes the virtual zombie node to look like original node. An algorithm is constructed for identifying the malicious node and the study outcome is found to excel better performance with respect to the existing system.

Keywords: Wireless adhoc network · Attack · Intrusion · Security
Routing protocol · Secure communication

1 Introduction

Wireless Adhoc Network is gaining pace in terms of commercial application owing to its simplicity in deployment and communication [1]. Owing to absence of any routers, the nodes in adhoc network undertakes routing decision by itself. This will imply the presence of decentralized architecture that calls for highest range of security threats [2]. Various forms of attacks e.g. black hole attack, wormhole attack, Sybil attack, Denial-of-Service attack etc. has been reported to be intruding the adhoc network since last decade [3]. Different forms of wireless adhoc network includes wireless sensor network, mesh network, mobile and vehicular adhoc network where such forms of attacks are already studied [4, 5]. The potential threat in such forms of network is to actually identify the form of attack owing to either dynamic topology or because of decentralization-based communication system. There are various form of security-based routing protocol designed to offer resistance against the potential threats but there is less report of any potential secure routing technique in this regards [6–10]. The biggest problem of existing routing protocol is that it is applicable only on a

© Springer International Publishing AG, part of Springer Nature 2019
R. Silhavy (Ed.): CSOC 2018, AISC 765, pp. 346–355, 2019.
https://doi.org/10.1007/978-3-319-91192-2_34

specific scenario of application in wireless adhoc network. This will mean that security solution of wireless sensor network is less likely is applicable on mobile adhoc network and vice-versa. Moreover, existing system further narrow down its implementation towards solving specific forms of attack. Therefore, this research trade-off leads to develop an urgency of investigation in order to evolve up with a superior and efficient security solution in wireless adhoc network.

Therefore, the proposed manuscript introduces a security solution in order to resists the security breaches in wireless adhoc network. Section 2 discusses about existing research techniques towards solving security problems in wireless adhoc network followed by brief discussion of the identified problem in Sect. 3. Adopted methodology to solve the problem is highlighted in Sect. 4 and implemented algorithm for proposed methodology is exhibited in Sect. 5. Section 6 exhibits the accomplished results of the proposed study followed by conclusion in Sect. 7.

2 Related Work

In present times, there have been various types of solutions towards securing wireless adhoc network. The security features can be boosted up if the nodes were incorporated with fault tolerance. The research carried out by Wu et al. [11] showed that enhanced fault tolerance could also enhance the node retention capability using AODV/DSR protocol. Restricting the overhearing process further leverages the security features in adhoc environment that can actually increase communication performance too as seen in work of Zamani et al. [12]. Adoption of watchdog schemes (e.g. Singh et al. [13]) and trust-based communication scheme (e.g. Gong et al. [14]) is also reported to offer increasing security level with good energy efficiency. Study towards securing hybrid networks has been presented by Babu et al. [15] which is claimed to be resistive against multiple lethal attacks. Jain and Tokekar [16] have revised conventional AODV to incorporate security measures to offer better communication performance. However, the inherent problems associated with dissemination of stale information are not addressed apart from various vulnerable threats in using AODV (e.g. Gupta and Rana [17]). Adoption of fixed rekeying and encryption using session-based key is also found to offer increased security in multicasting operation in adhoc network (e.g. Madhusudhanan et al. [18]). The study claims good data confidentiality that could compliment to secure routing scheme too (e.g. Wu and Liaw [19]). Constructing countermeasures on the basis of mobile beacons can assists in identifying the intrusion problems in adhoc network (e.g. Chen et al. [20]). However design requirements of such mobility-based adhoc network differs in vehicular network where infrastructural demands (e.g. road side unit) security apart from node security (e.g. Choi et al. [21]). There are various other techniques in existing approaches e.g. decentralized authentication [22], classification-based approach [23], multi-fold secure routing [24], Watchdog [25], routing with anonymity [26], approach for solving energy-based attacks [27], opportunistic routing using trust factor [28, 29], negotiation-based approach, etc. Hence, there are various security-based techniques designed for ensuring intrusion-free environment while performing communication in wireless adhoc network. An extensive survey has been conducted on the

same [30]. The associated problems of existing security approaches are briefly highlighted in next section.

3 Problem Description

The prominent research problem is that existing techniques are mainly based on secure
routing considering AODV without addressing the stale routing problems in it. Another
significant problem is increased vulnerability of compromised node owing to lack or
delay in updating mechanism for any topological changes in existing system. A control
message holds significant information about hops (number of nodes, time stamp, their
identity, communication ranges, energy), which is not found to be much used in
existing AODV-based secure routing schemes. At the same time, majority of the
existing solution is quite specific to a particular form of attacks and hence they are less
applicable if attack scenario is altered. Therefore, the problem statement is *"To design
a framework that could use simple and non-conventional security measures by identifying the degree of vulnerability in a node present in wireless adhoc network."*

4 Proposed Methodology

The prime objective of the proposed system is to introduce a framework that could offer
significant resistance from majority of the intrusion in wireless adhoc network using
analytical approach. The security scheme of the proposed system is shown in Fig. 1 as
follows:

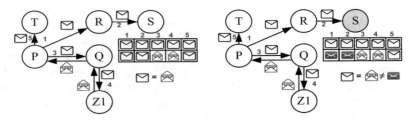

Fig. 1. Proposed security scheme

Every node (P) should be connected to a virtual node called as zombie node (Z1),
which is meant for capturing the attention of attacker A. The prime identification of
attacker is carried out using control messages relayed by the specific form of node
called as broker node. A broker node takes entire decision of packet transmission and
retransmission and hence it maintains a hop table. Every regular node has to be
associated at least with one single hop with the zombie node as the control message
relayed by the zombie node never matches with that of the attacker and hence the
source point of intruder could be easily explored. Say in Fig. 1, if S node is compromised, its response message will never be same as that mentioned by the hop table
retained by the broker node. Hence, zombie node don't assists in routing but it creates a

misleading communication scenario where it ensures that response control message relayed by the attacker never matches with the hop table and thereby protects from any form of topological changes. The next section outlines algorithm design.

5   Algorithm Implementation

The algorithm is mainly meant for evolving up with a solution that can not only identify the illegitimate node in the network but also resist them for further tampering the network. This mechanism of resistance is carried out by introducing a broker node followed by mitigating the attack. This section briefs about the algorithm responsible for this:

5.1   Algorithm for Constructing Broker Node

This algorithm takes the input of n (number of nodes), h (hop table) that after processing gives the output of τ (constructed broker node). The steps of the algorithm are as follow:

Algorithm for constructing broker node
Input: n, h
Output: τ
Start
init n, h
For all n
$h_2 \rightarrow$ explore(h=2)
   For j=1: size (h_2)
      $ix_2 \rightarrow [ix_2 ix_t]$, where ix_t=explore (h_2==1)
   End
      ix=int(ix_1, ix_2)
   For k=1: size (ix)
      $\tau \rightarrow$ int (h_2, h_1), where h_1=explore (h(ix)=1)
   End
      sort (size(τ))
End
End

After specifying the number of nodes with respective communication radius and network area, the network is deployed in random fashion (Line-1). A hop table h is created which retains all the neighborhood information within it and is stored within the node. The next task is to construct a broker node. For all the nodes (Line-2), the algorithm obtains the row of the hop matrix and explores only 2 hop nodes (Line-3). It is then followed by computation of common 1-hop nodes (Line-5) that allows the algorithm to compute all the 1-hop nodes for 2-hop neighbors (Line-4–5). The common elements of ix1 (stores 1-hop nodes) and ix2 are found using intersection method int (Line-7). Similar task is carried out for hops h1 and h2 (Line-9). Finally, the system computes the common value between

2-hop nodes and 1 hop node candidate and sorts it accordingly (Line-11). The algorithm introduces broker node in order to minimize the degree of duplicate packets during broadcasting and hence the primary responsibility of broker node is to select some efficient nodes to perform retransmission irrespective of all the nodes. These selections of nodes are carried out from its 2-hop adjacent nodes. Hence, only the subjects of the nodes chosen as broker node are eligible to perform retransmission thereby eliminating any feasibility of introducing any malicious message of duplicated origin.

5.2   Algorithm for Mitigating Attacks

The complete concept of attack mitigation is implemented with a novel concept of a specific form of a node called as zombie node. This algorithm takes the input of n (number of nodes), h (hop table), z (zombie) that after processing gives the output of mitigation and identification of attack. The steps of the algorithm are as follows:

Algorithm for mitigating attacks
Input: n, h (hop table), z (zombie)
Output: identification of attacker node
Start
init n, h, BN, z
Set ϕ (denial protocol)
    $x_{node} \rightarrow z_{node}|$ conf[advertise(n) $\notin \theta(c)$]
    For all y_{node}
        $c \in \theta(y_{node})$
        $c \rightarrow assign(x_{node}, \{d \in \theta(x_{node})\} \rightarrow BN$
    End
    $mal_{node} \rightarrow C-\theta(c)$
    $c \rightarrow assign\{x_{node} \rightarrow BN\}$
If c adds z_{node}
    c add z_{node} s.t. dist(y_{node} and $z_{node} < 3$)
Else
    remove BN
    Identify Attacker node
End

A normal node n will designate zombie node z. A denial protocol (Line-2) is constructed that allows the resistance against malicious node. The first protocol is that the beacon forwarded by the node xnode consists of information related to 1-hop adjacent node as well as the susceptible node znode should confirm that all the n nodes advertised should not be a part of θ, where θ is a set of all the unit hop adjacent nodes of susceptive nodes c(Line-3). Any node can be susceptive node but potential factors to identify susceptive node is its increase connectivity with many number of hops. As chances of attack increases with more connectivity. The second protocol is that the susceptive node c must investigate if there is any presence of zombie znode such that znode belong to $\theta(y_{node})$ for all the ynode tagged within the beacon (Line-5). More the cases of many non-tagged messages from the source node than more are the feasibility of the node to be acting as intruder (Line-11). If the presence of message is found in multiple place (>3 hops) than it confirm that susceptive node c to be an intruder. The algorithm than evaluates the possibilities of selecting xnode as broker node that could resume communication with the znode (Line-9). One of the potential steps of the algorithm is that it doesn't allow intruder to tag or represent any broker node (Line-8). Hence, there is no possibility that control message relayed by the broker node or intruder node to be equivalent. By this process, the system could easily identify the presence of malicious node. Too much presence of broker node is not required in the system and hence after the identification process, the algorithm removes extra broker node. Hence, only broker node responsible for communication is retained and rest are removed which maintains a good balance between security and communication performance. The hop table is instantly updated after this to ensure that all the routes could perform error-free and secure transmission of data. The next section outlines outcomes obtained.

6 Result Discussion

The study outcome is assessed with 100–500 MEMSIC nodes located in simulation area of 1000×1000 m^2 with 40 units of transmission range. The assessment of the proposed system is carried out using unconventional performance parameters e.g. average message size, percentage of required broker node, and percentage of zombies. The outcomes were compared with the secured AODV as majority of the security approaches in wireless adhoc network have used AODV routing.

The outcome had shown in Fig. 2 exhibits that size of transmitting message increases with increase in node density. However, proposed system offers better outcome in contrast to existing system. These messages are basically used for controlling topologies on the basis of the outcome of identification of malicious node. For effective analysis, the number of attacker nodes is increased to observe the detection capability. It is found that proposed system offers almost instantaneous response (processing time < 0.0265 s) against any malicious node by increasing message size. Considering 1000 number of hypothetical topologies in random order in order to represent adhoc network, the occurrences of false positive identification is found to be less than 3% (found during the trials of simulation). The proposed system offers security by rejecting selection of any broker node that is found or suspected to be adversary. However, conventional AODV based secure routing is incapable of performing these changes as they retain the stale information of the routing where the occurrences of the update of the hops are comparatively less in contrast to proposed system.

Similarly, Fig. 3 showcased the minimization of the dependencies of broker node. Although, both the system has similar trend of decline of broker node dependencies but proposed system offers better outcomes. It also shows that cost of overhead owing to communication is reduced with the increment in network size (i.e. density of the nodes). Hence, broker node is used wisely for both communication and security purpose in wireless adhoc network.

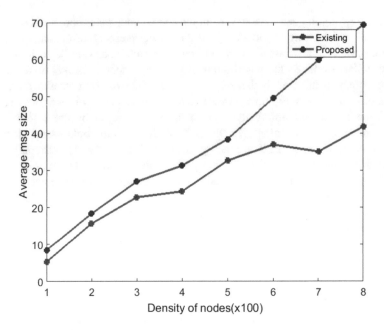

Fig. 2. Comparative analysis of average message size

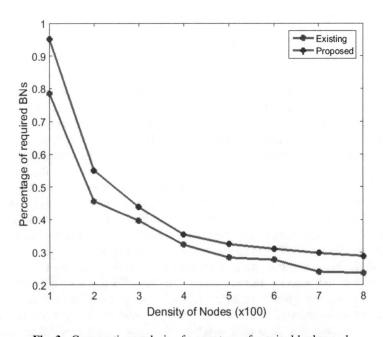

Fig. 3. Comparative analysis of percentage of required broker node

The proposed system uses zombies in order to capture the identification of the malicious node. Although, incorporation of zombie node is good for security but it requires frequent updating operation with other neighbor nodes. However, the proposed assessment shows that proposed system do not require increasing number of zombie node with increasing traffic in order to meet the security demands (Fig. 4). The percentage of reduction in zombie is approximately 30% (found during the trials of simulation) that shows that proposed system is not only secure but also offer cost effective communication performance. With reduced dependencies of increasing resources, the proposed system offers significant performance of security against any lethal threats over wireless adhoc network as its associated applications.

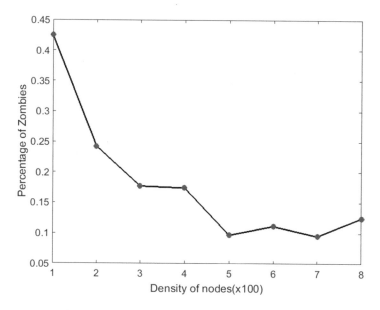

Fig. 4. Comparative analysis of percentage of zombies

7 Conclusion

The contribution of the proposed study are multi-folds. The first contribution of the proposed study is its broker node concept. Any node could act as a broker node but a specific selection mechanism for broker node is constructed on the basis of safety measures only. The study outcome has finally proven that dependency of the broker node over a period of time with increasing traffic minimizes and hence it don't offer any significant overhead to wireless adhoc network. The second contribution is to introduce a virtual node called as zombie node. The zombie node just offers certain forged response message which attracts the attacker and results in effective localization of attack event. Hence, it increases chances of identifying the higher probability of attack and can even stop it. It is because period assessment of the network allows more updated information about the intruder node. The third contribution of the study is that proposed system has not used any of the conventional security measures e.g. encryption-based, trust-based, reputation-based, optimization-based, but is more effective than this.

References

1. Sarkar, S.K., Basavaraju, T.G., Puttamadappa, C.: Ad Hoc Mobile Wireless Networks: Principles, Protocols, and Applications, 2nd edn., p. 349. CRC Press (2016)
2. Anjum, F., Mouchtaris, P.: Security for Wireless Ad Hoc Networks, p. 316. Wiley, New York (2007)
3. Bhatia, S.K., Mishra, K.K., Tiwari, S., Singh, V.K.: Advances in Computer and Computational Sciences: Proceedings of ICCCCS 2016, vol. 1, p. 775. Springer, Singapore (2017)
4. Khan, S., Pathan, A.-S.K.: Wireless Networks and Security: Issues, Challenges and Research Trends, p. 512. Springer, Heidelberg (2013)
5. Rawat, D.B.: Security, Privacy, Trust, and Resource Management in Mobile and Wireless Communications, p. 577. IGI Global, Hershey (2013)
6. Alcaraz, C., Lopez, J.: A security analysis for wireless sensor mesh networks in highly critical systems. IEEE Trans. Syst. Man Cybern. Part C (Appl. Rev.) **40**(4), 419–428 (2010)
7. Grover, J., Sharma, S.: Security Issues in Wireless Sensor Network-A Review. IEEE Xplore (2016)
8. Alani, M.M.: MANET security: a survey. In: IEEE International Conference on Control System, Computing and Engineering (ICCSCE 2014), Batu Ferringhi, pp. 559–564 (2014)
9. Qu, F., Wu, Z., Wang, F.Y., Cho, W.: A security and privacy review of VANETs. IEEE Trans. Intell. Transp. Syst. **16**(6), 2985–2996 (2015)
10. Barskar, R., Ahirwar, M., Vishwakarma, R.: Secure key management in vehicular ad-hoc network: a review. In: International Conference on Signal Processing, Communication, Power and Embedded System (SCOPES), Paralakhemundi, pp. 1688–1694 (2016)
11. Wu, W., Huang, N., Sun, L., Zheng, X.: Research article measurement and analysis of MANET resilience with fault tolerance strategies. Math. Prob. Eng. **2017**, 10 (2017)
12. Zamani, E., Soltanaghaei, M.: The improved overhearing backup AODV protocol in MANET. J. Comput. Netw. Commun. **2016**, 8 (2016)
13. Singh, R., Singh, J., Singh, R.: WRHT: a hybrid technique for detection of wormhole attack in wireless sensor networks. Mob. Inf. Syst. **2016**, 13 (2016)
14. Gong, P., Chen, T.M., Xu, Q.: ETARP: an energy efficient trust-aware routing protocol for wireless sensor networks. J. Sens. **2015**, 10 (2015)
15. Babu, M.R., Dian, S.M., Chelladurai, S., Palaniappan, M.: Proactive alleviation procedure to handle black hole attack and its version. Sci. World J. **2015**, 11 (2015)
16. Jain, A.K., Tokekar, V.: Mitigating the effects of black hole attacks on AODV routing protocol in mobile ad hoc networks. In: International Conference on Pervasive Computing (ICPC), Pune, pp. 1–6 (2015)
17. Gupta, A., Rana, K.: Assessment of various attacks on AODV in malicious environment. In: 1st International Conference on Next Generation Computing Technologies (NGCT), Dehradun, pp. 153–157 (2015)
18. Madhusudhanan, B., Chitra, S., Rajan, C.: Mobility based key management technique for multicast security in mobile ad hoc networks. Sci. World J. **2015**, 10 (2015)
19. Wu, W.-C., Liaw, H.-T.: A study on high secure and efficient MANET routing scheme. J. Sens. **2015**, 10 (2015)

20. Chen, H., Chen, W., Wang, Z., Wang, Z., Li, Y.: Mobile beacon based wormhole attackers detection and positioning in wireless sensor networks. Int. J. Distrib. Sens. Netw. **2014**, 10 (2014)
21. Choi, H.-K., Kim, I.-H., Yoo, J.-C.: Secure and efficient protocol for vehicular ad hoc network with privacy preservation. EURASIP J. Wirel. Commun. Netw. **2011**, 15 (2011)
22. Chuang, M.C., Lee, J.F.: TEAM: trust-extended authentication mechanism for vehicular ad hoc networks. IEEE Syst. J. **8**(3), 749–758 (2014)
23. Li, W., Yi, P., Wu, Y., Pan, L., Li, J.: A new intrusion detection system based on KNN classification algorithm in wireless sensor network. J. Electr. Comput. Eng. **2014**, 8 (2014)
24. Sekaran, R., Parasuraman, G.K.: A secure 3-way routing protocols for intermittently connected mobile ad hoc networks. Sci. World J. **2014**, 13 (2014)
25. Varshney, T., Sharma, T., Sharma, P.: Implementation of watchdog protocol with AODV in mobile ad hoc network. In: Fourth International Conference on Communication Systems and Network Technologies, Bhopal, pp. 217–221 (2014)
26. Yuan, W.: An anonymous routing protocol with authenticated key establishment in wireless ad hoc networks. Int. J. Distrib. Sens. Netw. **2014**, 10 (2014)
27. Vasserman, E.Y., Hopper, N.: Vampire attacks: draining life from wireless ad hoc sensor networks. IEEE Trans. Mob. Comput. **12**(2), 318–332 (2013)
28. Wang, B., Chen, X., Chang, W.: An efficient trust-based opportunistic routing for ad hoc networks. In: International Conference on Wireless Communications and Signal Processing (WCSP), Huangshan, pp. 1–7 (2012)
29. Gonzalez, J.M., Anwar, M., Joshi, J.B.D.: Trust-based approaches to solve routing issues in ad-hoc wireless networks: a survey. In: IEEE 10th International Conference on Trust, Security and Privacy in Computing and Communications, Changsha, pp. 556–563 (2011)
30. Neeli, J., Cauvery, N.K.: Insight to research progress on secure routing in wireless ad hoc network. Int. J. Adv. Comput. Sci. Appl. (IJACSA) **8**(6), 68–76 (2017)

SDQE: Sensor Data Quality Enhancement in Reconfigurable Network for Optimal Reliability

B. Prathiba[1(✉)], K. Jaya Sankar[2], and V. Sumalatha[2]

[1] Department of ECE, Jawaharlal Nehru Technological University, Anantapur, India
balireddyprathibha@gmail.com
[2] Department of ECE, Vasavi College of Engineering, Hyderabad, India
kottareddyjs@gmail.com, sumaatp@yahoo.com

Abstract. The future applications include a multi-technology based paradigm, which includes wireless sensor network, Internet of things and cloud computing to be synchronized for the accurate and real-time analytics. The applications client will be a smart phone user who will request the wireless sensor network data or key data points of sensors through the universal data centers. This paper highlights the problem identification of the sensor data quality and reliability aspects by proposing a model for sensor data quality enhancement (SDQE) by synchronizing the priority of critical data with time factor. The data request prediction based optimization is proposed to maximize the usefulness factor which is the measure of sensor data quality as reliability and minimize the energy consumption. The model is simulated in numerical computing platform and found acceptable response.

Keywords: Wireless sensor network · Internet of Things (IoT)
Sensor data quality

1 Introduction

The applications based on the wireless sensor network (WSN) is beyond the independent network of sensors, rather it is now scalable to a high extend with the event of Internet of things and ubiquitous computing [1]. The paradigm of such applications includes many WSNs as a sub-network of a larger system [2], where many sensors in a WSN and many such WSNs along with the suitable gateway synchronized with the step-up to the data centers to stream the data. The applications interface in such system will be in the user's portable devices where the data point of importance is acquired directly from the universal data centers. The timeliness, and data quality is the measure of reliability [3]. The vision of unique and ubiquitous applications solely depends on the data quality in terms of reliability. The typical architecture of the model is illustrated in Fig. 1, where typical WSNs for respective applications acquires real-world data and pushes it to nearest gateway using suitable communication protocols (6LoWPAN). Further the gateways pushes data through a dedicated channel to the data centers via fog/edge computing mechanism. This way the

© Springer International Publishing AG, part of Springer Nature 2019
R. Silhavy (Ed.): CSOC 2018, AISC 765, pp. 356–363, 2019.
https://doi.org/10.1007/978-3-319-91192-2_35

wireless sensor networks (WSNs) is the new trend given by Moore's Law for the future of computing device [4]. With growing deployment of the sensor networks the issue of data quality is an open research problem for designing the future generation sensor systems. A detailed survey has been conducted towards the requirements, issues, current trend and research gaps in enhancing the data quality in WSNs [5].

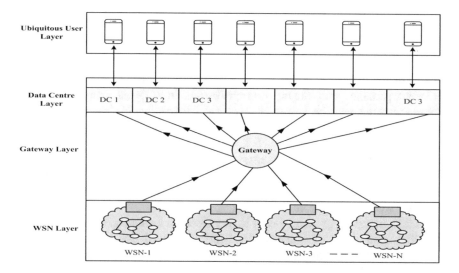

Fig. 1. Architectural representation of reconfigurable network with sensor network, gateway and date centre

In various ubiquitous applications, user's applications can drive the needs of specific information with desired format or particular quality whereas quality may depend upon various operating conditions too. The traditional approach of the offline outlier identification ceases its utility when the applications requires multi-layered approach of computing as similar context of the architecture proposed [6]. Achieving data purity is very contextual to the applications which need to be achieved in highly distributed and up-scaled WSN with high accuracy with optimal energy usage in the real-time scenario. The reconfigurable network is referred to a context where WSN is jointly studied with Internet-of-Things (IoT), Optical Network, Cloud computing etc. Such joint implications of heterogeneous network and technologies are also called as reconfigurable network (RCN). This paper aims to propose a novel algorithm to ensure reliability of sensory data processing in reconfigurable network of WSN. It is a very challenging task to ensure data quality in RCN as individual network operates on varied protocols to each other, so in such cases the data quality is checked using reliability factor as the data need to be processed from one to other layers in reconfigurable network. Many middleware's based services were proposed for this task [7, 8]. The proposed study does not consider designing (or enhancing any middleware approach), as basically such approaches are prominently hardware based approach so it cannot be validated effectively owing to different standardization, and most importantly, it belongs to third party applications. Hence the prime problems identified are mainly (1) design of a universal networking interface that supports reconfigurable network, (2) The effect of energy consumption, mobility factor and uncertainty of the user behaviour degrades the quality of the aggregated data from RCN. These two tasks are very changing to handle. Section 2 explains state of art related works, Sect. 3 discusses the methodology adopted for proposed solutions along with critical algorithm, and finally Sect. 4 illustrates results and analysis followed by conclusion.

2 Review of Literature

With the increasing demand of extension of wireless sensor network based applications there was a paradigm shift into the research direction in WSN and many reconfigurable networks were evolved, the work of Ghorbel et al. [9] describes the importance of having very high-level scalability as well as performance of real-time processing and storage of the data generated from the WSN. These can be achieved by proper synchronous mechanism with the cloud where sensor -cloud architecture may provision low cost solutions. The work of Maddar et al. [10] clearly describes the limitations of the restriction due to constraint resource in WSN and advocates that the integration of WSN with the cloud makes these networks scalable.

Apart from designing such RCN, it is very essential to validate the reliability of the WSN by measuring its extended lifetime. In the work of Baccarelli et al. [11], the convergence aspect of research from complex sensor data along with self-configurable platform to process data generated from the resource constrained sub-systems of WSN is described along with real-time energy efficiency optimization for RCNs. They validated the network lifetime performance of RCN with traditional resource.

The work carried out for resource optimization for achieving higher network lifetime does not consider the aspect of usefulness of sensory data which affect the quality for many analytics requirements. Zhu et al. [12], in their work, have investigated the synchronization among the ubiquitous users with certain mobility, WSNs and many non-context situations which demand some quality data to operate the application in order to provide accurate results. They mainly focus on the issues like recommendations of data, prediction of data along with feature information obtained from the data as per the real-time scenario.

Further the work towards the data quality and reliability, one of the state of art work by Zhu et al. [13], elaborates that the authors motivation is to facilitate the focus on various ubiquitous applications for pro-healthcare, real-time monitoring of the environment for managing the disasters many such applications for agriculture, structural health monitoring etc. where the application is sensitive enough to the reliability of the sensor data obtained from the RCNs. Then further many other works as a trail to this is being done by Yang et al. [14], Zhu et al. [15], Wan et al. [16].

In the context of RCNs, the data travel from various sources of sensor networks and due to some system failure there will be missing data which characterises this data with veracity. The uncertainty of the data needs intelligent prediction mechanism and one of the works by Lin et al. [17], in their work has handled the issue of veracity from the view point of trust. The literature available lack the issue of provisioning the data quality issues, challenges and techniques in RCNs. This paper introduces with a system model SDQE to handle the data reliability in terms of the usefulness as well as optimization of the energy usages.

3   System Model for SDQE

The system model of SDQE adopts an analytical modelling approach by developing a system performance functions to measure the reliability factor as usefulness for system dynamics of the proposed architecture of SDQE. The design constraints consider essentials components such as sensor networks (SN), number of nodes (N), number of ubiquitous clients (UC), along with essentials parameters such as iterations of communication cycles, slots for time (Ts), energy balance (E) etc. The design integrates all these components of WSN and pervasive computing to generate RCN. With the created RCN an algorithm is formulated for message prioritization by recording sensor data points with slots of a fixed time interval. The challenge of handling heterogeneity of devices and protocols involved in RCN is normalized with a filed bias factor (Ω). There exists a limitation to validate all possible forms of the RCN; therefore a probabilistic approach is adopted for data request probability based on slots of time.

3.1   Schema of SDQE

The schema of the sensor data quality enhancement in reconfigurable network for optimal reliability (SDQE) is shown in Fig. 2, which describes the building blocks of the model. The components of the schema include (1) reconfigurable network, (2) Characteristics modelling, (3) wireless sensor network, (4) pervasive computing, (5) Algorithm for (5a) message prioritization, (5b) Time-based prioritization, (5c) minimization of resource, 5d) energy model all using probability theory and finally, (6) computation of reliability factor of aggregated data.

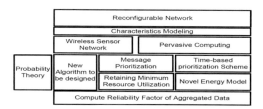

Fig. 2.   Schema of DQ Enhancement of Reconfigurable Network (SDQE)

3.2   Data Request Probability (DRP)

The data request probability requires, random localization of each WSN as (x, y) in a given deployment area. The system model is designed to check the scalability efficiency by taking independent variable of Number of sensor network as Ns with flexible number of nodes (N) in each sensor network as well as the number of ubiquitous clients (uc).

The model can be initialized with initial energy Eo which could be 100% in the beginning. As per the architecture the data centre is placed with defined location of (DCx, DCy) with its servers' units.

Algorithm to create data request probability

Input: x,y, Ns, N, uc, Eo , DCx, DCy
Output: Point of interest(POI), service Request (of S_R)
        Divide D in 24 hr time slot(Ts)
        Ts = {1, 2, 3…24}
            for all uc
                  generate request(Ruc)
                        for different time(Ts)
                                create time priority table
                                      for time instance
                                          calculate probability
                                          Calculate (P) of S_R
                                          Convert to probability
                                          Create POI table
                                end
                  end
        end

The point of interest table based on the probability factors of service request from various ubiquitous applications provides a base for developing mechanism of the missing data predictions, which is planned in our future work for achieving data purity where some machine learning algorithms will be explored on the same framework of SDQE. The discussion of the data veracity handling to meet data quality requirements in RCNs is beyond the scope of this paper. One of the instances of data request probability is shown in the Fig. 3 for different time slots and it can be generated from N numbers of ubiquitous users (UCs).

In Fig. 3, the time priority table for the data request probability for time slot is illustrated for 5 ubiquitous users.

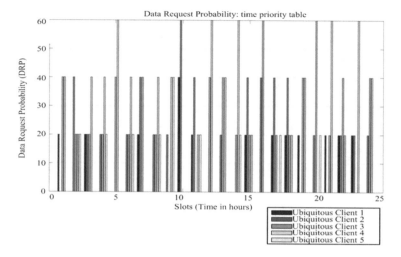

Fig. 3. Time priority table: Data Request Probability (DRP) vs time slot

4 Results and Analysis

Figure 4 include three performance graphs where the first column of the table shown the energy remains at every iteration of the communication cycle and it is shown that there is uniformity and consistency, which is also being justified by getting the consistent trend of number of nodes with available energy (Alive Node) in each communication cycle. Most important trend observed in the proposed model is the very uniform reliability for each node against the number of communication cycle.

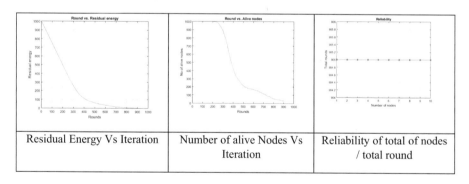

Residual Energy Vs Iteration	Number of alive Nodes Vs Iteration	Reliability of total of nodes / total round

Fig. 4. Performance graphs for residual energy, trend of live load, reliability

Figure 5 illustrates the pattern of the usefulness as a reliability factor against each round of the communication iteration and validates its consistency.

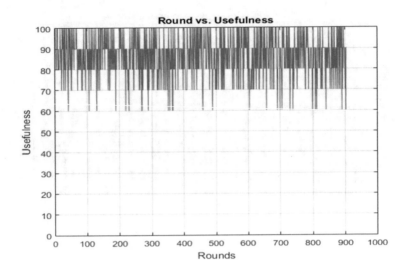

Fig. 5. Usefulness verses iteration

5   Conclusion and Future Work

This paper proposes a unified framework to create a scalable WSNs synchronized with the gateways and data centres namely SDQE for building a bed for introducing a reliability for ubiquitous applications with optimization of energy consumption and reliability in terms of user data demand with time slots by using probabilistic model. In future direction to extend this work the aspect of data veracity with some machine learning approach will be initiated to normalize the effect of mobility factor and other filed parameters which degrades the data quality in such RCNs.

References

1. Kocakulak, M., Butun, I.: An overview of wireless sensor networks towards Internet of things. In: IEEE 7th Annual Computing and Communication Workshop and Conference (CCWC), Las Vegas, pp. 1–6 (2017)
2. Çetin, R., Kadioğlu, T., Cesur, E., Ayten, U.E.: Design of wireless sensor network for Internet of Things structure. In: National Conference on Electrical, Electronics and Biomedical Engineering (ELECO), Bursa, pp. 402–405 (2016)
3. Zhu, C., Sheng, Z., Leung, V.C.M., Shu, L., Yang, L.T.: Toward offering more useful data reliably to mobile cloud from wireless sensor network. IEEE Trans. Emerg. Top. Comput. 3(1), 84–94 (2015)
4. Sen, S.: Invited: context-aware energy-efficient communication for IoT sensor nodes. In: 53rd ACM/EDAC/IEEE Design Automation Conference (DAC), Austin, pp. 1–6 (2016)
5. Prathiba, B., Sankar, K.J., Sumalatha, V.: Enhancing the data quality in wireless sensor networks — a review. In: International Conference on Automatic Control and Dynamic Optimization Techniques (ICACDOT), Pune, pp. 448–454 (2016)

6. Ghorbel, O., Ayedi, W., Snoussi, H., Abid, M.: Fast and efficient outlier detection method in wireless sensor networks. IEEE Sens. J. **15**(6), 3403–3411 (2015)
7. Cecílio, J., Furtado, P.: Existing middleware solutions for wireless sensor networks. In: Wireless Sensors in Heterogeneous Networked Systems. Computer Communications and Networks. Springer, Cham (2014)
8. Ajana, M.E.K., Harroud, H., Boulmalf, M., Elkoutbi, M.: Middleware Architecture in WSN. In: Benhaddou, D., Al-Fuqaha, A. (eds.) Wireless Sensor and Mobile Ad-Hoc Networks. Springer, New York (2015)
9. Alamri, A., Ansari, W.S., Hassan, M.M., Hossain, M.S., Alelaiwi, A., Hossain, M.A.: A survey on sensor-cloud: architecture, applications, and approaches. Int. J. Distrib. Sens. Netw. **9**(2) (2013). Article ID 917923
10. Maddar, H., Kammoun, W., Youssef, H.: Cloudlets architecture for wireless sensor network. In: Madureira, A., Abraham, A., Gamboa, D., Novais, P. (eds.) Intelligent Systems Design and Applications. ISDA 2016. Advances in Intelligent Systems and Computing, vol. 557. Springer, Cham (2017)
11. Baccarelli, E., Cordeschi, N., Mei, A., Panella, M., Shojafar, M., Stefa, J.: Energy-efficient dynamic traffic offloading and reconfiguration of networked data centers for big data stream mobile computing: review, challenges, and a case study. IEEE Netw. **30**(2), 54–61 (2016)
12. Zhu, C., Leung, V.C.M., Wang, H., Chen, W., Liu, X.: Providing desirable data to users when integrating wireless sensor networks with mobile cloud. In: Proceedings of the IEEE 5th International Conference on Cloud Computing Technology and Science (CloudCom), pp. 607–614 (2013)
13. Zhu, C., Sheng, Z., Leung, V.C.M., Shu, L., Yang, L.T.: Toward offering more useful data reliably to mobile cloud from wireless sensor network. IEEE Trans. Emerg. Top. Comput. **3**, 84–94 (2015)
14. Zhu, C., Leung, V.C., Shu, L., Ngai, E.C.H.: Green Internet of Things for smart world. IEEE Access **3**, 2151–2162 (2015)
15. Zhu, C., Leung, V.C., Yang, L.T., Shu, L., Rodrigues, J.J., Li, X.: Trust assistance in sensor-cloud. In: IEEE Conference on Computer Communications Workshops (INFOCOM WKSHPS), pp. 342–347. IEEE, April 2015
16. Wan, L., Han, G., Shu, L., Feng, N., Zhu, C., Lloret, J.: Distributed parameter estimation for mobile wireless sensor network based on cloud computing in battlefield surveillance system. IEEE Access **3**, 1729–1739 (2015)
17. Lin, H., Hu, J., Tian, Y., Yang, L., Xu, L.: Toward better data veracity in mobile cloud computing: a context-aware and incentive-based reputation mechanism. Inf. Sci. **387**, 238–253 (2017)

Relaxed Greedy-Based Approach
for Enhancing of Resource Allocation
for Future Cellular Network

Chanda V. Reddy[1([⊠])] and K. V. Padmaja[2]

[1] Department of Telecommunication Engineering, K.S. Institute of Technology, Bangalore, India
cvr.badami@gmail.com
[2] Department of Electronics and Instrumentation Engineering, R.V. College of Engineering,
Bangalore, India
padmajakv@rvce.edu.in

Abstract. The study considers the resource allocation (R_{Al}) for Orthogonal Frequency Division Multiple Access (OFDMA) future cellular network (i.e., Cloud-RAN), where multiple mobile operators can distribute the Cloud-RAN infrastructure as well as network resources possessed by infrastructure providers. We have designed the resource allocation system by solving the dual-coupled problems at two distinct levels (i.e., Upper Level and Lower Level). The first level problem responsible for slicing the front haul capacity (F_{cap}) and computation of cloud resources for all operators (O_p's). This would indeed tend to increase the overall profits for each O_p as well as infrastructure provider by accounting the numerical constraints on F_{cap} and computational resources. The study introduces a dual-level algorithmic approach to solve this two level R_{Al} problem. At first-level, system considers both U_{level} and L_{level} problems by relaxing discrete values with continuous ones. While in the second-level, we introduce two rounding methods to solve the optimal relaxed problems and attain a practical solution for the proposed problem. Finally, simulation results show that the designed algorithms efficiently perform the greedy approach to resource allocation and attain the discrete value very near to the total rate of upper bound acquired by solving resource allocation relaxed problems.

Keywords: Cloud-RANs · Discrete rate · OFDMA · Resource allocation

1 Introduction

With the increasing demand for wireless network and multimedia communication services needs more reliable and high-speed data transmission over wireless systems. Utilizing the wireless network temporal variations, the access point or base station can achieve multi user diversity through dynamic-scheduling of data transmission to users with acceptance network conditions. This mechanism is suitable for packet delay tolerant network traffic. Though the gain of multi-user diversity increases the throughput of the entire system, this may result in declined the system performance regarding transmission delay. To stabilize the delay and efficiency, a dynamic wireless network is

© Springer International Publishing AG, part of Springer Nature 2019
R. Silhavy (Ed.): CSOC 2018, AISC 765, pp. 364–373, 2019.
https://doi.org/10.1007/978-3-319-91192-2_36

considered for designing of cross-layer framework [1]. Nevertheless, the next generation wireless communication networks (e.g., 4-G/5-G) are expected to results the higher capacity regarding cost-effective manner which supports the tremendous growth in network traffic and wireless services [2, 3]. Some existing methods have shown that the conventional system of single state network architecture could not be efficient since the average of bit loading demand is typically lesser than the intended peak demand [3, 4]. Hence, the new accessing techniques for Cloud-based Radio Access Network (Cloud-RAN) and Virtualized Wireless Network (VWN) have to pay more attention to both academic and industry.

By perceiving several communications and processing functions in the cloud, Cloud-RAN allows more efficient utilization of resources, which provides the high network throughput and minimizing the network operation and costs. With the help of VWNs and multiple operators can efficiently share the several network resources like as Radio-Spectrum, computing resources, and backhaul capacity many more; therefore to attain a possible profit of the Cloud RAN and VWN technologies, has to address numerous technical challenges [4]. The resource allocation determines the allotment of centralized computation resource [5, 6], radio spectrum and energy allocation in Cloud-RAN and slicing of infrastructure resources for various operators to optimize the desired model objectives in VWN which are the primary research challenges [7–9].

Therefore, this paper represents a novel of OFDMA based virtualized future cellular network approach to provide an end to end network slicing. The Sect. 2 provides an existing research work followed by problem definition in Sect. 3. Section 4 illustrates the proposed research methodology and algorithm implementation will discuss in the Sect. 5. The comparative analysis of accomplished result is discussed in the Sect. 6. Finally, the research conclusion is presented in Sect. 7.

2   Related Work

In this section, the author discusses some existing research study being carried out toward OFDMA techniques adopted for enhancing bit allocation performance. The prior study [10] has discussed existing spectrum sharing techniques towards 5-G networks. The aim was to exploit radio spectrums very efficiently, to minimize the future spectrum requirements and maximize the small cells network capacity. Especially, the paper work mainly focuses on to achieve more throughput during operators have the same traffic patterns. In [11] Heuveldop, have addressed the virtual resource slicing problem, which increases the system capacity regarding function to accomplished a cumulative rates and allocated the resource slots when meeting the requirement of every slice is studied. In [12] author introduced a dynamic resource allocation mechanism which addressed the network slicing problems for various operators. Also represented the new mechanism, which provides a flexible definition of requirements for various service providers. The system performance was calculated regarding time complexity concerning the normalized sum rate for various service providers. Reddy and Padmaja [13] have presented a mechanism of packet-prioritization to maximize the performance of both bit-loading and power-allocation on OFDM access exploiting a group of users. The system evaluation was carried out by measuring of system performance with conventional resource-allocation techniques.

Wang et al. [14] have proposed the virtualized network resource (VNR) scheme and leveraged the virtual base station mechanism which is dynamically designed for small cells by assigning of VNR, i.e., virtual front haul link connecting the digital-unit and radio-unit and formulated the system optimization problem and traffic patterns. The simulation results showed that proposed Cloud-RAN with virtual base station scheme achieves the significant energy consumption compared with conventional RAN with Cloud-RAN. In [15] Luo et al. has introduced the uplink and down-link energy minimization strategy to support the mobile users and increases the mobile data rates. The study also addressed the two joint algorithms for uplink and downlink transmission and designed a beam forming model whose performance was calculated and analyzed by comparing with other existing techniques. Shi et al. [16] have introduced a novel design of cloud radio access network and computed a low cost low powered remote radio access which is responsible to energy minimization beam forming problem. For a solution, authors have proposed a Greedy algorithm which gives the near optimum performance and showed that designed algorithm efficiently reduced the network energy consumption and illustrated the significance of taking the transmission link energy consumption. In [17] Ha et al. illustrated the combined multipoint transmission scheme for downlink C-RANs. The prime objective was to optimize the remote radio heads helping each user and reduce the total power transmission during supervision of front haul capacity and quality of service constraints.

Additionally, the data compression problem for minimization of front haul capacity constraint has been introduced [18] where focused on reducing the capacity of data transmission over the front-haul network. Whereas in [19] Lui and Zhang have proposed uplink transmission of OFDMA based Cloud RAN model, where the mobile operators are allotted with orthogonal sub carriers for data transmission. Particularly, this study addressed the joint wireless power management and front haul quantization model for sub carriers which increase the throughput. The designed algorithm solved the joint optimization problem while front haul quantization scheme was applied. In [20] authors considered the uplink cloud-based radio access network model and have introduced distributed front haul compressed model which receives the signals from remote radio heads and incorporated the multi-access fading scheme in cloud RANs into compressive sensing computation and manage end to end recovery of the transmitted signal from the mobile users. Finally, from the simulation results, authors evaluated the probability of the accuracy of active user detection and measured the relationship tradeoff between uplink-capacity and front haul compression in cloud RAN. Werthmann et al. [21] introduced a cloud base-station (BS) concept for 4-G communication networks. In this study author presented a multi-layer framework which illustrates the network traffic, user distribution, a mobile radio network (MRN) and formulates resource requirements. By implementing this system, author measured the simulation results regarding user load distribution concerning data traffic.

3   Problem Description

To the best of our insight, the future cellular networking model is considering to restricts on limited front haul capacity and computational resources has not yet considered in the existing studies and also not consider the virtual Cloud-based Radio Access Network (Cloud-RAN). Though spectrum-allocation in centralized networks also illustrated in existing work; it is merely considered as a simple approach. Additionally, compared with general C-RAN architecture proposed virtualized OFDMA model for wireless networks is more efficient to enhance the resource allocation problem.

4   Proposed Methodology

An analytical research technology is exploited on proposed research methodology by referring the study [22]. The primary aim of the proposed system is illustrating a novel dual-level network slicing algorithm for improving the resource computation and transmission rate by considering the optimal rate and quantization bit-allocation in virtualized cloud-based radio access networks.

The Fig. 1 highlights the schematic view of the dual-level network slicing approach. The system considering the virtual resource allocation model for OFDMA based Cloud RAN where service provider leases the resources to various operators which support their corresponding mobile operators. The system implementation carryout by dual level coupled problem (i.e., upper-level and lower level). The first level (U_{Level}) determine the resource slicing (Sl) method for resource computation and front haul capacity which improve the benefits of service provider and operators. While on the second level (L_{level}) system determine the resource allocation for each mobile operator. Particularly, each lower level problem is solved by the corresponding mobile operator which increases the

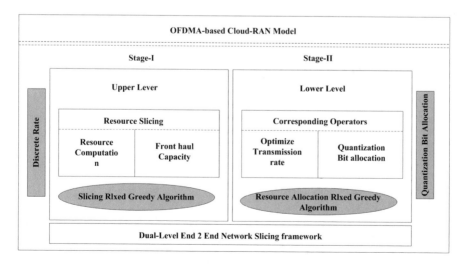

Fig. 1. Schematic view of dual-level network slicing model

discrete rate of its operators by considering the quantization bit-allocation and transmission rate for resource slicing approach.

The proposed system develops a dual stage network slicing model to solve the dual-
based coupled problem. In the first stage, the system determines the Resource Allocation
Relaxed (R_A-Rlxed) problem and in the second stage system deals with optimum rate
and bit allocation problem for two given points, which allows the system to formulate
an efficient algorithm to solve the R_A-Rlxed problem. For computation performance
system evaluates the designed algorithm regarding number of resource blocks
concerning running time (sec).

5   Algorithm Implementation

The prime reason for the proposed algorithm is to solve a dual-level resource allocation
(R_A) computation where algorithm solves the relaxed (R_{lxed}) problem by applying
Greedy R_A algorithm. The extensive simulation studies are conducted where system
evaluate the efficiency of the developed algorithm along with the impacts of various
system parameters on the system running time. The below algorithm illustrates the
implementation process of the proposed system.

To obtain a discrete and feasible solution for the lower level problem, the system
considers the rounding method to generate the corresponding discrete values. The
rounding method must be applied carefully since the generating discrete results may not
satisfy the actual cloud computation and front haul constraints. In the following section
system defining two rounding algorithms:

Algorithm-II: (IR) Rounding Algorithm

Input: Number of O_p
Output: Solve the R_{lxed} Lower-Level Problem
Start
    1. **Initialize** $D_r \rightarrow R_{mini}$ for all (k,s)
    2. Select the tolerance variable τ for convergence
    3. **Repeat**
        a. **for** k $\in$ K **do**
            i. Repeat
                1. Fix D_r and update F_{cap}
                2. Fix F_{cap} and update D_r
            ii. Until convergence
        b. **End for**
        c. Evaluate all X_k^s with new D_r and F_{cap}
        d. **Update** sub gradient variable
    4. **End**
End

The above algorithm iteratively performs the convergence operations for each O_p.
Initially, the algorithm considers the number of O_p and selects the D_r (data rate) for

corresponding cell (k), which is nearer to the value in R and set it to that value for following iterations. This algorithm impalement step by step rounding process which will repeat until convergence.

Algorithm-III: Rounding and Adjusting (RA) Algorithm

Input: Number of O_p

Output: Solve the R_{lxed} Upper-Level Problem

Start

    1. **Initialize** $C_{eff} \rightarrow C_{cloud}$ and $F_{cap} \rightarrow F'_{cap}$

    2. **Repeat**

        a. Calculate front haul capacity and computational effort

    3. **Update** C_{eff} and F_{cap}

    4. **End**

End

The above algorithm represents the single-time rounding and adjusting process which implements in two stages for each O_p. Initially system round all $[D_r]$ and $[F_{cap}]$ to their corresponding rates in R. Then, if any quantization constrains are desecrated, system round down the relating variables step by step where variable that affects the desecrated constraints the majorly selected for each rounding process.

Algorithm-IV: Slicing Relaxed (Sl-R_{lxed}) Greedy Algorithm

Input: Number of Resource blocks

Output: resource slicing set of O_p's

Start

    1. **Initialize** number of network resource for all operator (O_p)

    2. **for** O_p o do

        a. Evaluate the front haul capacity (F_{cap}) for remote radio heads

        b. Set $(F_{cap}) \leftarrow [F_{cap}]'$, $\forall(k,s) -> K \times S_k$

        c. Set $R_k \rightarrow \max[R, s, t \leq t(F_{cap}), \forall(k,s)$

    3. **End of for**

    4. Evaluate the upper bound of the discrete rate of all O_p

    5. Find (K, S) $\rightarrow$ max $[C_k]$

    6. Minimize the (R_k)

End

The implementation of the above-mentioned resource slicing algorithm was carried out over a specific number of *resource blocks* deployed randomly over a *simulation area*. Initially, this algorithm processes in two stages. In the first stage, the network resources of infrastructure providers are sliced into operators based on its upper-bound rates. Later, every operator allocates the sum rate and bits quantization to its mobile users. Specifically, the system initially estimates the upper-bound of sum rate on each resource block (K) and cell (S). Then system allocates the front haul capacity and resource computation for distinct operators (O_p) based on the upper-bound of the rate of particular O_p. while in the second stage, the system solves the Lower-level problem for each O_p. Particularly, the system optimizes the bit-allocation to increase the discrete rate of all mobile users by relaxing the corresponding variables to continual ones and resolving the problem.

Algorithm-I: Resource Allocation (R_{Al}) Relaxed (R_{lxed}) Greedy Algorithm

Input: Number of Resource blocks

Output: Optimize resource allocation

Start

1. **Initialize** number of network resource for all operator (O_p)

2. Select data rate (D_r) on resource block as the higher value in R   where D_r $\leq \min(S_k)$

3. **Repeat**

    a. Evaluate F_{cap} and get feasible vector 'b' by using flooring operation to (b)'

    b. If vector 'b' is feasible then

        i.   Select D_r ($D_r > R_{mini}$) equivalent to the largest value of resource blocks assigned to O_p in cell 'k'

        ii. Minimize its total rate to lower in R

    c. **End if**

4. **End**

In the above algorithm (R_{Al} R_{lxed}-Greedy.Algo) system alleviate the difficulty of R_{Al} for each O_p by setting the sum rate and quantization bits for every mobile user from O_p in each cell (k). Then sum rate required by the cloud to precisely decode the signals for each user from O_p in (k) cell. Particularly system considers the Lower-level problem, and it can be referred as a dual-level problem for given variables of D_r's into a single level problem. Therefore, this single-level relaxed problem can be resolved smartly to obtain the network blocks for all O_p. Mainly, the system finds the data rates (D_r) from higher to lower which describe the effectiveness of the proposed problem.

Therefore, the optimization and bits quantization problem are carried out by solving the dual-level problem for all users from a set of O_p's. Based on the implementation results system evaluates the accuracy of the proposed algorithm. The following section illustrates the results accomplished from the proposed study (Table 1).

Table 1. Summery of notations

Symbole	Description
O_p	Operator
D_r	Data rate
K	Number of cells (blocks)
C_{cloud}	Cloud Computational resource
C_{eff}	Computational effort
F_{cap}	Front haul capacity
K	Number of resource blocks
S	Set of cells
S_k	Set of resource blocks
Ck	Required C_{eff} for signal S_k in cell k
R_k	Data rate on resource block in cell k
R_{Al}	Resource allocation
R_{lxed}	Resource allocation relaxed

6   Results Discussion

What this section discusses simulation study experimented in MatLab environment. For result analysis, the system considered the three set of O_p's and assigned some resource blocks to each operator. Also, the system randomly allocates the resource blocks to all users in each cell. During simulation process, system set the limit of the F_{cap} for various cells (k). The simulation process evaluates the complexity of the proposed algorithms where Fig. 2 illustrates the number of resource blocks vs. average processing time. The system processing time values are generated by averaging over (1000) runs. It can be noticed that proposed first two algorithms (i.e., prop. IR.Algo, prop RA.Algo) need more running time by comparing with other two algorithms (i.e., SI R_{lxed} Greedy.Algo and R_A R_{lxed} Greedy.Algo). Additionally, the proposed Rounding algorithms contain the longest processing time whereas the Relaxed-Greedy algorithm implements in a short period. Thus the computation running time maximizes as the number of resource blocks in each cell maximizes as expected.

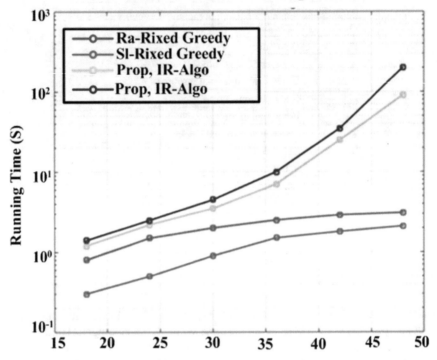

Fig. 2. Simulation outcome of comparative performance analysis

7   Conclusion

The proposed system introduced a new OFDMA virtualized framework for future cellular networks (Cloud-RAN) which supports the multiple operators through sum rate and bit allocation for mobile users distributed by each operator. The proposed methodology aims to improve the total profit of infrastructure provider as well as operators by applying quantization constraints on front haul capacity and cloud-computation limit. The simulation results have discussed that proposed algorithms greatly perform the greedy R_{Al} algorithm and attain the discrete rate very closer to the upper bound sum rate acquired by resolving the relaxed problems.

References

1. Cisco Visual Networking Index: Global mobile data traffic forecast update, 2016–2021, Cisco, White Paper, February 2017
2. Cerwall, P., et al.: Ericsson mobility report, Ericsson, Stockholm, Sweden, Technical report, June 2017
3. China Mobile Research Institute, "C-RAN: The road towards green RAN," China Mobile, White Paper (2011)

4. Checko, A., et al.: Cloud RAN for mobile networks - a technology overview. IEEE Commun. Surv. Tutor. **17**(1), 405–426 (2015)
5. Wubben, D., et al.: Benefits and impact of cloud computing on 5G signal processing: flexible centralization through cloud-RAN. IEEE Signal Process. Mag. **31**(6), 35–44 (2014)
6. Suryaprakash, V., Rost, P., Fettweis, G.: Are heterogeneous cloud-based radio access networks cost-effective? IEEE J. Sel. Areas Commun. **33**(10), 2239–2251 (2015)
7. Costa-Perez, X., Swetina, J., Guo, T., Mahindra, R., Rangarajan, S.: Radio access network virtualization for future mobile carrier networks. IEEE Commun. Mag. **51**(7), 27–35 (2013)
8. Liang, C., Yu, F.R.: Wireless network virtualization: a survey, some research issues, and challenges. IEEE Commun. Surv. Tutor. **17**(1), 358–380 (2015)
9. Peng, M., Wang, C., Lau, V., Poor, H.V.: Fronthaul-constrained cloud radio access networks: insights and challenges. IEEE Wirel. Commun. **22**(2), 152–160 (2015)
10. Luoto, P., Pirinen, P., Bennis, M., Samarakoon, S., Scott, S., Latva-Aho, M.: Co-primary multi-operator resource sharing for small cell networks. IEEE Trans. Wireless Commun. **14**(6), 3120–3130 (2015)
11. Kokku, R., Mahindra, R., Zhang, H., Rangarajan, S.: NVS: a substrate for virtualizing wireless resources in cellular networks. IEEE/ACM Trans. Netw. **20**(5), 1333–1346 (2012)
12. Kamel, M.I., Le, L.B., Girard, A.: LTE wireless network virtualization: dynamic slicing via flexible scheduling. In: Proceedings of the IEEE VTC Fall, pp. 1–5, September 2014
13. Reddy, C.V., Padmaja, K.V.: Leveraging communication performance for OFDMA using novel bit loading and allocation of power. In: International Conference on Wireless Communications, Signal Processing and Networking (WiSPNET), Chennai, pp. 1606–1610 (2016)
14. Wang, X., et al.: Energy-efficient virtual base station formation in optical-access-enabled cloud-RAN. IEEE J. Sel. Areas Commun. **34**(5), 1130–1139 (2016)
15. Luo, S., Zhang, R., Lim, T.J.: Downlink and uplink energy minimization through user association and beamforming in C-RAN. IEEE Trans. Wirel. Commun. **14**(1), 494–508 (2015)
16. Shi, Y., Zhang, J., Letaief, K.B.: Group sparse beamforming for green cloud-RAN. IEEE Trans. Wirel. Commun. **13**(5), 2809–2823 (2014)
17. Ha, V.N., Le, L.B., Đào, N.-D.: Coordinated multipoint transmission design for cloud-RANs with limited fronthaul capacity constraints. IEEE Trans. Veh. Technol. **65**(9), 7432–7447 (2015)
18. Park, S.-H., Simeone, O., Sahin, O., Shamai (Shitz), S.: Robust layered transmission and compression for distributed uplink reception in cloud radio access networks. IEEE Trans. Veh. Technol. **63**(1), 204–216 (2014)
19. Liu, L., Bi, S., Zhang, R.: Joint power control and fronthaul rate allocation for throughput maximization in OFDMA-based cloud radio access network. IEEE Trans. Commun. **63**(11), 4097–4110 (2015)
20. Rao, X., Lau, V.K.N.: Distributed fronthaul compression and joint signal recovery in cloud-RAN. IEEE Trans. Signal Process. **63**(4), 1056–1065 (2015)
21. Werthmann, T., Grob-Lipski, H., Proebster, M.: Multiplexing gains achieved in pools of baseband computation units in 4G cellular networks. In: Proceedings of the IEEE 24th International Symposium on Personal, Indoor, and Mobile Radio Communications (PIMRC), pp. 3328–3333, September 2013
22. Necoara, I., Patrascu, A.: Iteration complexity analysis of dual first-order methods for conic convex programming. Optim. Methods Softw. **31**(3), 645–678 (2016)

ITM-CLD: Intelligent Traffic Management to Handling Cloudlets of the Large Data

Chetana Tukkoji[1(✉)] and K. Seetharam[2]

[1] Visveswaraya Technological University, Belagavi, India
chetana.rnd@gmail.com
[2] Department of CSE, NMAM-Nitte, Udupi, India

Abstract. The Cloud computing environment is the ultimate infrastructure for almost every kind of application missioned as a smart city or smart application concept, where heterogeneous network generating large data file of varied characteristics of increasing volume on time line which data at move as velocity etc. need to be stored, processed and analyzed. This paradigm shift from proprietary infrastructure to cloud infrastructures leads sudden and random loads as cloudlet to it. The conventional traffic management methods lack the robustness in terms of handling synchronization of heterogeneous network generated large data system. This paper proposes an intelligent traffic management namely ITM-CLD to provision a mechanism of varied traffic load in cloud environment for large data stream. The model ITM-CLD is simulated on numerical computation platform and computes performance metrics such as (1) Cloudlet handling time, (2) Unused resource, (3) Unused memory and finally, (4) Resource cost. Theses metrics are compared for different kinds of traffic load as job category with state of art work and it exhibits better performance.

Keywords: Traffic management · Resource allocation · Cloud computing

1 Introduction

The availability of sensor networks as Internet of Things (IoT), collaborative platforms as social networks generates a continuous time series data. These generated data may include images, text, video, audios, sensor readings, weblogs etc and having some unique characteristics such as its large size (volume), heterogeneity (variety) of data as well as uncertainty (veracity) into it along with a dynamic data (velocity) which is defined as big data. The storage of big data is not possible in the form of NTFS or UNIX file system. The Big data is stored in a distributed file system (DFS), where many nodes are clustered and data is stored on various nodes as DFS. Initially organizations adopted to have their own servers' rooms and cluster of nodes. But the proprietary setup of clusters is not cost effective as well as having many functional constraints. To overcome these constraints cloud computing service providers initiated to provision clusters of nodes to store, access and process the Big data on the cloud itself. There is many future smart city or smart custom applications where the heterogeneous network generates heterogeneous data through the mobile smart devices that forms a large stream of data

© Springer International Publishing AG, part of Springer Nature 2019
R. Silhavy (Ed.): CSOC 2018, AISC 765, pp. 374–381, 2019.
https://doi.org/10.1007/978-3-319-91192-2_37

as well as characterized as Big Data. There are many research pertaining to the data acquisition, storages, processes and handling distinguished characteristics of big data such as volume, variety, velocity, veracity etc. along with analytics but all these efforts will be fruitful if the cloud environment provided a better and optimal resource provisioning.

Though, the cloud computing infrastructures are very large in its resources but yet limited as well as there is growing pattern of shift from proprietary cluster of nodes to the cloud cluster. This paradigm shift leads the requirement of resource optimization by optimal traffic management processes. The cloud setups consisting data centers which claims higher I/O performance but due to the increasing pattern of setting up cluster-nodes on a cloud in order to store and process larger data there are cases of traffic blockage because of server blockage due to low processing constraints which leads lower performance to the data centers. Appropriate traffic architecture on random process especially when large data is to be considered as well it needs to be realized.

2   Related Work

In order to conduct detail survey for in-depth study of traffic management techniques or methods for handling larger files in the cloud computing environment, the digital library IEEE Xplore is consider as a primary source of data. For initial data collection a keyword cloud computing is used which gives 39,411 publication articles, which is further narrowed down by giving sub-search using a keyword traffic management that gives 986 articles including 775 conferences and 181 journals. Assuming that journal papers are state of art work, so finally papers from 2011 till date (13th–Jan 2018) is taken who having more emphasis on traffic management in context of large data.

There will be a human population explosion in the near future which paves the foundation to build a concept of smart city so that large population in the urban area can be facilitated with fast services of applications. These smart applications will be a collaboration of smart heterogeneous networks with sensors and communication synchronization. The work of Mazza et al. mainly handles the issue of handling the data transfer, data processing of the data generated from this heterogeneous network to cloud which is of large data type or even called big data [1]. In their work, they have presented a framework namely urban mobile cloud computing(UMCC) for offload mechanism using load balancing of large traffic coming from different networks synchronous to power and computation optimization. This work is a good contribution towards mobility and traffic control domain of large data handling on cloud. The smart phones of future will have a real-time synchronization with the cloud that leads robust and efficient mobile cloud computing requirements. Cloudlet is the emerging technology to balance between demand and the resource. In the work of Liu et al. issue of the balance between computational capacity of cloudlet and available bandwidth is handled by formulating a multi-resource allocation problem for latency sensitive mobile applications. The model utilizes semi Markov decision process (SMDP) with cost function minimization as linear optimization problem. To validate the model, they simulated it to check its robustness and found it outperforming for greedy admission control [2]. The Table 1 tabulates few important works towards review of the related literature. Feng et al. have considered the problem of cloud traffic management and provided the solution approach as a framework [3]. The problem of CPU resources management and handling uncertainty of the user request along with data center operation expenditure is being handle by means of decomposition and virtual machine scheduling by [5] Guan et al., [4] Assi et al., [5], Wanis et al. [6], Kantarci et al. [7] respectively. The resource allocation approach per handling sensitive job and internet data center is proposed by Xu et al. [8], Xu et al. [9] respectively. The work of Doyle et al. proposes load balancing approach for the big data traffic management on cloud [10]. The Quality of service for multimedia is studied with video transcoding approach by Wei et al. [11]. An extensive survey on big data traffic management on cloud is conducted by Chetana et al. to explain the issues trend and technologies [12]. Multitenancy is the one another problem which is handle using approach of virtualization computing by Duan et al. [13]. Literature witness the use of genetic algorithm for data center management, one of such work is proposed by Rankothge et al. [14].

Table 1. Node cluster (N_c) resource structure

Node cluster (Nc) type	Storage capacity (S_c)	Total VMs(VM_{tot})
Cloudlet (S): T1, T2, T3, T4	X, 2X, 4X, 8X	4Y, 3Y, 2Y, Y
Cloudlet (M): T1, T2, T3, T4	X, 2X, 4X, 8X	8Y, 6Y, 4Y, 2Y
Cloudlet (H): T1, T2, T3, T4	X, 2X, 4X, 8X	12Y, 9Y, 6Y, 3Y

Literature reveals that there has been many research related to the traffic management with the problems including CPU resource management, handling issues of uncertainty of cloudlet request, handling critical jobs, load balancing, resource allocation etc. with mechanism of framework, decompositions, genetic algorithm etc. There literature lacks a frame work to evaluate more synchronized architectures to compute various perform-ance metrics such as time take to handle the cloudlets, amount of unused resources, unused memory and value of resource cost. Computation of these values may provide better vision of traffic management for large traffic stream in cloud context.

3   ITM-CLD System Architecture

The typical system architecture of the proposed ITM-CLD is modeled as a mathematical model where the cloudlet Ci ∈ ({C1, C2, C3…Cn}, where i > 0 any natural number. There are three kinds of cloudlet traffic for the large data stream is considered namely as J ∈ {Small(S), Medium (M), High (H)}, which might be mapped for both the notion of the size as well as priority as an assumption to simplify the model. The basic archi-tectural diagram of the ITM-CLD is shown in the Fig. 1, where it shown that different application sources or networks is generating the data having characteristics of Big Data. These generated data are pushed via dedicated or shared channel to the upper layer of architecture called ITM-CLD cloudlet unit, which handle the arrived cloudlets. They arrived cloudlets are pushed to the cloud infrastructure through the virtualization layers of node clustered setup on the cloud (T1, T2…Tn) with different storage capacity and server or resource size, then the ITM-CLD protocol initiates to have a priory process of resource management to handle the traffic more intelligently. The Sect. 4 describes in details the scheme as an algorithm approach of ITM-CLD.

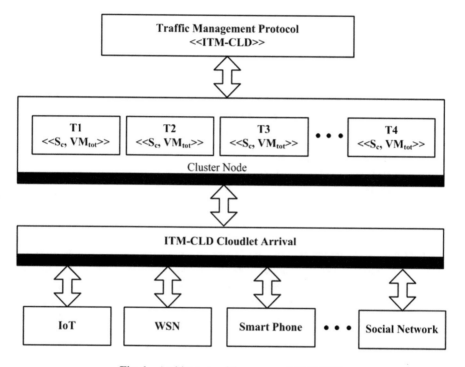

Fig. 1. Architecture of the proposed ITM-CLD

Setup of virtual machines (VM) as a node clusters (Nc), with different capacities is modelled as type (T), where T ∈ {Small Capacity (T1), Medium Capacity (T2), large capacity (T3), very large capacity (T4)}. The capacity of cluster node is designated with its storage capacity (Sc) as well as the total number of VMs (VMtot) in the Nc. Table 1 illustrates the structure of these four node clusters.

In Fig. 1 a layered architecture of ITM-CLD is shown where the ITM-CLD cloudlet arrival unit receives and manages different cloudlet of type S, M and H.

4   ITM-CLD Algorithm

The solution space for proposed ITM-CLD is realized with analytical objective functions for the dynamic system of the ITM-CLD as minimization of cloudlet handling time, optimization of unused memory as well as resource along with overall cost minimization function. The scheme utilizes alliance algorithm (IAA).

Intelligent Alliance Algorithm: **ITM-CLD**

Input: T, C, VM, Host Machine, Memory
Output:Ct, Ru, Mu, Cr
  Start
  Cv←[Xn]; Cv is host machine capacity
    Check
      Cv ≥ Memory
        Check
          Host machine available (un available)
            Update availability of host machine
              Update alliance vector
        Apply greedy approach by providing incentive
        Update incentive of host machine
        [Value, Host machine]←Max (Incentive)
        Update best alliance vector
        Compute Ct, Ru, Mu, Cr
  End

The intelligent alliance algorithm (IAA), is based on the conception of pulling the resource on the basis of certain network resources. A capacity vector of the entire host machine is maintained for structure of availability and non-availability of VM's as well resource in terms of memory. Depending upon the type of traffic request and incentive is provisioned to early allocate the resources, which makes ITM-CLD adoptable to the uncertain as well as balanced resource provisioning to overcome the trade off resource versus task as well as under and over provisioning to make it efficient traffic management scheme for handling large data on cloud. The IAA can be further optimized to be synchronically with large data components to make it adoptable in real time scenario. The future work is to extend it to benchmark with more state of art methods to overcome many assumptions by considering more granular components while building mathematical model and improvise performances.

5   Results and Analysis

In order to evaluate the ITM-CLD, the model is simulated on the numerical computing platform, the observations of the proposed method for three different kinds of traffic patterns is tabulated for an instance (Tables 2 and 3).

Table 2. Performance metric for small, medium and high traffic pattern for proposed ITM-CLD

Traffic pattern	Cloudlet handling time (msec)	Unused resource	Unused memory	Resource cost
Small (S)	1.1536	1.5429	2.4434	4.2725
Medium (M)	0.86474	1.6132	2.6499	4.6663
High (H)	1.1057	1.8651	2.9718	4.7852

Table 3. Performance metric for small, medium and high traffic pattern for proposed ITM-CLD for existing system

Traffic pattern	Cloudlet handling time (msec)	Unused resource	Unused memory	Resource cost
Small (S)	1.9378	2.6501	4.3151	4.7288
Medium (M)	3.0587	3.0779	4.8478	5.1251
High (H)	4.0687	3.0644	4.9582	4.9826

The tabulated observed value is shown into bar graph of Figs. 2, 3, 4 and 5 for cloud handling time, unused resource, unused memory and resource cost respectively.

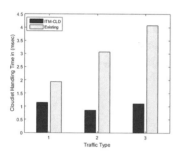

Fig. 2. Cloud handing time in mini second for traffic type (1) ITM-CLD, (2) Existing

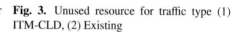

Fig. 3. Unused resource for traffic type (1) ITM-CLD, (2) Existing

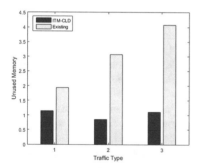

Fig. 4. Unused memory for traffic type (1) ITM-CLD, (2) Existing

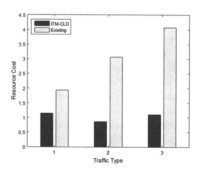

Fig. 5. Resource cost for traffic type (1) ITM-CLD, (2) Existing

Figure 2 illustrates that the cloudlet handling time for all three types of traffic is lower in case of ITM-CLD as compare to the existing method. The theoretical average computation is taken as existing values for all the performance benchmarking. The unused resource is comparatively lower in ITM-CLD, which shows the effectiveness of the ITM-CLD as compared to the existing mechanisms.

In Fig. 4, the performance comparison of the unused memory exhibits the very low memory wastage as compare to existing methods, whereas the resource cost is comparatively very high for the existing methods as compared to the ITM-CLD.

6 Conclusion

The fast adoption of the various applications which generates large or big data requires a flexible, effective and fast as well as cost effective platform for the data managements from view point of storage, access as well processing. Cloud computing the paradigm which qualifies these requirements. But the under provisioning as well as over provisioning method of resource allocations brings traffic bottleneck for handing random as well as growing traffic demand. This paper introduces a simple yet effective system model for handing traffic management for the large data sources like big data stream on cloud infrastructure. The performance metrics demonstrates the effectiveness of the proposed ITM-CLD. In future the ITM-CLD will be optimized for real time ecosystem variable consideration while formulating analytical model.

References

1. Mazza, D., Tarchi, D., Corazza, G.E.: A unified urban mobile cloud computing offloading mechanism for smart cities. IEEE Commun. Mag. **55**(3), 30–37 (2017)
2. Liu, Y., Lee, M.J., Zheng, Y.: Adaptive multi-resource allocation for cloudlet-based mobile cloud computing system. IEEE Trans. Mob. Comput. **15**(10), 2398–2410 (2016)
3. Feng, C., Xu, H., Li, B.: An alternating direction method approach to cloud traffic management. IEEE Trans. Parallel Distrib. Syst. **28**(8), 2145–2158 (2017)
4. Assi, C., Ayoubi, S., Sebbah, S., Shaban, K.: Towards scalable traffic management in cloud data centers. IEEE Trans. Commun. **62**(3), 1033–1045 (2014)
5. Guan, B., Wu, J., Wang, Y., Khan, S.U.: CIVSched: a communication-aware Inter-VM scheduling technique for decreased network latency between co-located VMs. IEEE Trans. Cloud Comput. **2**(3), 320–332 (2014)
6. Wanis, B., Samaan, N., Karmouch, A.: Efficient modeling and demand allocation for differentiated cloud virtual-network as-a service offerings. IEEE Trans. Cloud Comput. **4**(4), 376–391 (2016)
7. Kantarci, B., Mouftah, H.T.: Inter-data center network dimensioning under time-of-use pricing. IEEE Trans. Cloud Comput. **4**(4), 402–414 (2016)
8. Xu, D., Liu, X., Niu, Z.: Joint resource provisioning for internet datacenters with diverse and dynamic traffic. IEEE Trans. Cloud Comput. **5**(1), 71–84 (2017)
9. Xu, D., Liu, X., Vasilakos, A.V.: Traffic-aware resource provisioning for distributed clouds. IEEE Cloud Comput. **2**(1), 30–39 (2015)
10. Doyle, J., Shorten, R., O'Mahony, D.: Stratus: load balancing the cloud for carbon emissions control. IEEE Trans. Cloud Comput. **1**(1), 1 (2013)
11. Wei, L., Cai, J., Foh, C.H., He, B.: QoS-aware resource allocation for video transcoding in clouds. IEEE Trans. Circ. Syst. Video Technol. **27**(1), 49–61 (2017)
12. Tukkoji, C.D., Seetharam, K.: A survey on big data traffic management with the help of cloud computing. Int. J. Eng. Res. Dev. **12**(10), 27–34 (2016)

13. Duan, J., Yang, Y.: A load balancing and multi-tenancy oriented data center virtualization framework. IEEE Trans. Parallel Distrib. Syst. **28**(8), 2131–2144 (2017)
14. Rankothge, W., Le, F., Russo, A., Lobo, J.: Optimizing resource allocation for virtualized network functions in a cloud center using genetic algorithms. IEEE Trans. Netw. Serv. Manag. **14**(2), 343–356 (2017)

A Novel Computational Modelling to Optimize the Utilization of Intrusion Detection Paradigm in a Large-Scale MANET

Najiya Sultana[✉]

Department of Computer Science and Engineering, TAIBAH University, Medina, Saudi Arabia
najiya.research@gmail.com

Abstract. Over the past decade, mobile ad-hoc networks (MANET) gained the attention of researchers and became a key technology in many aspects owing to its potential applicability and increased usage in providing efficient wireless networking. The ability of enabling an instant temporary wireless networking scenario in situations like flooding and defense made MANET a prominent domain of research. Although, it has been extensively studied with respect to different means of issues including network security, power consumption issues etc. but the core findings in the area of security were found mostly limited to theoretical contributions. Moreover, An intrusion detection systems (IDS) enable different procedures involved into monitoring the activities being exercised in a MANET whether; it poses any suspicious or malicious events that could be harmful for the entire system. The conventional IDS models are more likely to consume higher level of energy which minimizes the network lifetime owing to rapid depletion of node's battery power. The study thereby primarily addressed this issue and come up with an efficient scheme which targets to optimize the time period in which IDS remain busy in a large-scale MANET. It also incorporated a technique which relates probabilistic theory of optimization to bring an effective cooperation among IDSs and neighbor nodes which leads to reduce their individual busy time. The proposed approach aims to reduce busy time of individual IDS while maintaining their effectiveness towards achieving defined tasks. To support the performance efficiency the proposed study developed an algorithm and simulated it over a numerical computing tool in terms of different performance parameters.

Keywords: Mobile ad-hoc networks · Intrusion detection systems
Energy consumption optimization

1 Introduction

In last two decades the concept of MANET came into practice in order to simplify the infrastructure less, dynamic wireless communication and routing among self-configuring, automated mobile devices [1]. However, these types of networks mostly consider deployment of mobile nodes in a temporary situation by establishing communication and routing among them without taking any infrastructure assisted support [2]. Nodes

© Springer International Publishing AG, part of Springer Nature 2019
R. Silhavy (Ed.): CSOC 2018, AISC 765, pp. 382–392, 2019.
https://doi.org/10.1007/978-3-319-91192-2_38

associated into these types of communications play a crucial role by behaving like a host or router at different time instances based on the network requirements. As MANET do not employ any additional infrastructure to provide routing oriented support, hence each node deployed takes help of its neighbor node to transmit its respective data to a destination node which is not within its communication radius (C_r). The prime advantage of deployment of MANET over conventional wireless networking is its inherent characteristics that include dynamicity of topology where they can be easily set up and dissembled [3]. The dynamicity of topology formation in this regards complies the requirement of few nodes to be moved out while other nodes to join the network which makes MANET a standalone network. Along with these features MANET can be interfaced with internet or other networks which require extended routing with higher degree of connectivity and coverage if there exist no infrastructure based support [4, 5]. In the recent times MANET covers a wide area of applications where one of the most prominent areas is vehicular ad-hoc networks (VANET) which initiates a self-configuring network of moving vehicles well-capable of communicating with each other in order to avoid road accidents and other inconveniences. It has become a promising technology to enhance the road safety and traffic efficiency [6]. However, the conventional MANET lacks security aspects due to several factors including *(i) Mobility of nodes, (ii) Poor wireless radio links, (iii) No intervention of centralized unit* etc. which makes the system vulnerable to various potential network attacks. To provide security solutions a lot of research effort has been expanded towards improving the MANET system and its operations from any violations against security preservations. However, in order to adequate the convention of security approaches against multi-level threats, the current researchers provided a way of tracking down the suspicious activities being exercised within a MANET namely IDS [7, 8]. It is basically a network activity monitoring process in addition with anomaly detection procedure which ensures whether any attacks are currently being performed to interrupt the system regulations. Usually in a resource constrained network like MANET, IDS usually get implemented over the member mobile nodes at each time instance with the purpose of continuous network monitoring. Hence, it results cost-overhead to the battery operated MANET in terms of power consumption and computational resource utilization [9]. The study thereby proposed a probabilistic novel scheme well capable of analyzing and optimizing the busy time period consumed by IDS over MANET during continuous monitoring period. It also aims to determine an optimal trade-off between the performance effectiveness of IDS and their minimized busy time with an addition of multi-model cooperative communications among different IDSs. The cooperative communication model has been designed and experimented over a numerical computing environment to validate the proposed system. Through numerical simulation the study obtained superior outcomes in terms of power consumption and computational cost which is further validated with a comparative performance analysis. The pattern of the overall paper is planned with different segments such as Sect. 2: which discusses the most recent convention of works carried out with respect to the problem of usage of IDS systems in a MANET. On the other hand Sect. 3: briefs the problems addressed followed by Sect. 4: where the system modelling of the proposed scheme has been discussed along with algorithm design and implementation. The extensive simulation outcomes supported by a comparative performance analysis is discussed under Sect. 5. Finally, Sect. 6: concluded the overall contribution of the research.

2   Related Work

This section presents the conventional techniques being adopted into practice with the aim of achieving energy efficient MANET performance without compromising the security requirements and effectiveness of IDS by optimizing their active period of executional state. The study of Dong et al. [10] presented an extensive analysis subjected to optimize the network topology that concerns edge-self monitoring with the purpose of expanding network lifetime. The study adopted optimization problem in a similar direction where they intended to reduce the number of communication links and monitor nodes as well to conserve the energy. Our proposed study also works with the aim of energy conservation of MANET but in a different context. The study of Khail et al. [11] presented a protocol namely Sleep-Wake Aware Local Monitoring (SLAM) deployed exclusively to monitor the special guard nodes of a sensor network. It basically reduces the time interval in which a special guard node remains awake for the purpose of monitoring if there is any suspicious activity happening within the system. Similarly in the study of Hoang Hai et al. [7] a scheme enabling optimal selection of intrusion detection agents (IDA) for WSN is presented with both theoretical and analytical viewpoint. A game-theoretical probabilistic convention has adopted in the study of Fitaci et al. [8] which ensures optimal usage of IDAs to achieve higher security level in a, MANET. Clegg et al. [9] presented a novel approach which optimally selects a set of eligible nodes in which each node monitors other set of nodes within the network to optimize the incoming traffic load. A problem formulation has been taken into consideration by the study of Zheng et al. [12] where the proposed approach determines m out of M sniffers with optimal decision. The selected each sniffer further employed into one of the K channels for monitoring the data transmission activities. The experimental analysis clearly demonstrated the performance efficiency of the proposed system where evaluation of trade-off between computations cost and rate of learning is empirically derived. The study of Tsikoudis et al. [13] modelled a novel architecture namely LEoNIDS which is subjected to perform network level intrusion detection by means of providing an optimal trade-off between power consumption and detection latency. Muradore and Quagila [14] introduced a packet based encryption which is incorporated to minimize the energy consumption during intrusion detection among networked control systems. The extensive analysis of various studies also reveals the matter of fact that game theory is widely adopted into various studies for the realization of optimization of intrusion detection system in terms of energy consumption and detection accuracy. The literatures of Shen [15], Afgah et al. [16], Alpcan and Basar [17], Liu et al. [18] introduced several game theory based solution approaches for modeling of energy efficient IDS while taking care of different issues like cooperation and selfishness of nodes within a wireless network. The next segment of the study formulates the problem description on the basis of conventional research issues unaddressed.

3 Problem Description

A thorough investigation of above highlighted literatures show various research attempts towards optimizing the active time period of conventional IDSs but most of the literatures are found limited to only large scale WSNs not MANETs. Although different attempts have been taken towards energy efficient intrusion detection but at the same time various techniques witnessed lacking computational efficiency along with detection accuracy. These factors further leads to generating communication overhead within a resource constrained system. Therefore designing energy efficient IDS while achieving its effectiveness with negligible computational complexity has become utmost concern of researchers. Along with this one more fact that kept the research timeline active is that most of the IDS run over every set of battery constrained node all the time. These nodes are also appeared to have very limited computational resources. Therefore, optimizing the executional active time period when the IDS monitors each and every activity happening inside the network has become challenging from design and development perspective.

4 Proposed Design Methodology

The study aims to formulate a novel system design pertaining to solve the problems associated with efficient usage of conventional IDSs over MANET. The design and development of the proposed analytical model is explicitly carried out in a numerical computing environment by taking various simulation parameters into consideration. It is basically designed with a probabilistic perspective where the active time duration of IDAs are reduced up to a great extent without compromising its effectiveness. The design

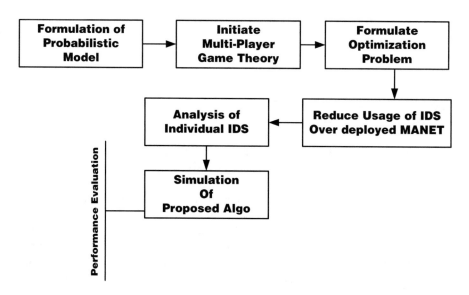

Fig. 1. Proposed design methodology

of the proposed technique has been carried out based on a multi-player game theory based cooperative approach where the effects of individual IDSs on different reduced network activity are analyzed. The following Fig. 1 exhibits the design approach adopted into the proposed system to attain efficient usage of IDS within a cooperative communication of MANET. The proposed technique on individual IDS utilizes local information and processes it to distribute among other IDSs with more scalability.

The study basically formulated two different approaches to solve the problem associated with usage of IDS theoretically.

- The study initially formulated an empirical modeling where the approach of solving the problem associated with minimization of the active period of IDS has been taken into consideration.
- Firstly, it applied an optimization approach with a problem from a view point where the case study analyzes the communication paradigm happening between the nodes being monitored by its one-hope-adjacent nodes using a game theory based analysis.
- Secondly, another viewpoint of problem has been taken into consideration where a node continuously monitors its adjacent nodes.

The study also arrived at the solutions obtained from the optimization problem where a distributed analytical algorithm has been designed to be integrated with every node in the network.

This section also generalizes the design aspects associated with the proposed system. Suppose, a wireless network consisting 6 numbers of wireless nodes where Node 3 resides into the network with a central tendency as below:

Each node runs IDS which target to monitor its adjacent nodes for detecting malicious activates. The above Fig. 2 depicts pictorially an assumption where an individual node is being monitored by its adjacent nodes IDS for detecting suspicious events taking place inside that particular node. The study exclusively considered 1-hop neighbor. Suppose a node n_i where $i = 3$ having α number of neighbors within a particular time period as mentioned in the above Fig. 2 where node 3 having 4 neighbors at a particular instant of network traffic movement i.e. $\alpha = 4$ in this context. It clearly denotes that each of the adjacent nodes continuously monitors n_3 to detect malicious activity. However, if at any instant of time all or a subset of adjacent nodes have detected malicious activities happening within n_3, it actually depends on the detection rate being computed from each IDS agents running on the top of them. However, a closer observation of the above figure reveals a scenario where n_3 is being monitored by all its 4 adjacent nodes where each of the 4 nodes is running IDS components all the time. Therefore, it can be said that it is not required to keep IDS components active in all the adjacent nodes as IDS components consume most of the computational resources and processing elements while tracking down the activities associated with n_3. Therefore, the study attempted to minimize the computational resource consumption by employing an efficient algorithm which incorporates a probabilistic approach to minimize the usage of IDS components while running in active time by addressing the aforementioned problem. As each node set up comprises IDS component running over it therefore, the study intended to optimize the involvement of adjacent nodes all the time during the neighbor node monitoring phase. The proposed work also emphasized to consider and incorporate efficient IDS which do not poses higher false detection rate. Therefore, the design constraints associated with IDS component more often affect the efficiency of the intrusion recognition procedure. The study focuses more about enhancing the performance of IDS in terms of efficient uses rather not concern about its design characteristics. An assumption also taken into consideration which states that the incorporation of number of active IDSs required monitoring adjacent nodes on the basis of the security requirements imposed there. The consideration of this aforementioned assumption leads to generate a trade-off between security level and energy consumption. However, it clearly depicts that more level of security requirements demands deployment of more IDSs over a particular adjacent node at a time. Hence, the more number of computing processes involved into each IDS component for continuous monitoring of a particular node result in higher energy consumption. The study formulated an optimization problem where it mostly emphasized on optimizing the probability of active time $\phi(l/\alpha)$ of a node n_3 is being monitored by its adjacent nodes. Considering the above highlighted scenario in Fig. 2, suppose node n_3 is being monitored by its individual adjacent nodes where each node separately runs its respective IDS component with a probability (P_t). The study also imposed a security level l to the node getting monitored at every instance of time Here security level denotes the active participation of at least S_L nodes for monitoring an adjacent node at any instant of time. The following mathematical expression in (Eq. 1) computes the probability of a particular node n_i being monitored by its adjacent nodes as follows:

$$\phi(l/\alpha) = \sum_{j=1}^{\alpha} \binom{\alpha}{j} p_t^j (1 - p_t)^{\alpha - j} \tag{1}$$

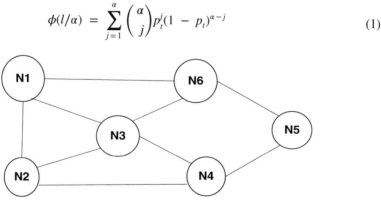

**Problem from point of View-1: Node 3 is being monitored by
node 1,2,4,5,6**

Fig. 2. A MANET where each node lies within the communication range of others

The proposed optimization problem formulated by means of minimizing the value of probability $\phi(l/\alpha)$ which is stated as follows:

Minimize $\phi(l/\alpha)$ by subjecting

$$\phi(l/\alpha) = \sum_{j=1}^{\alpha} \binom{\alpha}{j} p_t^j (1 - p_t)^{\alpha-j} \leq \Delta T \tag{2}$$

Where $\Delta T + \varepsilon$ denotes a minimum positive optimal threshold value (Probabilistic Value) defined with the context where it achieves the desired security level at any instant of time period. The value of threshold can depend upon different application scenario. The optimal solution of the above stated Eq. (2) thereby suggest that the monitored node is required by be watched by at least number of running IDSs where each one will have a probability of $\Delta T + \varepsilon$. The system computes monitoring probability and based on that a cooperative communication happens between nodes in terms of broadcasted rely and send messages. However, while a node is monitored with probability of 1 by at least α neighbor the optimization problem formulation happens towards optimizing Eq. 2 towards 1 with respect to $p = 1$ and irrespective of considering the values of α. The aforementioned condition formulation for optimization of usage of IDS further assisted into developing a novel algorithm for the effective management of IDS towards balancing the trade-off between energy efficient computation and accuracy of suspicious activity detection. The optimization problem formulation determines and computes the ideal probability of monitoring for each individual node and its associated IDSs. Each node enabled IDS therefore remains active for the rest of the time with the minimum monitoring probability allocated over a desired security level. The study considered p_{min_i} as minimum probability of node IDS component i to keep tracking down the events taking place among its adjacent nodes within a desired level of security. The following algorithm thereby designed and employed in order to achieve minimum usage of IDS components within a desired level of security. In this proposed algorithm degree of node represents its associate adjacent nodes at any point of time instance.

Proposed Algorithm Design for effective usage of IDS agents

---

start

1. $n_i \rightarrow broadcast(m(send\_degree))$ where $i \in R$
2. $\alpha_i (m(reply\_degree)) \rightarrow n_i$ where $i \in R$
3.   **once the reply obtained by n_i it does the following:**
4.   *for* $i \leftarrow 1:m_i$
5. $K_{Degree} / \alpha = min(reply\_degree);$
6.   **Check condition if $1 > \alpha$**
7.   **then** $p_{\min i} = 1$
8.   **else**
9.   **min** $p_t \leftarrow p_{\min i}$
10.   **Such that**
11.   **Apply eq. 2**
12. $\phi(l / \alpha) = \sum_{j=1}^{\alpha} \binom{\alpha}{j} p_t^{\ j} (1 - p_t)^{\alpha - j} \leq \Delta T$

13. *end*
14. *end*

end

The above algorithm design and developed for the purpose of attaining energy efficient usage of IDS component running on the top of adjacent nodes without compromising their effectiveness. This mechanism is actually deployed on each node to make their IDS component more effective in terms of energy efficiency and computational efficiency. The algorithm basically computes the minimum monitoring probability for each node enabled IDS considering optimal degree. Basically the monitored node computes the monitoring probability during the communication/intrusion detection process.

The algorithm considers that a node n_i is being monitored from its adjacent one hop neighbor nodes and vice versa where number of adjacent nodes here considered $\alpha = 4$. The algorithm initially performs computation of probability associated with each node for monitoring its adjacent nodes. In (step-1) n_i broadcast a data-packet consisting of a message seeks degree of its neighbor nodes. The message name is annotated here with a keyword name send_degree. The prime consideration here taken is message should get transmitted within one-hop neighbors. On receiving the receiver nodes (i.e. adjacent neighbors) reply to the sender node its degree of associativity termed with a keyword called reply_degree (step-2). Finally the algorithm checks the least value of α (step-5) and checks the condition whether $1 > \alpha$ in (step-6). If it is so then it assigns minimum monitoring probability of n_i to 1 (step-7) else $p_{\min}$ is assigned to p_t. Further the minimum monitoring probability of n_i computed using Eq. 2 with which it has to monitor its adjacent nodes. In (step-2) a malicious neighbor may perform attack by sending false degree of information to n_i and try to disrupt the system. The nest segment of the study further performs an extensive analysis to validate the performance efficiency of the proposed algorithm.

5   Results Discussion

The algorithm has been designed and simulated on a numerical computing environment supported by a 64-bit system where the performance efficiency of the algorithm has been validated with respect to processing time in sec. The extensive simulation outcomes shows that our proposed algorithm consumes very less amount of computation time while determining the minimum monitoring probability of each node deployed within a MANET. The following Fig. 3 clearly exhibits that the proposed technique which is integrated with each IDS running over nodes consumes very less amount of execution time which has a direct impact on energy efficient usage of computational resources.

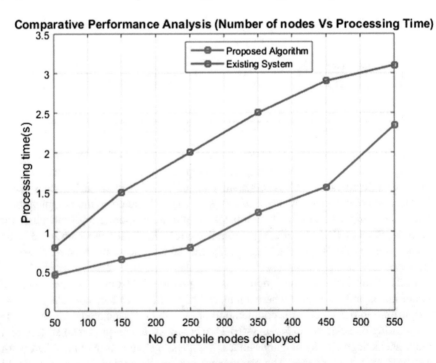

Fig. 3. Comparative performance analysis

A closer interpretation of the above Fig. 3 clearly shows that the processing time curve orientation in case of proposed system is gradually increases with respect to the number of deployed nodes enabled IDS in a MANET. On the other hand the existing system exhibits processing time up to 3 s and consumes more number of computational resources while running IDS component which is not at all desirable within a resource constrained network like MANET. Therefore, the comparative performance analysis clearly depicts that the proposed algorithm outperforms the conventional system by depicting negligible computational complexity.

6   Conclusion

The study presented a novel approach subjected to optimize the usage of IDS component for energy efficient monitoring of MANET nodes. The study developed an empirical modeling on the basis of a game theory based probabilistic approach to optimize the probability of monitoring time (active period) of IDSs, which has a direct impact into energy efficient communication and routing. The study validated the proposed approach considering a experimental analysis where the performance efficiency of the proposed approach has been exhibited in terms of algorithm processing time. It clearly illustrated that our proposed solution approach takes lesser processing time hence poses less computational complexity as compared to the existing system.

References

1. Zeadally, S., Hunt, R., Chen, Y.-S., Irwin, A., Hassan, A.: Vehicular ad hoc networks (VANETS): status, results, and challenges. Telecommun. Syst. 50(4), 217–241 (2012)
2. Bhoi, S.K., Khilar, P.M.: Vehicular communication: a survey. IET Netw. 3(3), 204–217 (2014)
3. Marti, S., Giuli, T.J., La, K., Baker, M.: Mitigating routing misbehavior in a mobile ad-hoc environment. In: Proceedings of the 6th Annual ACM/IEEE International Conference on Mobile Computing and Networking, pp. 255–265, August 2000
4. Bernsen, J., Manivannan, D.: Unicast routing protocols for vehicular ad hoc networks: a critical comparison and classification. Pervasive Mob. Comput. 5(1), 1–18 (2009)
5. Kumar, R., Dave, M.: A comparative study of various routing protocols in VANET, CoRR, vol. abs/1108.2094 (2011)
6. Zhang, M., Wolff, R.S.: A border node based routing protocol for partially connected vehicular ad hoc networks. J. Commun. 5(2), 130–143 (2010)
7. Hoang Hai, T., Huh, E.-N.: Optimal selection and activation of intrusion detection agents for wireless sensor networks. In: Proceedings of the Future Generation Communication and Networking (FGCN 2007), vol. 1, pp. 350–355, 6–8 December 2007
8. Fitaci, S.M., Jaffres-Runser, K., Comaniciu, C.: On modeling energy-security trade-offs for distributed monitoring in wireless ad hoc networks. In: Proceedings of the Military Communications Conference, MILCOM 2008, pp. 1–7. IEEE, 16–19 November 2008
9. Clegg, R.G., Clayman, S., Pavlou, G., Mamatas, L., Galis, A.: On the selection of management/monitoring nodes in highly dynamic networks. IEEE Trans. Comput. 62(6), 1207–1220 (2013)
10. Dong, D., Liao, X., Liu, Y., Shen, C., Wang, X.: Edge self-monitoring for wireless sensor networks. IEEE Trans. Parallel Distrib. Syst. 22(3), 514–527 (2011)
11. Khalil, I., Bagchi, S., Shroff, N.B.: SLAM: sleep-wake aware local monitoring in sensor networks. In: Proceedings of the 37th Annual IEEE/IFIP International Conference on Dependable Systems and Networks (DSN 2007), pp. 565–574 (2007)

12. Zheng, R., Le, T., Han, Z.: Approximate online learning algorithms for optimal monitoring in multi-channel wireless networks. IEEE Trans. Wirel. Commun. **13**(2), 1023–1033 (2014)
13. Tsikoudis, N., Papadogiannakis, A., Markatos, E.P.: LEoNIDS: a low-latency and energy-efficient network-level intrusion detection system. IEEE Trans. Emerg. Top. Comput. **PP**(99) (2014)
14. Muradore, R., Quaglia, D.: Energy-efficient intrusion detection and mitigation for networked control systems security. IEEE Trans. Ind. Inform. **11**(3), 830–840 (2015)
15. Shen, S.: A game-theoretic approach for optimizing intrusion detection strategy in WSNs. In: Proceedings of the 2011 2nd International Conference on Artificial Intelligence, Management Science and Electronic Commerce (AIMSEC), pp. 4510–4513, 8–10 August 2011
16. Afgah, A., Das, S.K., Basu, K.: A non-cooperative game approach for intrusion detection in sensor networks. In: Proceedings of the VTC 2004, Fall 2004
17. Alpcan, T., Basar, T.: A game theoretic approach to decision and analysis in network intrusion detection. In: Proceedings of the 43rd IEEE Conference on Decision and Control, December 2004
18. Liu, Y., Man, H., Comaniciu, C.: A game theoretic approach to efficient mixed strategies for intrusion detection. In: Proceedings of the IEEE International Conference on Communications (ICC 2006) (2006)

DSP-IR: Delay Sensitive Protocol for Intelligent Routing with Medium Access Control

A. C. Yogeesh[1]([⊠]), Shantakumar B. Patil[1], Premajyothi Patil[1], and H. R. Roopashree[2]

[1] Department of Computer Science and Engineering,
Nagarjuna College of Engineering and Technology, VTU, Bengaluru, India
yogeesh13@gmail.com
[2] Publicis Sapient Consulting Pvt. Ltd., Bengaluru, India
roopashree.r@res.christuniversity.in

Abstract. The wireless sensor network has been positioned itself from a complete network system to a sub-net of future internet namely Internet of Things, where a communications among anything to anything is possible. The extensible use of wireless sensor network makes it more risky if the security threats are not handled wisely. The conventional methods adopted for securing the WSN vulnerability based attacks introduces delay, which brings congestion in the routing flow as well as influence the quality of service. The proposed DSP-IR is a secure routing algorithm to handle security with delay sensitivity. The DSP-IR framework evaluates MAC protocols including S-MAC, Q-MAC and IH-MAC along with RSA, AES and DSP-IR encrypt process. For all the three combinations performance evaluation is done by simulating the model to know the behavior of residual energy, energy consumption and average packet delay with varying interval of message arrival time and it is found that the proposed IH-MAC with DSP-IR exhibits better performance.

Keywords: Wireless sensor network · Secure routing
MAC for sensor network

1 Introduction

Wireless Sensor Network (WSN) either as an independent network or sub-net of internet of things (IoT) along with pervasive and ubiquitous future generation applications in Industry 4.0 vision may provision many intelligent applications in the wide variety of domains of IBM vision of smart planet as well as many cyber physical world based any-to-any communication based applications [1]. In future, all these applications will be critical applications, which will be treated as a lifeline. There are two requirements that ensures success of this vision (1) optimal quality of service

© Springer International Publishing AG, part of Springer Nature 2019
R. Silhavy (Ed.): CSOC 2018, AISC 765, pp. 393–402, 2019.
https://doi.org/10.1007/978-3-319-91192-2_39

(QoS) such as maximization of packet delivery, reliability in terms of timely delivery, higher throughput etc., in optimal use of resources like energy, computational resources (memory, processing) and bandwidth, (2) Security against any kind of adversary due to vulnerability of unique characteristic of WSN which includes data integrity, energy harassment, compromising of routing path etc. [2–4]. The focus of this paper is on security aspects in WSN. There are various kind of attacks is witnessed in literature such as cross layer attack [5], node capture attack [6], wormhole attack etc. [7]. The solution methods developed for combating these attacks provisions adoptable level of security but at the same time it introduces delay. This paper aims to provide a solution framework for security adopting one traditional method without considering delay factor and proposing DSP-IR as a new security algorithm considering delay as a critical factor. In Sect. 2, review of related work is mentioned. The description of DSP-IR system model is described in Sect. 3. Section 4 describes solution as algorithm for DSP-IR process mechanism and finally, Sect. 5 describes behavior of DSP-IR through graphs of performance metrics followed by conclusion in Sect. 6.

2   Review of Related Work

Wireless Sensor networks (WSN) applicability becomes useful once it ensures the security whereas on implementation of security protocols some amount of delay is introduced. This section describes the recent work in the field of security especially who has considered delay as one of the importance parameters while protocol development for routing, secure routing else optimization of energy for achieving optimal quality of service. In WSN, it is very easy to implant additional nodes from adversaries. Due to the self-configurability characteristics of ad-hoc technology, the implanted nodes can be part of network. In this case node-to-node authentication becomes mandatory for secure routing. Many trust based secure routing protocols have been proposed in past [8–11]. The security of the WSN for application confidentiality is prime concern because the attacker can inject the intrusion to the applications to hack the confidential data. Under this scenario, the intrusions need to be identified before it harms the network system [12]. In the work of Yavuz et al. [13] performed the investigation over the k-connectivity in WSN security under random key distribution scheme with un-reliable links. Message authentication over the wireless sensor network is more important as it can be corrupted or accessed by unauthorized users. Towards this problem of WSN author Li et al. [14] discussed about the communication overhead, scalability issues associated with node attacks and introduced an scalable scheme for authentication based on elliptic curve cryptography (ECC). In WSN, the power efficiency can be obtained by using the clustering routing protocol but the application of these protocols without any proper security may leads to security vulnerabilities in the system [15]. In order to tackle this problem the work of [16] proposed an efficient scheme which adopts sensor based authentication in sensor nodes and establishes a symmetrical key and it will be updated periodically to bring effectiveness in network lifetime with high level of security. Similarly, the work of [17] described a key management scheme for static WSN by adding nodes, where the pair of keys was generated which provides high level of security. A detailed survey is being conducted on energy efficient secure routing protocols [18] and on the basic of research gap found an energy efficient secure routing protocol is proposed as EEESR for the heterogeneous WSNs [19]. This paper introduces a delay sensitive secure routing mechanism with different MAC-protocols.

3   System Model of DSP-IR

The system model of DSP-IR initiates with independent variable of N sensor nodes deployed in geographical area of A. The algorithm of random localization of N node is applied to distribute the node (N_x, N_y) across area A to achieve optimal coverage for cost effectiveness and proper connectivity. The Graph G(v) represents the node deployment. The model is designed robustly to validate for any positioning (SNx, SNy) of the sink node as contrast to many conventional methods where the sink nodes are place outside the deployment area, where as the DSP-IR can be evaluated by placing sink node anywhere. Figure 1 shows the initial deployment stage of G(v, SN) as a WSN.

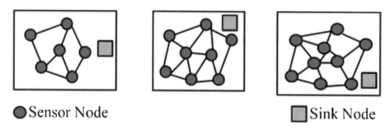

● Sensor Node   ■ Sink Node

Fig. 1. Deployment of sensor and sink node: G(v, SN)

To compute the Euclidian distance(d_{ij}), among the paired N_{ij}, where, i = N ∈ {1, 2, ...N}, j = N ∈ {1, 2, ...N − 1}, by incrementing value of a radius(R), if the $d_{ij} \leq R$ and covering M hop neighbor, set it as communication range. Figure 2 shows the process computations of communication range setup for optimal coverage requirements.

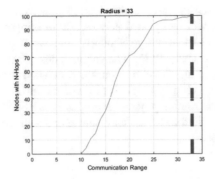

Fig. 2. N-hop neighborhood based communication range optimization

The system model of DSP-IP is modelled to test it for varied inter-arrival period (T_{arr}) for the data in such a way that $T_{arr} \in \{1, 2...K\}$, where $K = 10$. The very objective of the DSP-IP is to design a scalable mechanism of the data encryption process which can compensate the introduced delay due to encryption process. Traditionally block cipher based encryption algorithms such as skipjack encrypt (SJE) is used in the context of WSN, further to meet the reliability goal of data transfer the AES based encrypt is used with different key size of 256, 192 and 128 bits but it poses computational and memory overhead which is not suitable for the delay sensitive routings [20, 21].

4   DSP-IR Process Mechanism

The system model described in Sect. 3 is solved with a solution approach of a cryptographic protocol using RSA, AES and DSP-IR with scalable encrypt (EP) with different MAC-protocol such as S-MAC, Q-MAC and IH-MAC. Section 4.1 describes the tree based routing mechanism adopted for the DSP-IR.

4.1   Tree-Based Routing for DSP-IR

With the deployment of G(v, SN) at communication range (C_r) and inter-arrival period (T_{arr}), a tree base routing protocol is adopted. The description of the algorithms is shown as below:

Tree based routing protocol for DSP-IR

Input: N_x, N_y , N , A , SNx, SNy, C_r,
   Output:One-Hop
   Start
   $C[] \leftarrow d_{ij} \le C_r$, where C = connection matrix
   For each Node (N)

$$D_{N \rightarrow BS} \leftarrow \sqrt{((Nxi - SNx)2} + \sqrt{((Nyi - SNy)2}$$, where $D_{N \rightarrow BS}$ is distance of

node to BS
   If $D_{N \rightarrow BS} \le C_r$
   Update flag = 1 in C []
   CT $\leftarrow$ [N_x, N_y& SNx, SNy] , CT= Connection tree
   [cost, path]$\leftarrow$Routing function(C[] , CT, Start(N), End(N))
   One-Hop$\leftarrow$Path(i)
                          End
   End

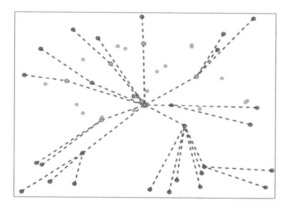

Fig. 3. Tree based routing topology for DSP-IR

The typical deployment of the tree based routing topology is shown in the Fig. 3 with sink node at the center.

4.2   Scalable Encrypt for DSP-IR

The description of the algorithm scale encrypt (SE) for DSP-IR is given, where the massage size is considered of 64 bits as well as the key size if of 64 bit and if key size taken is 56 then it is padded with additional 8-bits.

Scalable Encrypt/Decrypt for DSP-IR

Input: a) Massage of 64 Bit, b) key = 56/64 bit
Output: a) 64-Bit Massage/Key after encryption/Decryption
Encryption Validation Check
$K \leftarrow f_{rand}(S)$, k is key, S = seed
    Check key parity for 64 bits
        If K = 56
            $K \leftarrow K + 8$
        End
    Check massage Length(M) for 64 Bit
        If M < 64 Bits
            $M \leftarrow M(64)$
        End
Encryption Primitives
    $[M_l, M_r] \leftarrow f_{split}(M)$
        Do, key mixing with 8- Subscription table
            Create 8-Binary S-boxes
                $SBox \leftarrow f_{sbox}(M_l, M_r, K)$, Create Substitution function
                $PBox \leftarrow, f_{pbox}(M_l, M_r, K)$, Create permutation function
    Initialize ,$P_{initial}$, P_{final}
Key Scheduling
$P_{choice1 \& 2}(L/R) \leftarrow$ Permuted Choice One with $K \leftarrow 64/56$
    Do Rotation in Key Schedule (R_{ks})
    $K_s \leftarrow f(K, Shift)$, K_s : Key shift
        [L-subkey, R-subkey]$\leftarrow P_{choice1 \& 2}(L/R, K)$
    For 16 cipher rounds
    half key: 32-bit
            expended half key: 32-bit to 48-bit
      if encryption, apply sub-keys in the original order
      mixed with sub-key: 48-bit
      if decryption, apply sub-keys in the reverse order
      mixed with sub-key: 48-bit
        substitution: 48-bit to 32-bit
        permutation: 32-bit
      XOR by Feistel function: 32-bit
           final permutation
                C=[L, R], if encrypt
                C = [R,L];if decrpt
64-Bit (M/K: E/D) $\leftarrow$ Final Permutation (C)

---

5 Performance Analysis

The system model described in Sect. 3, for DSP-IR with scalable encrypt (EP) along with RSA and AES to knows simulated on numerical computing platform for various performance metrics. Figures 4, 5 and 6 shows its pattern. The default medium access protocol (MAC) for WSNs is s-Mac which is different as compared to the conventional one in IEEE 802.11 as it requires energy saving by means of sleep scheduling across event and non-event occurring situations. The DSP-IR is validated with RSA and AES along with S-MAC, Q-MAC and IH-MAC [22–24]. The trade off among the overall cost performance and the reliability in terms of timely delivery as an important QoS parameter in WSN, the results obtained from the simulation of proposed DSP-IR can be adoptable in real time application with further consideration of issues like fault-tolerant etc.

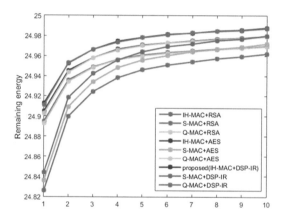

Fig. 4. Remaining energy vs Packet arrival interval with S-MAC: RSA/AES/DSP-IR, Q-MAC: RSA/AES/DSP-IR, IH-MAC: RSA/AES/DSP-IR

The Fig. 4 shows that the remaining energy is having higher pattern for the proposed IH-MAC - DSP-IR as compared to the other combinations, which makes it more adoptable from the energy optimization view point.

The Fig. 5 shows that the pattern of the average packet delay which is having a lower pattern for the proposed IH-MAC - DSP-IR as compared to the other combinations, which makes it more adoptable from the energy optimization view point.

The Fig. 6 shows that the pattern of the energy consumption per bit against increasing value of the packet arrival time and the energy used per bit is having lower pattern for the proposed IH-MAC - DSP-IR as compared to the other combinations, which makes it more consistent protocol from the energy view point.

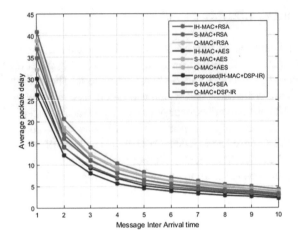

Fig. 5. Average packet delay vs Packet arrival interval with S-MAC: RSA/AES/DSP-IR, Q-MAC: RSA/AES/DSP-IR, IH-MAC: RSA/AES/DSP-IR

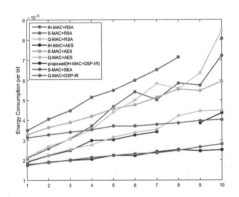

Fig. 6. Energy consumption vs Packet arrival interval with S-MAC: RSA/AES/DSP-IR, Q-MAC:RSA/AES/DSP-IR, IH-MAC: RSA/AES/DSP-IR

6 Conclusion

The proposed DSP-IR with scalable encrypt synchronized with a IH-MAC protocol provides a delay optimized routing in a secure manner by balancing the optimality of overall energy consumption as well as per bit energy utilization to ensure it more scalable. This achieved due to low computational overhead as it uses stretchy key size as per the requirements of robustness. This scheme is very much suitable for many of the wireless sensor network based delay critical emergency application where the requirement is to have delay sensitive secure and stable routing protocol. The framework of IH-MAC is also evaluated with different encrypt process of RSA and AES customized for WSN along with other MACs as S and Q-MAC. It realizes the objective function aiming to maintain a good balance between minimization of energy consumption and maximization for security strength.

References

1. Nukala, R., Panduru, K., Shields, A., Riordan, D., Doody, P., Walsh, J.: Internet of Things: a review from 'Farm to Fork'. In: 2016 27th Irish Signals and Systems Conference (ISSC), Londonderry, pp. 1–6 (2016). https://doi.org/10.1109/issc.2016.7528456
2. Karlof, C., Wagner, D.: Secure routing in wireless sensor networks: attacks and countermeasures. In: Proceedings of the First IEEE International Workshop on Sensor Network Protocols and Applications, pp. 113–127 (2003). https://doi.org/10.1109/snpa.2003.1203362
3. Bai, Y.: Industrial Internet of Things over tactile internet in the context of intelligent manufacturing. Cluster Comput., 1–9 (2017). https://doi.org/10.1007/s10586-017-0925-1
4. Badarinath, R., Prabhu, V.V.: Advances in Internet of Things (IoT) in manufacturing. In: IFIP International Conference on Advances in Production Management Systems, APMS, pp. 111–118 (2017). https://doi.org/10.1007/978-3-319-66923-6_13
5. Tayebi, A., Berber, S., Swain, A.: Wireless sensor network attacks: an overview and critical analysis. In: 2013 Seventh International Conference on Sensing Technology (ICST), Wellington, pp. 97–102 (2013). https://doi.org/10.1109/icsenst.2013.6727623
6. Lin, C., Wu, G., Yu, C.W., Yao, L.: Maximizing destructiveness of node capture attack in wireless sensor networks. J. Supercomput. **71**(8), 3181–3212 (2015)
7. Buch, D., Jinwala, D.: Detection of wormhole attacks in wireless sensor network. In: 3rd International Conference on Advances in Recent Technologies in Communication and Computing (ARTCom 2011), Bangalore, pp. 7–14 (2011)
8. Basan, A., Basan, E., Makarevich, O.: A trust evaluation method for active attack counteraction in wireless sensor networks. In: 2017 International Conference on Cyber-Enabled Distributed Computing and Knowledge Discovery (CyberC), Nanjing, China, pp. 369–372 (2017)
9. Tomić, I., McCann, J.A.: A survey of potential security issues in existing wireless sensor network protocols. IEEE Internet Things J. **4**(6), 1910–1923 (2017). https://doi.org/10.1109/jiot.2017.2749883
10. Jedidi, A., Mohammad, A.: History trust routing algorithm to improve efficiency and security in wireless sensor network. In: 2017 14th International Multi-Conference on Systems, Signals & Devices (SSD), Marrakech, pp. 750–754 (2017)
11. Qin, D., Yang, S., Jia, S., Zhang, Y., Ma, J., Ding, Q.: Research on trust sensing based secure routing mechanism for wireless sensor network. IEEE Access **5**, 9599–9609 (2017)
12. Butun, I., Morgera, S.D., Sankar, R.: A survey of intrusion detection systems in wireless sensor networks. IEEE Commun. Surv. Tutor. **16**(1), 266–282 (2014). https://doi.org/10.1109/surv.2013.050113.00191
13. Yavuz, F., Zhao, J., Yağan, O., Gligor, V.: k-connectivity in random K-out graphs intersecting Erdős-Rényi graphs. IEEE Trans. Inf. Theory **63**(3), 1677–1692 (2017). https://doi.org/10.1109/TIT.2016.2634422
14. Li, J., Li, Y., Ren, J., Wu, J.: Hop-by-Hop message authentication and source privacy in wireless sensor networks. IEEE Trans. Parallel Distrib. Syst. **25**(5), 1223–1232 (2014). https://doi.org/10.1109/TPDS.2013.119

15. Shen, J., Tan, H., Moh, S., Chung, I., Liu, Q., Sun, X.: Enhanced secure sensor association and key management in wireless body area networks. J. Commun. Netw. **17**(5), 453–462 (2015). https://doi.org/10.1109/JCN.2015.000083

16. Lee, S., Kim, K.: Key renewal scheme with sensor authentication under clustered wireless sensor networks. Electron. Lett. **51**(4), 368–369 (2015). https://doi.org/10.1049/el.2014.3327

17. Gandino, F., Montrucchio, B., Rebaudengo, M.: Key management for static wireless sensor networks with node adding. IEEE Trans. Ind. Inform. **10**(2), 1133–1143 (2014). https://doi.org/10.1109/TII.2013.2288063

18. Yogeesh, A.C., Patil, S.B., Patil, P.: A survey on energy efficient, secure routing protocols for wireless sensor networks. Int. J. Eng. Comput. Sci. **5**(8), 17702–17709 (2016). ISSN 2319-7242

19. Yogeesh, A.C., Patil, S.B., Patil, P.: FEESR: framework for energy efficient secured routing in heterogeneous sensor network. In: 2016 International Conference on Circuits, Controls, Communications and Computing (I4C), Bangalore, pp. 1–7 (2016)

20. Biswas, K., Muthukkumarasamy, V., Wu, X.W., Singh, K.: Performance evaluation of block ciphers for wireless sensor networks. In: Choudhary, R., Mandal, J., Auluck, N., Nagarajaram, H. (eds.) Advanced Computing and Communication Technologies. Advances in Intelligent Systems and Computing, vol. 452. Springer, Singapore (2016)

21. Bandirmali, N., Erturk, I., Ceken, C.: Securing data transfer in delay-sensitive and energy-aware WSNs using the scalable encryption algorithm. In: 2009 4th International Symposium on Wireless Pervasive Computing, Melbourne, VIC, pp. 1–6 (2009). https://doi.org/10.1109/iswpc.2009.4800606

22. Shi, Y., Wang, J., Fang, X., Gu, S., Dong, L.: Modelling and control of S-MAC based wireless sensor networks control system with network-induced delay in industrial. In: 2017 4th International Conference on Information Science and Control Engineering (ICISCE), Changsha, pp. 1539–1544 (2017). https://doi.org/10.1109/icisce.2017.321

23. Liu, Y., Elhanany, I., Qi, H.: An energy-efficient QoS-aware media access control protocol for wireless sensor networks. In: IEEE International Conference on Mobile Adhoc and Sensor Systems, Washington, DC, pp. 189–191 (2005). https://doi.org/10.1109/mahss.2005.1542798

24. Ceken, C.: An energy efficient and delay sensitive centralized MAC protocol for wireless sensor networks. Comput. Stand. Interfaces **30**(1), 20–31 (2008)

A Novel, Lightweight, and Cost-Effective Mechanism to Secure the Sensor-Gateway Communication in IoT

Shamshekhar S. Patil[1]([⊠]) and N. R. Sunitha[2]

[1] Department of Computer Science and Engineering, Dr. AIT, Bengaluru, India
shamshekhar.patil@dr-ait.org
[2] Department of Computer Science and Engineering,
SIT, Tumkur, Karnataka, India
nrsunitha@gmail.com

Abstract. Ensuring highest degree of resistance against potential adversaries in Internet-of-Thing (IoT) is still an open challenge, and the prime reason behind this is the computational complexities associated with designing security algorithms and its corresponding transformation between the cloud and Wireless Sensor Network (WSN) via gateway node. Hence, the proposed system develops a lightweight and highly responsive encryption technique to offer minimal resource consumption from the resource-constraint sensor nodes. The significant contribution is also to introduce a novel bootstrapping key mechanism with the unique generation of the secret key to maintain both forward and backward secrecy. The study outcome shows that proposed system is highly practical to offer reduced resource consumption and faster algorithm processing time in the presence of dynamic scenario of IoT.

Keywords: Internet-of-Things (IoT) · Wireless Sensor Network
Security · Privacy · Encryption · Secret key · Key management

1 Introduction

Internet-of-Things (IoT) is a formed by integrating cloud services with Wireless Sensor Network (WSN) [1]. However, the present concept of IoT is not in line with this as it offers a gateway which is connected to the sensor nodes to accumulate the data and process as per the demand [2]. It will mean that there is no emphasis on the actual principle of WSN where nodes form a network and use its routing scheme to collect data, process and forward it to sink [3]. At the same time, the type and intensity of threats on the internet as well as on WSN are very different from each other [4, 5]. Although there have been various studies carried out toward cloud security [6], WSN security [7], and IoT security [8], there was no discussion about any single security protocol that offers uniform security solution to all above problems. This is a major research gap in the literature of the present time. Although the area of cryptography has witnessed a tremendous revolution in the form of public key cryptography [9, 10], there is no evidence till date that a single encryption scheme could offer same security resistance performance for both internet and WSN together. There is a higher possibility of

© Springer International Publishing AG, part of Springer Nature 2019
R. Silhavy (Ed.): CSOC 2018, AISC 765, pp. 403–412, 2019.
https://doi.org/10.1007/978-3-319-91192-2_40

spreading the malicious codes from the internet to WSN as well as vice versa with being even detected. Hence, there is an emergent need of evolving up with a secure protocol that offers uniform security strength to both clouds as well as WSN. Therefore, the proposed study aims to represent a very new security scheme that offers uniform security performance to both cloud resources and WSN nodes. While designing the proposed system, the core emphasis was laid to develop a very lightweight encryption technique that doesn't use too many resources, especially from the resource constraint sensor nodes. Section 2 discusses the existing research work followed by problem identification after reviewing existing literature in Sect. 3. Section 4 discusses proposed methodology followed by an elaborated discussion of algorithm implementation for secret key bootstrapping and generation in Sect. 5. Comparative analysis of accomplished result is discussed under Sect. 6 followed by conclusion in Sect. 7.

2   Related Work

This section briefs about different methods implemented in recent times for securing IoT-based communication system. The work carried out by Wang et al. [11] has used the policy-based technique for securing the IoT terminals using prototyping-based approach. Dao et al. [12] have addressed the problem of key agreement using unique features from social networks. Issues of biometric security are addressed by Hossain et al. [13] with the aid of encryption using the pairing of keys. Strengthening of the middle-ware system in IoT has been emphasized in the work of Tiburski et al. [14]. Security issues related to eavesdropping has been addressed in the work of Xu et al. [15] where analytical modeling was used for computing probability of security breach. Literature has also seen the discussion over the capillary network of IoT where Guiliano et al. [16] have presented secure access to the devices connected to IoT. Cheng et al. [17] have introduced a technique where malware prevention scheme has been highlighted to be solved using patching based security technique. Adoption of game theory is also found to be an approach for identifying anomalies on the network as seen in the work of Sedjelmaci et al. [18]. Discussion about security threats in smart grid has been put forward by Chin and Chen [19] while the correlation between energy and security has been put forward Trappe et al. [20] and Kumar [21]. Wolf [22] and Burg et al. [23] have briefed about threat models for the cyber physical intrusion. Ruan et al. [24] have presented a framework for a key exchanging mechanism for leveraging the authentication process in IoT. The hardware-based approach for generating pseudo-random number is designed by Lopez et al. [25] for strengthening encryption process. A unique study of analyzing network threats has been put forward by Mohsin et al. [26] using the probability-based framework. The similar direction of work has also been carried out by Xu et al. [27] using ontology and semantic-based approach for network security. Adoption of attribute-based cryptography has been advocated by Ambrosin et al. [28] over hardware model. Premnath and Haas [29] have investigated the impact of the size of the secret key on the IoT security. Ko and Keoh [30] have presented the recursive process of service virtualization for leveraging the security during relaying the services. The next section briefs about the research problems extracted after reviewing the literature.

3 Problem Description

Majority of the existing security techniques considers IoT as a combined scenario without much considering the security complexities to be different on the internet and in sensor nodes. Majority of the existing mechanisms are highly recursive that is not feasible for a resource-constraint node to execute for a longer run. Therefore, it will lead to an excessive shortening of the resource from the sensor node thereby degrading both communication and security standard. The adversaries act very differently in the wireless sensor network (WSN), and it has never been checked how they behaved when they slipped to internet communication channel. Hence, a robust and cost-effective security scheme is required for securing both gateway node as well as sensor node from being prone to an ever increasing level of threats in IoT network and its respective applications.

4 Proposed Methodology

The proposed system aims to design a unique and lightweight framework that offers secure and seamless communication between the wireless sensor node and IoT gateway node. The implementation is carried out using analytical research methodology and a schematic diagram of it is highlighted in Fig. 1.

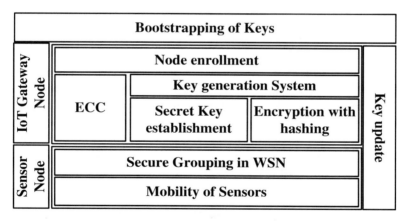

Fig. 1. Schematic diagram of proposed system

The proposed system introduces a novel mechanism of bootstrapping public keys followed by the process of sensor node enrollment within the gateway nodes. The proposed system uses Elliptical Curve Cryptography (ECC) for generating keys followed by the secret key establishment as performing encryption using simplified hashing techniques. Once the secure key has been established with gateway node and all its member sensor nodes than a secure grouping is applied to allocate legitimate nodes to the IoT gateway nodes. The construction of the proposed system also considers the mobility of the sensor node to assess the dynamic scenarios with potential supportability of a faster secret key update. The next section briefs about the significant algorithm constructed for this purpose.

5   Algorithm Implementation

This section discusses the prominent algorithm that has been implemented in the proposed system. The proposed system uses mainly two algorithms, i.e., (i) algorithm for bootstrapping the key and (ii) algorithm to generate a secret key. An elaborated discussion of this algorithm is as below:

(i)   Algorithm for key Bootstrapping

This algorithm is executed within the gateway node which is responsible for generating a new form of preliminary secret. The algorithm considers input as α (IoT Attribute) and e (ECC tuple parameters) that after processing yields β (Bootstrapped key). The steps involved in the proposed algorithm are as follows:

Algorithm for key bootstrapping
Input: α, e
Output: β
Start
1. *init* α
2. *compute* H^e | $H = \{a_1, a_2, a_3, a_4\}$
3. $\beta \rightarrow (e \| k_{pu} \| H^e)$
End

The algorithm assumes an IoT attribute α as a variable that contains a set of a different number of protocols executed by a gateway node (Line-1). As the proposed system uses ECC, hence, it computes the set of prime numbers that start from 2^{α} till 0. The algorithm also considers ECC tuple parameters e which is a set of many unit finite field parameters over ECC curve. Using private key K_{pr}, the proposed system uses following mathematical expression to calculate public key K_{pu},

$$K_{pu} = K_{pr} \times e_4 \tag{1}$$

We assume $e = [e_1, e_2, e_3, e_4]$ are parameters that constructs e in Line-2. The next part of the study is to form an encrypted pass keys H considering a_1, a_2, a_3, and a_4 to be the elements of matrix H (Line-2). The computation is carried out as follows:

$$\begin{aligned} a_1 &\rightarrow r \cdot e_3^2 \\ a_2 &\rightarrow e_3^2 \cdot r \cdot e_3 \\ [a_3 a_4] &\rightarrow [e_3 \cdot r]^3 \cdot e_3 \end{aligned} \tag{2}$$

Finally, the proposed system obtained bootstrapped key β using e, K_{pu}, and H_e (Line-3). It will also mean that a gateway node will now publish the bootstrap key β to all the sensors. One of the significant contributions of this algorithm is it offers extensive data integrity as well as a significant level of privacy. The performance of the algorithm is unaffected eventually after passing from a different number of gateway nodes. The algorithm not only protects the key information of the gateway node but also protects the secret key which is generated in the next algorithm by mutual authentication between the gateway node and sensor node.

(ii) **Algorithm for secret key generation**

This algorithm is responsible for generating a final secret key that takes input arguments of T_{range} (Transmission Range), k_{pub} (public secret key), and g (prime number). The processing yields the output of γ_2 (Secret Key). The step of the algorithm is described below:

Algorithm for secret key generation
Input: T_{range}, k_{pub}, g
Output:γ_2
Start
1. **For** i=1: N
2. $n_{par} \rightarrow find(d \leq T_{range})$
3.   **For** j=1:size(n_{par})
4. $\gamma_1 \rightarrow c_1.K_{pub} + |c_2, g|$
5. $\gamma_2 \rightarrow c_3.g$
6. $[\gamma_3\,\gamma_4] \rightarrow [e_2.nb]$ & $\gamma_5 \rightarrow [r_1 + f(\gamma_3,\gamma_4)]$
7.   st=$f(e_1)$
8. ...... **If** $\gamma_5.e == (h(x))$
9. $\gamma_2 \rightarrow a_2 *\ id *r*\ id*g$
10.   **End**
11. **End**
12. **End**
End

The algorithm considers all the sensor nodes N (Line-1) followed by obtaining the location information of the nodes as well as computing distance between all the other respective sensors d. All the nodes that are out of transmission range T_{range} of the gateway are termed as partitioned node n_{par} (Line-2). The prime idea is to protect such node and hence security is incorporated for such nodes (Line-3). The algorithm computes a temporary variable id by multiplying unique identifier ui with e_4. The first security token generated is γ_1 (Line-4), where the variable c_1 represents the scalar product of unique node identifier, a_1, e_4, and K_{pu}, while the variable c_2 represents scalar product of unique node an identifier and identifier of the node with low processing capability g (line-4). The next security token is generated by the scalar product of c_3 and node with low processing capability (Line-5). The variable c_3 is obtained by multiplying

$a_2 * id * \gamma_1 * ui * e_4$ and an identifier of the node with low processing capability g. The algorithm also computes other three security tokens (Line-6) as follow:

$$\gamma_3 = a_3 * id * r_2 * \gamma_1 * r_3 * r_4 * g;$$
$$\gamma_4 = a_4 * id * r_2 * \gamma_1 * r_3 * r_4 * g;$$
$$\gamma_5 = r_1 + ui * st + r_5 * \gamma_4;$$

Where the variable r_2 and r_1 are random number. A final encryption pass key st is generated after applying a function f(x), where $f(x) \rightarrow a_3 * id * r_2 * \gamma_1 * r_3 * r_4 * g$ (Line-7). A condition is stated to find if the two different forms of keys generated by gateway node and sensor node matches (Line-8) then only it will call for computing the final secret key γ_2 (Line-9). The algorithm potentially maintains the highest number of layers of security and hence can significantly offer both forward and backward secrecy. The algorithm offers better privacy as well as non-repudiation. The next section discusses the results being obtained from the proposed algorithm.

6   Results Discussion

As the proposed study introduces a novel mechanism to generate a secret key in IoT devices hence, it is essential to check the performance of the key generation mechanism concerning its resource consumption. We have also seen that work carried out by Zhao et al. [31] also keeps the similar motive of IoT security has been increasingly cited among the research community. Hence, we compare the outcome of proposed study with that of Zhao et al. work concerning resource consumption.

We implement group key update for the entire gateway to obtain a global view of the performance. The existing algorithm of Zhao et al. [31] is implemented in our test bed. We use a new evaluation factor called Time Instance 1 or TI1 that represents how many times the group key has been updated. It can be seen that there is no significant difference between the proposed and existing system on different scales of velocity (v = 1 m/s, 2 m/s, 4 m/s, 8 m/s, 16 m/s). This trend makes clear that usage of ECC in both the system offers the similar performance of resource consumption. It also shows that increased of node velocity also increases resource drainage. An interesting observation is that existing system has implemented 4 level of iterative steps (initialization, verification of certificate authority, mutual authentication, and hashing) to obtain resource consumption trend shown in Fig. 2(a). On the other hand, proposed system doesn't use any form of certificates and doesn't use manifold iterative steps, unlike any existing system. Hence, the effectiveness of the proposed system is more compared to the existing system to offer similar resource dependencies but on a different scale of internal operative function. Another distinct difference is that the approach of Zhao et al. [31] doesn't offer any form of a grouping mechanism for performing updating operation of the secret key and such update operation is carried out by the node, which also has a chance to bear any form of rogue identity. However, the proposed system offers a grouping mechanism where all the secret keys of gateway node are grouped to formed update, and thereby it not only increases the security features but also offers faster operation while performing the update. This feature of the proposed system will assist maximum security strength compared to existing approach.

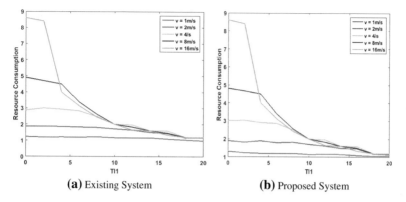

(a) Existing System   **(b)** Proposed System

Fig. 2. Comparative analysis of resource consumption for group key update

The updating mechanism of the group key is carried out when any sensor either join a new gateway during their mobility or vice-versa (i.e., departing from the old gateway node). While doing so, only the low processing sensors are priorities so that they can fulfill their task of data forwarding within a stipulated period. The sensor with higher processing capabilities is allowed to wait until TI1 duration. Hence, we consider another variable called Time Instance 2 or TI2 to represent such waiting time. Such form of scheduling while performing the key update on a global scale of gateway node has not been seen in any existing authentication approach of IoT. This is quite a practical scenario that is implemented in the proposed system to show that it excels less resource consumption while performing the establishment of the secret key as shown in Fig. 3.

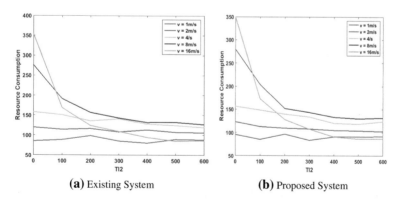

(a) Existing System   **(b)** Proposed System

Fig. 3. Comparative analysis of resource consumption for secret key establishment

The proposed algorithm also offers approximately 49% of improvement of existing approaches on IoT security as shown in Fig. 4. Hence, the proposed algorithm can be stated to be cost-effective security solution with faster response time.

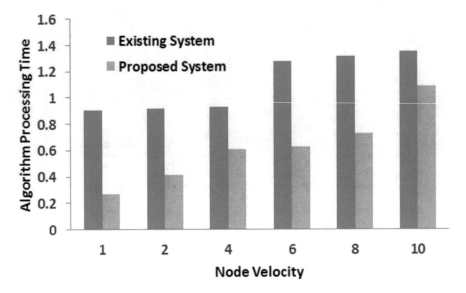

Fig. 4. Comparative analysis of algorithm processing time

7 Conclusion

The proposed study has presented a technique where ECC has been used completely in a different way to offer security. It is because the conventional approach of implementing ECC is not resistive against message tampering-based attack which is very often in cloud computing. Hence, the proposed system has offered the solution in the form of the following contribution –(i) it creates a unique communication model to include both cloud and WSN environment retaining all their legacy operations connected by a gateway node, (ii) a novel bootstrapping mechanism to generate public key to be used the gateway node for its allocated enrolled member sensor nodes, (iii) A novel secret key generation scheme has been introduced using simple hashing mechanism. Our future study will be to further optimize the security strength.

References

1. Dawson, M., Eltayeb, M., Omar, M.: Security Solutions for Hyperconnectivity and the Internet of Things. IGI Global, Hershey (2016)
2. Gravina, R., Palau, C.E., Manso, M., Liotta, A., Fortino, G.: Integration, Interconnection, and Interoperability of IoT Systems. Springer (2017)
3. Khan, S., Pathan, A.S.K., Alrajeh, N.A.: Wireless Sensor Networks: Current Status and Future Trends. CRC Press, Boca Raton (2016)
4. Selmic, R.R., Phoha, V.V., Serwadda, A.: Wireless Sensor Networks: Security, Coverage, and Localization. Springer (2016)
5. Gilchrist, A.: IoT Security Issues. Walter de Gruyter GmbH (2016)

6. Riddle, A.R., Chung, S.M.: A survey on the security of hypervisors in cloud computing. In: 2015 IEEE 35th International Conference on Distributed Computing Systems Workshops, Columbus, OH, pp. 100–104 (2015)
7. He, D., Zeadally, S., Kumar, N., Wu, W.: Efficient and anonymous mobile user authentication protocol using self-certified public key cryptography for multi-server architectures. IEEE Trans. Inf. Forensics Secur. 11(9), 2052–2064 (2016)
8. Krejčí, R., Hujňák, O., Švepeš, M.: Security survey of the IoT wireless protocols. In: 2017 25th Telecommunication Forum (TELFOR), Belgrade, Serbia, pp. 1–4 (2017)
9. Shim, K.A.: A survey of public-key cryptographic primitives in wireless sensor networks. IEEE Commun. Surv. Tutor. 18(1), 577–601 (2016)
10. Kiviharju, M.: On the fog of RSA key lengths: verifying public key cryptography strength recommendations. In: 2017 International Conference on Military Communications and Information Systems (ICMCIS), Oulu, pp. 1–8 (2017)
11. Wang, J., Hong, Z., Zhang, Y., Jin, Y.: Enabling security-enhanced attestation with Intel SGX for remote terminal and IoT. IEEE Trans. Comput. Aided Des. Integr. Circ. Syst. 37(1), 88–96 (2018)
12. Dao, N.N., Kim, Y., Jeong, S., Park, M., Cho, S.: Achievable multi-security levels for lightweight IoT-enabled devices in infrastructureless peer-aware communications. IEEE Access 5, 26743–26753 (2017)
13. Hossain, M.S., Muhammad, G., Rahman, S.M.M., Abdul, W., Alelaiwi, A., Alamri, A.: Toward end-to-end biomet rics-based security for IoT infrastructure. IEEE Wirel. Commun. 23(5), 44–51 (2016)
14. Tiburski, R.T., Amaral, L.A., de Matos, E., de Azevedo, D.F.G., Hessel, F.: The role of lightweight approaches towards the standardization of a security architecture for IoT middleware systems. IEEE Commun. Mag. 54(12), 56–62 (2016)
15. Xu, Q., Ren, P., Song, H., Du, Q.: Security enhancement for IoT communications exposed to eavesdroppers with uncertain locations. IEEE Access 4, 2840–2853 (2016)
16. Giuliano, R., Mazzenga, F., Neri, A., Vegni, A.M.: Security access protocols in IoT capillary networks. IEEE Internet Things J. 4(3), 645–657 (2017)
17. Cheng, S.M., Chen, P.Y., Lin, C.C., Hsiao, H.C.: Traffic-aware patching for cyber security in mobile IoT. IEEE Commun. Mag. 55(7), 29–35 (2017)
18. Sedjelmaci, H., Senouci, S.M., Taleb, T.: An accurate security game for low-resource IoT devices. IEEE Trans. Veh. Technol. 66(10), 9381–9393 (2017)
19. Chin, W.L., Li, W., Chen, H.H.: Energy big data security threats in IoT-based smart grid communications. IEEE Commun. Mag. 55(10), 70–75 (2017)
20. Trappe, W., Howard, R., Moore, R.S.: Low-energy security: limits and opportunities in the Internet of Things. IEEE Secur. Priv. 13(1), 14–21 (2015)
21. Kumar, S.D., Thapliyal, H., Mohammad, A.: FinSAL: FinFET-based secure adiabatic logic for energy-efficient and DPA resistant IoT devices. IEEE Trans. Comput. Aided Des. Integr. Circ. Syst. 37(1), 110–122 (2018)
22. Wolf, M., Serpanos, D.: Safety and security in cyber-physical systems and Internet-of-Things systems. Proc. IEEE 106(1), 9–20 (2018)
23. Burg, A., Chattopadhyay, A., Lam, K.Y.: Wireless communication and security issues for cyber-physical systems and the Internet-of-Things. Proc. IEEE 106(1), 38–60 (2018)
24. Ruan, O., Chen, J., Zhang, M.: Provably leakage-resilient password-based authenticated key exchange in the standard model. IEEE Access 5, 26832–26841 (2017)
25. López, A.B.O., Encinas, L.H., Muñoz, A.M., Vitini, F.M.: A lightweight pseudorandom number generator for securing the Internet of Things. IEEE Access 5, 27800–27806 (2017)

26. Mohsin, M., Sardar, M.U., Hasan, O., Anwar, Z.: IoTRiskAnalyzer: a probabilistic model checking based framework for formal risk analytics of the Internet of Things. IEEE Access **5**, 5494–5505 (2017)
27. Xu, G., Cao, Y., Ren, Y., Li, X., Feng, Z.: Network security situation awareness based on semantic ontology and user-defined rules for Internet of Things. IEEE Access **5**, 21046–21056 (2017)
28. Ambrosin, M., et al.: On the feasibility of attribute-based encryption on Internet of Things devices. IEEE Micro **36**(6), 25–35 (2016)
29. Premnath, S.N., Haas, Z.J.: Security and privacy in the Internet-of-Things under time-and-budget-limited adversary model. IEEE Wirel. Commun. Lett. **4**(3), 277–280 (2015)
30. Ko, H., Jin, J., Keoh, S.L.: Secure service virtualization in IoT by dynamic service dependency verification. IEEE Internet Things J. **3**(6), 1006–1014 (2016)
31. Zhao, G., Si, X., Wang, J., Long, X., Hu, T.: A novel mutual authentication scheme for Internet of Things. In: Proceedings of 2011 International Conference on Modelling, Identification and Control, Shanghai, pp. 563–566 (2011)

Quality of Service (QoS) Aware Reconfigurable Optical Add/Drop Multiplexers (ROADM) Model with Minimizing the Blocking Rate

G. R. Kavitha[1(✉)] and T. S. Indumathi[2]

[1] Department of Telecommunication Engineering,
Visvesvaraiah Technological University, Bangalore, India
kavigr@gmail.com
[2] Department of Digital Electronics and Communication Systems,
Visvesvaraiah Technological University, Bangalore Region, Bangalore, India

Abstract. The development of the Wavelength Switched (WS) Optical Networks (ONet) is introduced to extract the significances of entire Optical Switching (OS) fabrics pertaining to high efficiency and automation. However the selective switches of wavelength (SW) indicates the prior switching elements subjected with Reconfigurable Optical Add/Drop Multiplexers (ROADM) frameworks exhibiting the different color-less, direction-less and also with contention-less (Col-Dl-Conl) switching features. With the implementation of the Col-Dl-Conl based ROADM, the WSw ONet nodes will be asymmetrical subjected to its switching abilities. In the process of Quality of Service (QoS) enhancement, choosing switching abilities plays a major role subjected to number of ports (Np) and the asymmetry depends over SSw port size. This paper introduces an efficient model of ROADM switching connectivity (SwC) to achieve Qos for given Np. The performance analysis is performed by considering QoS and found that it depends on SwC.

Keywords: Quality of Service (QoS) · ROADM
Selective switches of wavelength (SSw) · Switching connectivity
Wavelength Switched (WSw) Optical networks (ONet)

1 Introduction

The current development in the WS technology enables the multi-degree ROADM architectures with Col-Dl-Conl switching. The WS has become a suitable enabler in upcoming Wavelength Division Multiplexing (WDM) mesh networks [1]. The WS chooses the separate wavelengths from multiple ingress switches and ports to a common egress port and a key feature of WSw based RAODM. The ROADM exhibits the highly asymmetric switching ability, i.e., a switch in which a signal over ingress port will reach to egress ports subset [2]. These kinds of restrictions were mainly considered in researches subjected to the Routing and Wavelength Assignment (RWA) problem. Most of the existing works in RWA algorithm either assumed a network is having a physical layer or having physical layer impairment, with node architectures which are completely flexible. The Sect. 3 some of the researches meant for RWA algorithm

© Springer International Publishing AG, part of Springer Nature 2019
R. Silhavy (Ed.): CSOC 2018, AISC 765, pp. 413–422, 2019.
https://doi.org/10.1007/978-3-319-91192-2_41

design by considering the nodes having architectural constraints like asymmetric switching (AsySw) [3]. This paper introduces an Integer Linear Programming (ILP) model, i.e., RWA with Asymmetric Nodes (AsyN) for existing RWA issues. The optimized large-scale ILP model that provides the accurate solution for large RWA architectures with a wavelength of 670 considering all AsyN and given SwC matrix. In this, the RWA_AN model is modified to evaluate the SwC matrix for the given number of switch connections and ports concerning QoS and compare with existing model [4].

This paper is organized into following sections. Section 2 gives the ROADM architecture, and Sect. 3 discusses the review of the literature. Section 4 describes the problem formation while Sect. 5 explains the proposed system. The results analysis and conclusion were discussed the significant contribution of the paper.

2 ROADM Architectures

The notion of ROADMs is the chief components in optical networks that can be built to reconfigure the future generation networks [5]. ROADMs enable the dynamically add/drop operation else ensure that it passes through an identical Wavelength Division Multiplexing (WDM) channel group in the network nodes eliminating the requirement of transformations through Optical-Electrical-Optical conceptualization (O-E-O). Therefore, it creates an impact on the optical performance, cost-effectiveness and configuration adaptability. Initially, the design of ROADMs supported line and ring architectures, functioning at a degree value 2, whereas the ROADMs being currently developed support the higher degree node values. This is necessary for designing and developing the upcoming generation's optical networks. To enable this, the dominating technologies of WSS-based ROADMs which are having multiple degrees are studied and emerge as an extremely promising technique that is easily flexible with the function of degree updation [1–4].

The Fig. 1(a) illustrates the operation of a 4th degree WSS based ROADM that is positioned at a node within the network. This type of node can be configured to reach its neighbor accordingly starting from the South, North, West and East directions and also vice-versa. The optical signal which is received by all directions individually is split via a Power Splitter (PS) and is further directed towards the WSS positioning on the outer side of rest three directions and the module of local DROP. The WSS individually chooses and integrates the wavelengths attained by the rest three directions, received by the module of local ADD and will be later on directed in the direction which is desired. Hence, incorporating this architecture, a wavelength of any value can be that is entering in the network node will be routed at the output of either one or more directions. Moreover, it is not mandatory the service providers would always target with the investigating approach of pay as one grows. This indicates that the preferred practice method is asymmetric switching in nodes. The Fig. 1(b) demonstrates the asymmetric possible architecture of ROADM. Its degree is four, and the configuration utilized for it is 1 × 4 PSs and 1 × 4 WSSs, precisely ensuring that a subset from their port travels the other directions and the left out ports are given for reservation to conduct scaling further [5].

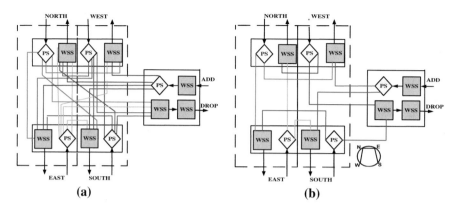

Fig. 1. (a). 4-degree ROADM (Symmetric Architecture) **(b).** A 4-degree ROADM (Asymmetric Architecture)

The switching connectivity between the four directions in the mentioned asymmetric ROADM can be represented as a small circular shape in the lower most right corner in Fig. 1(b). The three highlighted lines in the circular shape interconnect the directions which are easy to reach and are termed as the Internal Port connection (IPCs).

3 Related Works

This section prioritizes the review of existing researches in WDM networks focusing on RWA issues. In this regard, various concepts and techniques were introduced under different level traffic patterns. The review works for Integer Linear Programming (ILP) were found in Jaumard et al. [6] where Symmetric and Asymmetric traffic patterns are discussed. The extended study of Jaumard et al. [7] expressed the comparative analysis of all the improvements in ILP is discussed. However, these above researches were not meant for internal switching design of optical nodes. Chen et al. [8] introduced two significant mechanisms subjected to dynamic light path provisioning with distance vector (DisV) and link state (LkS) mechanisms in optical WDM mesh networks exhibiting Asymmetric Nodes (AsyN). The Lks mechanism introduces algorithms like K-shortest path (Dijkstra), Asymmetric switching aware (AsySA) and complete path searching (CPS). The outcomes replicate that the ASySA Dijkstra's algorithm attains high blocking probability and the CPS algorithm attained factorial computational complexity and was non-polynomial. Thus, all these algorithms cannot be able to scale with increased network size. The author presented an information diffusion based routing solution for the DisV mechanism. Hashiguchi et al. [9] expressed the idea of providing the resilience towards node failures in WDM networks with AsyN. This gives the search for node disjoints paths pair (in which one is working path and other is for backup path). Similarly, the Bhandari [10] and Surballe and Tarjan's [11] mechanisms can also compute the optimal disjoint paths in WDM networks with Symmetrical Nodes (SyN), but these algorithms fail with the network with

AsyN. Author implemented [11] algorithm to avoid the AsyN related issues and achieves better time complexity and lags with optimality proof.

4 Problem Formation

To formulate a problem wavelength division multiplexing (WDM) optical network is considered, and the network is represented by directed multi-graphs (dMG) as:

$$dMG = (N_{set}, A_{set}) \tag{1}$$

Where N_{set} represents the node set and A_{set} represents the arc set.

$$N_{set} = N_1, N_2, \ldots, N_n \tag{2}$$

Here, every node is associated with a node of the physical network.

$$A_{set} = A_1, A_2, \ldots, A_n \tag{3}$$

Here, every arc is subjected to a fiber link of the physical network.

The total number of the arc's from N to N' is equal to the number of fibers supporting traffic from N to N'.

In this an expanded dMG (EdMG) is considered and is noted as:

$$EdMG = (N_{Set}^E, A_{Set}^E) \tag{4}$$

Where N_{Set}^E represents the expanded N_{Set} and A_{Set}^E represents expanded A_{Set}.

$$N_{Set}^E = \underset{N \in N_{set}}{U} P^N \tag{5}$$

Where P^N indicates the set of the port of node N.

$$A_{Set}^E = \underset{N \in N_{set}}{U} A^N \tag{6}$$

Where A^N indicates the set of links connecting to ports of nodes.

The following Fig. 2 illustrates the general form of dMG indicating the multi-fiber optical network and Fig. 3 illustrates the extended form of dMG. The arcs are represented by a1, a2, a3... a8 and nodes as N_1, N_2, and N_3.

The set of available wavelengths can be represented as:

$$\lambda_{set} = \lambda_1, \lambda_2, \ldots, \lambda_n \tag{7}$$

Where $W = |\lambda_{set}|$

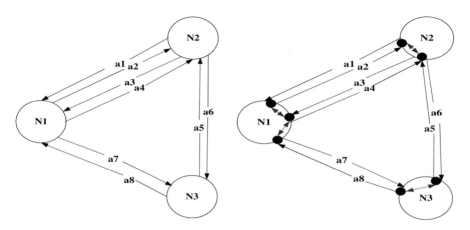

Fig. 2. General Form of dMG   **Fig. 3.** Extended form of dMG

Traffic is given as T and *Tsd* indicates the requested connections from the source node (*Ns*) to Destination Node (*Nd*).

$$\text{Consider } SD = \{(Ns, Nd) \in N \times N : Tsd > 0\} \tag{8}$$

In this, single-hop routing is considered, i.e., of the same wavelength from a source to destination connection requests. For RWA_asymmetric nodes (RWA_AN) an expanded dMG (*EdMG*) with respect to WDM Optical network with AN (for the given set of asymmetric connections) and a set of requested node connections to identify the desired short-path (*Sp*) for every permitted connections in which *Sp* is combination of routing path and wavelength (i.e, p & λ) which means no bi-directional paths sharing are allowed for same λ.

For RWA_OSA, the given EdMG relevant to WDM optical network consisting of less switching abilities like switch connections among the ports of node N represented as SwN. This RWA_OSA model is aimed to enhance the QoS by finding the switching node configuration. This paper aims to maximize the permitted connections counted with QoS and minimized blocking rate.

5   Proposed Model

This Section Describes The Proposed System Indicating Two Models Of RWA_AN And RWA With Optimal Asymmetric Switch (RWA_OAS).

5.1   Model of RWA_AN

The designed model of RWA_AN depends on the configuration concept. Consider, W_c set of wavelength configuration which is subjected to the maximal pair of link disjoint-paths, which can be considered to satisfy the given fraction of connections.

A W_c is represented by a vector, "b^c i.e., b_{sd}^c = connection requests from N_s and N_d which is supported with configuration c.

The W_c is maximal if there is no other W_c' is exist which means: $b_c' \geq b_c$

In this, two pairs of the variable exist. Consider Z_c as selected configuration occurrences each having a different wavelength. The variable "y_{sd}" indicates the accepted connections from N_s and N_d in SD.

The derivation of the objective function can be obtained as:

$$\max \sum_{Ns, Nd} y_{sd} \tag{9}$$

Associated with

$$\sum_{c \in W_c} Zc \leq W \tag{10}$$

$$\sum_{c \in W_c} b_{sd}^c Zc \geq y_{sd} \qquad (N_s, N_d) \in SD \tag{11}$$

$$y_{sd} \leq T_{sd} \quad (N_s, N_d) \in SD \tag{12}$$

$$(Zc \in N) \qquad c \in Wc \tag{13}$$

Equation (10) provides only the available wavelengths can be assigned and (11) indicates support towards requested connections. Equation (12) mentions that the accepted connections will not exceed the demand.

5.2   Model of RWA_OAS

In this model the wavelength configuration (W_c) is modified and in each configuration of W_c is indicated with two binary vectors of b_c and m_c in which $m\ell_c = 1$ if W_c utilizes the link (ℓ) internal port connection else it will be zero.

In this, one more variable set is introduced: i.e. $x\ell = 1$, if l is selected for internal port connection else, it will be zero.

The model of RWA_OAS has similar objective functions, and it exhibits additional functions like:

$$\sum_{c \in W_c} m_\ell^c Z_c \leq Wx\ell \qquad n \in N, \ell \in A_{set} \tag{14}$$

$$\sum_{\ell \in A_{set}} x_\ell \leq Sn \qquad n \in N \tag{15}$$

$$\sum_{\ell \in w+(n)} x_\ell \geq 1; \quad \sum_{\ell \in w-(n)} x_\ell \leq 1 \qquad n \in A_{set}^E \tag{16}$$

The Eq. (14) presents that the link (ℓ) is utilized to configure only if the internal port connection within the switching matrix $(x\ell = 1)$ is selected. The Eq. (15) indicates that the internal port connections of an asymmetric node are not exceeding the internal port connections for a particular node. The Eq. (16) assures that minimum one internal port connection per node in expanded dMG $(EdMG)$.

6   Results Discussion

The two RWA_AN and RWA_OAS models were implemented using the Open Programming Language and solved using CPLEX 12.5. The programs were run on a 2.2 GHz AMD Operated on a 64-bit processor with 4 GB RAM. The results analysis is done for Quality of service vs. switching connectivity for a given number of ports, under two switching scenarios. In thus, twenty traffic instances are considered. During 1st traffic (SD_0) instance, the directed traffic demand matrix (T = [Tsd]) is obtained by drawing the (integer) traffic demands (in Sp units) uniformly at random in {0, 1, 2, 3, 4, 5}. The following traffic instances correspond to incremental traffic: SD_i * SD_(i + 1) where SD_(i + 1) is developed on SD_i through random addition from 1 up to 5 more requests for every node pairs.

Table 1 gives request sets characteristics. For every traffic instance, the number of node pairs with requests ($|SD|$) is provided, and the overall traffic requests are formulated by:

$$\sum_{\{Ns, Nd\}\in SD} dsd$$

Observed that 182 are the highest node pair.

Table 1. Request sets characteristic

Traffic instances	$|SD|$	$\sum_{\{Ns, Nd\}\in SD} d_{sd}$	Traffic instances	$|SD|$	$\sum_{\{Ns, Nd\}\in SD} d_{sd}$
SD_0	142	346	SD_10	182	3973
SD_1	171	696	SD_11	182	4364
SD_2	179	1043	SD_12	182	4739
SD_3	180	1413	SD_13	182	5103
SD_4	182	1797	SD_14	182	5488
SD_5	182	2191	SD_15	182	5828
SD_6	182	2541	SD_16	182	6198
SD_7	182	2880	SD_17	182	6538
SD_8	182	3229	SD_18	182	6900
SD_9	182	3611	SD_19	182	7300

6.1   Quality of Service vs. Switching Connectivity

In Fig. 4, we investigate the impact of increasing the number of ports on the grade of service, and then the optimization of the switching connectivity on the grade of service. To do so, we run experiments on different sets of traffic requests, and for a different number of ports, for both models. Indeed, we increase the number of switching capabilities by one bidirectional IPC (two directional IPCs in opposite directions) for the following set of nodes: {v3, v4, v7, v8}. While this increased switching capability does not affect much the results of the RWA_OAS Model (see lines entitled NSF_OAS and NSF_OAS_ADD), it is different for the RWA AN Model (see lines entitled NSF_AN and NSF_AN_ADD): It is observed that this allows a 9.6% average increase of the grades of service for the traffic instances. However, if the switching connectivity matrix is optimized according to the solution of RWA_OAS model, no additional IPC is required for improving the grades of service. Note that even the improved GoS of model RWA AN, i.e., GoS2_AN, does not reach the values of the GoS1_OAS. Consequently, it is worth optimizing the switching capabilities of a ROADM to save on the number of ports, or on the switching connectivity requirements.

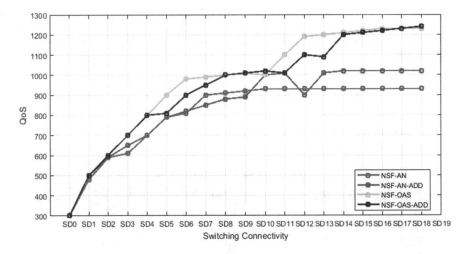

Fig. 4. Internal connections impacts

6.2   Analysis of the Number of Hops

In Fig. 5, wavelength paths percentage based on the number of hops (i.e., the number of links) in the optimal integer solutions for both models. Indeed, lines entitled "1-hop", "2-hops", "3-hops" and "4-hops" describe the proportion of wavelength paths with 1, 2, 3 or more than 4 links of the length respectively. When the number of traffic demands increases, going from SD 0 to SD 19, we can see that the number of "1-hop" paths goes up while the number of "4-hops" paths goes down, even to zero for almost traffic instances. The results show that: concerning the objective of maximizing the

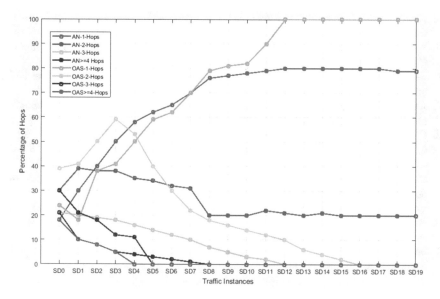

Fig. 5. Number of hops analysis

number of requests in a network with a fixed number of wavelengths, when traffic demands are increased, the solution consists essentially of single hop optical paths.

7 Conclusion

This paper introduced the significantly efficient, scalable optimization model to determine the better switching matrices for each ROADM. This RWA_AN and RWA_OSA model is aimed to enhance the QoS by finding the switching node configuration. The plots represented in Figs. 4 and 5 indicates the analysis of the QoS concerning Switching connectivity and number of hops analysis and is found that the models enhance the QoS with minimized blocking rate. This paper can be implemented in future research under dynamic traffic scenario to understand the ROADM flexibility in communication domain.

References

1. Zyskind, J., Srivastava, A. (eds.): Optically Amplified WDM Networks. Elsevier, San Diego (2011)
2. Zang, H., Jue, J.P., Mukherjee, B.: A review of routing and wavelength assignment approaches for wavelength-routed optical WDM networks. Opt. Netw. Mag. **1**, 47–60 (2000)
3. Kavitha, G.R., Indumathi, T.S.: Situational analysis of significant research contribution in optical network. Int. J. Comput. Appl. **112**(15) (2015)

4. Kavitha, G.R., Indumathi, T.S.: Enhanced constraint-based optical network for improving OSNR using ROADM. Int. J. Appl. Innov. Eng. Manag. **3**(3), 518–527 (2014)
5. Kavitha, G.R., Indumathi, T.S.: Novel ROADM modeling with WSS and OBS to Improve Routing Performance in Optical Network (2016)
6. Jaumard, B., Meyer, C., Thiongane, B.: Comparison of ILP formulations for the RWA problem. Opt. Switch. Netw. **4**, 157–172 (2007)
7. Jaumard, B., Meyer, C., Thiongane, B.: On column generation formulations for the RWA problem. Discret. Appl. Math. **157**, 1291–1308 (2009)
8. Chen, Y., Hua, N., Wan, X., Zhang, H., Zheng, X.: Dynamic lightpath provisioning in optical WDM mesh networks with asymmetric nodes. Photon Netw. Commun. **25**, 166–177 (2013)
9. Hashiguchi, T., Zhu, Y., Tajima, K., Takita, Y., Naito, T., Jue, J.P.: Iteration-free node-disjoint paths search in WDM networks with asymmetric nodes. In: International Conference on Optical Networking Design and Modeling - ONDM, pp. 1–6 (2012)
10. Bhandari, R. (ed.): Survivable Networks: Algorithms for Diverse Routing. Kluwer Academic Publishers, Massachusetts (1999)
11. Suurballe, J., Tarjan, R.: A quick method for finding shortest pairs of disjoint paths. Networks **14**, 325–336 (1984)

A Mixed Hybrid Conjugate Gradient Method for Unconstrained Engineering Optimization Problems

David A. Oladepo[1], Olawale J. Adeleke[2,3(✉)], and Churchill T. Ako[1]

[1] Department of Petroleum Engineering, College of Engineering,
Covenant University, PMB 1023, Ota, Nigeria
[2] Department of Mathematics, College of Science and Technology,
Covenant University, PMB 1023, Ota, Nigeria
wale.adeleke@covenantuniversity.edu.ng
[3] DST-NRF Centre of Excellence in Mathematical and Statistical Sciences
(CoE-MaSS), Johannesburg, South Africa

Abstract. A new hybrid of the conjugate gradient (CG) method that combines the features of five different CG methods is proposed. The corresponding CG algorithm generated descent directions independent of line search procedures. With the standard Wolfe line search conditions, the algorithm was shown to be globally convergent. Based on numerical experiments with selected large-scale benchmark test functions and comparison with classical methods, the method is very promising and competitive.

Keywords: Unconstrained optimization problems
Hybrid nonlinear conjugate gradient method · Descent direction
Global convergence · Standard Wolfe line search conditions
Numerical experiment

1 Introduction

Conjugate gradient (CG) method is an efficient gradient-based method that has been successfully applied to unconstrained problems in engineering. The general form of the problem which CG algorithms solve is given as follows:

$$\min[f(x) : x \in \mathbf{R}^n] \tag{1}$$

where $f \colon \mathbf{R}^n \to \mathbf{R}$ is a continuous and differentiable objective function whose gradient, $g(x)$, exists. The CG method solves (1) iteratively using the recurrence rule

$$x_{k+1} = x_k + \alpha_k d_k \tag{2}$$

© Springer International Publishing AG, part of Springer Nature 2019
R. Silhavy (Ed.): CSOC 2018, AISC 765, pp. 423–431, 2019.
https://doi.org/10.1007/978-3-319-91192-2_42

In (2), α_k is the step-length at the kth iteration and can be evaluated using a suitable line search technique (exact or inexact), while d_k is the search direction generated by the following rules

$$d_0 = -g_0 \quad \text{for} \quad k = 0; \quad d_k = -g_k + \beta_k d_{k-1} \quad \text{for} \quad k \geq 1 \qquad (3)$$

where β_k is the CG method update parameter and $g_k = \nabla f(x_k)$. Different values of β_k correspond to different CG methods. For instance, $\beta_k^{FR} = \frac{||g_k||^2}{||g_{k-1}||^2}$, $\beta_k^{HS} = \frac{g_k^T y_k}{d_{k-1}^T y_k}$, $\beta_k^{PRP} = \frac{g_k^T y_k}{||g_{k-1}||^2}$, $\beta_k^{LS} = \frac{g_k^T y_k}{-d_{k-1}^T g_{k-1}}$, $\beta_k^{DY} = \frac{||g_k||^2}{d_{k-1}^T y_k}$ are the Fletcher-Reeves [4], Hestenes-Stiefel [5], Polak, Ribire and Polyak [10,11], Liu-Storey [9] and Dai-Yuan [3] methods respectively.

In these methods, $|| \cdot ||$ is the Euclidean norm and $y_k = g_k - g_{k-1}$. For strict convex objective functions, the methods are equivalent. However, for non-convex functions, their behaviors differ. Although these methods represent the earliest CG methods, other variants have also been proposed recently with some exhibiting nice convergence and computational properties.

The FR and DY methods have been identified as having the best convergence results (see Al-Baali [1] and Dai and Yuan [3] for comprehensive proofs). However, for general objective functions, the two methods perform poorly in computations. Conversely, the HS and PRP methods have good computational strength even though they exhibit poor convergence results. This contrasting standpoint is the main motivation behind the development of hybrid methods which are constructed with the objective of overcoming any deficiencies in two or more methods. For instance, a well constructed hybrid method of FR and PRP should perform well computationally as well as yield good convergence properties.

In Wei et al. [12], a new nonlinear CG method was proposed where the value of β_k was given as

$$\beta_k^{WYL} = \frac{||g_k||^2 - \frac{||g_k||}{||g_{k-1}||} g_k^T g_{k-1}}{||g_{k-1}||^2}$$

Yao et al. [13] extended the work in [12] to the HS and LS methods and came up with a new version of the CG algorithm with the $\beta_k's$ proposed as

$$\beta_k^{MHS} = \frac{||g_k||^2 - \frac{||g_k||}{||g_{k-1}||} g_k^T g_{k-1}}{d_{k-1}^T (g_k - g_{k-1})}$$

and

$$\beta_k^{MLS} = \frac{||g_k||^2 - \frac{||g_k||}{||g_{k-1}||} g_k^T g_{k-1}}{-d_{k-1}^T g_{k-1}}$$

These two methods combined together give the YWH method. A number of hybrid methods have been constructed based on [12,13]. For instance, Li and Zhao in [8] proposed a hybrid algorithm featuring β_k^{PRP} and β_k^{WYL} and given as

$$\beta_k^{P-W} = max\{\beta_k^{PRP}, \beta_k^{WYL}\}$$

More recently, Jiang *et al.* [7] proposed a four term hybrid CG method with β_k obtained as

$$\beta_k^{JHJ} = \frac{\|g_k\|^2 - max\{0, \frac{\|g_k\|}{\|g_{k-1}\|}g_k^T d_{k-1}, \frac{\|g_k\|}{\|g_{k-1}\|}g_k^T g_{k-1}\}}{d_{k-1}^T(g_k - g_{k-1})}$$

Building on this idea, Jian *et al.* [6] introduced another four term hybrid method with the β_k in their case computed as

$$\beta_k^N = \frac{\|g_k\|^2 - max\{0, \frac{\|g_k\|}{\|g_{k-1}\|}g_k^T g_{k-1}\}}{max\{\|g_{k-1}\|^2, d_{k-1}^T(g_k - g_{k-1})\}}$$

This method is simply one of DY/FR/WYL/YWH as it can be reduced to any of these methods with various assumptions.

The efficiency of the hybrid methods constructed with the ideas in [12,13] above serves as a motivation to develop a new method with five terms with

$$\beta_k^{DAA} = \frac{\|g_k\|^2 - max\{0, \frac{\|g_k\|}{\|g_{k-1}\|}g_k^T g_{k-1}\}}{max\{\|g_{k-1}\|^2, d_{k-1}^T(g_k - g_{k-1}), -d_{k-1}^T g_{k-1}\}} \tag{4}$$

Independent of any line search process, this method will always satisfy the descent direction criterion, that is, $d_k^T g_k < 0$. This will be shown in a moment from now. Observe that the proposed method is an hybrid of the FR, DY, WYL, MHS and MLS methods. Hence, the method is capable of exhibiting the *nice* characteristics of these methods. A CG algorithm to implement the proposed method is presented in the next section.

2   Hybrid CG Algorithm and Descent Property

Hybrid CG Algorithm
Step 1: Initialize $x_0 \in \Re^n$, $\epsilon > 0$. Set $d_0 = -g_0$ and $k = 1$.
Step 2: While $\|g_k\| \leq \epsilon$, continue to Step 3; otherwise, stop.
Step 3: Obtain α_k by a suitable line search technique (in this case, the Standard Wolfe Line Search Procedure)
Step 4: Generate the sequences $\{x_k\}$, $\{g_k\}$ and $\{d_k\}$, where $\beta_k = \beta_K^{DAA}$.
Step 5: Set $k = k + 1$ and return to Step 2.

In what follows we establish the descent property of the algorithm.

Theorem 1: Let d_k and g_k be generated by the hybrid CG algorithm above. Then, the search direction d_k satisfies the descent condition

$$d_k^T g_k \leq 0 \quad \text{for each} \quad k \geq 0 \tag{5}$$

Proof: For $k = 0$, it can be shown that $d_0^T g_0 = -\|g_0\|^2$. Suppose we assume that $d_{k-1}^T g_{k-1} < 0$ for each $k-1$ and $k > 2$, and that $\beta_k^{DAA} = 0$ it will be obvious by the inner product of (3) with g_k that $d_k^T g_k = -\|g_k\|^2 + \beta_k d_{k-1}^T g_k = -\|g_k\|^2 < 0$.

Thus, we shall always assume that $\beta_k^{DAA} \neq 0$. Five different cases are of interest and considered as follow:

Case I: If $g_k^T g_{k-1} \leq 0$ and $\|g_{k-1}\|^2 \geq d_{k-1}^T (g_k - g_{k-1}) \geq -d_{k-1}^T g_{k-1}$

From (5) it follows that $\beta_k^{DAA} = \beta_k^{FR}$. Therefore, from (3) we obtain

$$
\begin{aligned}
d_k^T g_k = g_k \left(-g_k + \beta_k^{FR} d_{k-1}\right) &= -\|g_k\|^2 + \frac{\|g_k\|^2}{\|g_{k-1}\|^2} d_{k-1}^T g_k \\
&\leq \frac{d_{k-1}^T g_{k-1}}{\|g_{k-1}\|^2} \|g_k\|^2 = \frac{\|g_k\|^2}{\|g_{k-1}\|^2} d_{k-1}^T g_{k-1} < 0
\end{aligned}
\tag{6}
$$

The first inequality uses the fact that $-\|g_{k-1}\|^2 \geq d_{k-1}^T (g_k - g_{k-1})$, while the second inequality uses the assumption at the beginning of this proof and the fact that $\beta_k^{FR} > 0$.

Case II: If $g_k^T g_{k-1} \leq 0$ and $d_{k-1}^T (g_k - g_{k-1}) \geq \|g_{k-1}\|^2 \geq -d_{k-1}^T g_{k-1}$, then (5) reduces to $\beta_k^{DAA} = \beta_k^{DY}$. Therefore,

$$
\begin{aligned}
d_k^T g_k = g_k \left(-g_k + \beta_k^{DY} d_{k-1}\right) \\
= -\|g_k\|^2 + \frac{\|g_k\|^2}{d_{k-1}^T (g_k - g_{k-1})} d_{k-1}^T g_k \\
= \frac{d_{k-1}^T g_{k-1}}{d_{k-1}^T (g_k - g_{k-1})} \|g_k\|^2 < 0
\end{aligned}
\tag{7}
$$

since $\|g_{k-1}\|^2 \geq 0 \Rightarrow d_{k-1}^T (g_k - g_{k-1}) \geq 0$ and the assumption that $d_{k-1}^T g_{k-1} < 0$

Case III: If $g_k^T g_{k-1} > 0$ and $\|g_{k-1}\|^2 \geq d_{k-1}^T (g_k - g_{k-1}) \geq -d_{k-1}^T g_{k-1}$, then $\beta_k^{DAA} = \beta_k^{WYL}$. The fact that $g_k^T g_{k-1} > 0$ means we can get an inequality $0 < cos\theta_k < 1$ where θ_k is the angle between g_k and g_{k-1}. Therefore, following the same process as the above cases, we get

$$
\begin{aligned}
d_k^T g_k = g_k \left(-g_k + \beta_k^{WYL} d_{k-1}\right) \\
= -\|g_k\|^2 + \frac{\|g_k\|^2 - \frac{\|g_k\|}{\|g_{k-1}\|} g_k^T g_{k-1}}{\|g_{k-1}\|^2} g_k^T d_{k-1} \\
< \frac{(1 - cos\theta_k) \|g_k\|^2}{\|g_{k-1}\|^2} d_{k-1}^T g_{k-1} < 0
\end{aligned}
\tag{8}
$$

Case IV: If $g_k^T g_{k-1} > 0$ and $d_{k-1}^T (g_k - g_{k-1}) \geq \|g_{k-1}\|^2 \geq -d_{k-1}^T g_{k-1}$, then (5) becomes β_k^{MHS}. Therefore, from (3), case III above and the fact that $\beta_k^{DAA} \neq 0$, we obtain

$$
\begin{aligned}
d_k^T g_k = g_k \left(-g_k + \beta_k^{MHS} d_{k-1}\right) \\
= -\|g_k\|^2 + \frac{\|g_k\|^2 - \frac{\|g_k\|}{\|g_{k-1}\|} g_k^T g_{k-1}}{d_{k-1}^T (g_k - g_{k-1})} g_k^T d_{k-1} \\
< \frac{\|g_k\|^2 d_{k-1}^T g_{k-1}}{d_{k-1}^T (g_k - g_{k-1})} < 0
\end{aligned}
\tag{9}
$$

Case V: If $g_k^T g_{k-1} > 0$, $-d_{k-1}^T g_{k-1} \geq \|g_{k-1}\|^2$ and $-d_{k-1}^T g_{k-1} \geq d_{k-1}^T(g_k - g_{k-1})$, then $\beta_k^{DAA} = \beta_k^{MLS}$ and the following process holds.

$$d_k^T g_k = g_k\left(-g_k + \beta_k^{MLS} d_{k-1}\right)$$

$$= -\|g_k\|^2 + \frac{\|g_k\|^2 - \frac{\|g_k\|}{\|g_{k-1}\|} g_k^T g_{k-1}}{-d_{k-1}^T g_{k-1}} g_k^T d_{k-1} \tag{10}$$

$$< \frac{(1 - cos\theta_k)\|g_k\|^2 g_k^T d_{k-1}}{-d_{k-1}^T g_{k-1}} < 0$$

The second inequality follows the facts that $-d_{k-1}^T g_{k-1} \geq d_{k-1}^T(g_k - g_{k-1}) \Rightarrow d_{k-1}^T g_k < 0$ and $d_{k-1}^T g_{k-1} < 0 \Rightarrow -d_{k-1}^T g_{k-1} > 0$.

Results in (7)–(11) confirm that the new method satisfies the descent property in its approach to obtaining the optimal value of the objective function.

Another important property of our hybrid method is as stated in the following result. The process of proof in [10] may be adopted to establish this result.

Lemma 1: The relation $0 \leq \beta_k^{DAA} \leq \frac{d_k^T g_k}{d_{k-1}^T g_{k-1}}$ is true for any $k \geq 1$.

3   Global Convergence

We now show the global convergence result for the new hybrid method under the influence of the standard Wolfe line search conditions given as

$$f(x_k + \alpha_k d_k) \leq f(x_k) + \delta \alpha_k g_k^T d_k \tag{11}$$

and

$$g(x_k + \alpha_k d_k)^T d_k \geq \sigma g_k^T d_k \tag{12}$$

with $0 < \delta \leq \sigma_1$

To do this, the following common assumptions on the objective function are stated.

Assumption:
i. The level set $\Omega = \{x \in \mathbf{R}^n | f(x) \leq f(x_0)\}$ is bounded.
ii. In some neighborhood $N \in \Omega$, $f(x)$ is Lipschitz continuously differentiable, i.e., there exists a positive constant L such that

$$\|g(x) - g(y)\| \leq L\|x - y\|, \forall x, y \in \mathbf{N} \tag{13}$$

Assumptions (i)–(ii) imply that a positive constant m exists such that

$$\|g(x)\| \leq m, \quad \forall x \in \mathbf{N}$$

An important result for proving the global convergence properties of nonlinear CG algorithms is the famous Zoutendijk condition [14] which is stated as a lemma here.

Lemma 2: Suppose Assumptions (i)–(ii) hold. Given any iteration of the form (2), where d_k is a descent direction and α_k satisfies (12) and (13), then we have

$$\sum_{k=0}^{\infty} \frac{(g_k^T d_k)^2}{\|d_k\|^2} < +\infty \tag{14}$$

A straightforward proof of this lemma can be found in [3].
Next we state and prove the global convergence result for the five-term hybrid method.

Theorem 2: If Assumption 1 holds and $\{x\}$ is generated by the Hybrid CG Algorithm, then

$$\lim_{k \to \infty} inf \|g_k\| = 0 \tag{15}$$

Proof: Let $\lim_{k \to \infty} inf \|g_k\| \neq 0$. Then, since $\|g_k\| > 0$, there exists a constant $n > 0$ such that $\|g_k\|^2 \geq n, \forall k$. Squaring both sides of (3) we have

$$\|d_k\|^2 = \left(\beta_k^{DAA}\right)^2 \|d_{k-1}\|^2 - 2d_k^T g_k - \|g_k\|^2$$

Dividing through the above equation by $\left(d_k^T g_k\right)^2$ and using the result of Lemma 1, the following is obtained

$$\frac{\|d_k\|^2}{\left(d_k^T g_k\right)^2} \leq \frac{\|d_{k-1}\|^2}{\left(d_{k-1}^T g_{k-1}\right)^2} + \frac{1}{\|g_k\|^2} \tag{16}$$

Since $\frac{\|d_1\|^2}{\left(d_1^T g_1\right)^2} = \frac{1}{\|g_1\|^2}$, (16) gives

$$\frac{\|d_{k-1}\|^2}{\left(d_{k-1}^T g_{k-1}\right)^2} \leq \sum_{i=1}^{k} \frac{1}{\|g_i\|^2} \leq \frac{k}{n}, \quad \forall k$$

This implies

$$\sum_{k \geq 1} \frac{(g_k^T d_k)^2}{\|d_k\|^2} = \infty$$

which contradicts (14). Hence, the result is proved.

4   Numerical Test

In this section, a mild numerical experiment is presented to investigate the efficiency of the new hybrid method described above. The method with (7) as the CG parameter was tested numerically on Windows 7 with installed RAM of 2GB. The code for the algorithm was written with MatLab 7.10 using the Wolfe line search criteria (12) and (13) where $\delta = 0.0001$ and $\sigma = 0.9$. The initial value of α_k is 0, and the stopping condition is $\|g_k\| \leq 10^{-6}$. Since the hybrid method combines the structures of the FR, DY, WYL, MHS and MLS methods, it is

Table 1. Numerical results for DAA method

Function name	n	DAA
Ext. Rosenbrook	5000	70/4.0e−15/9.3e−8/0.54
	10000	93/1.3e−14/4.7e−7/1.11
Diagonal 4	5000	18/3.8e−17/1.8e−7/0.19
	10000	18/7.6e−17/2.6e−7/0.26
Ext. Himmelblau	5000	63/7.1e−16/2.5e−7/0.69
	10000	63/1.7e−15/3.9e−7/1.11
Ext. Beale	5000	69/4.4e−14/7.9e−7/1.45
	10000	63/2.8e−13/4.4e−7/2.35
Ext. Blk Dia BD1	5000	45/5.8e−14/5.6e−7/0.39
	10000	35/4.3e−15/3.7e−7/0.45

Table 2. Numerical results for FR method

Function name	n	FR
Ext. Rosenbrook	5000	107/8.7e−15/3.5e−7/0.83
	10000	520/3.1e−13/7.9e−7/7.90
Diagonal 4	5000	17/1.0e−16/2.3e−7/0.21
	10000	17/2.0e−16/3.3e−7/0.37
Ext. Himmelblau	5000	53/6.0e−15/5.8e−7/0.56
	10000	53/1.2e−14/8.0e−7/0.94
Ext. Beale	5000	102/1.8e−13/5.5e−7/2.94
	10000	178/3.5e−14/5.8e−7/9.22
Ext. Blk Dia BD1	5000	698/NaN/NaN/15.10
	10000	234/4.3e−14/9.3e−7/4.37

Table 3. Numerical results for DY method

Function name	n	DY
Ext. Rosenbrook	5000	68/2.3e−15/6.8e−8/0.56
	10000	68/4.6e−15/9.7e−8/0.85
Diagonal 4	5000	15/2.4e−17/1.4e−7/0.15
	10000	15/4.8e−17/2.0e−7/0.23
Ext. Himmelblau	5000	52/6.4e−15/6.5e−7/0.52
	10000	52/1.2e−14/9.1e−7/0.86
Ext. Beale	5000	242/2.7e−13/5.7e−7/6.33
	10000	242/5.5e−13/8.0e−7/11.63
Ext. Blk Dia BD1	5000	817/4.7e−14/9.8e−7/9.68
	10000	163/3.6e−14/9.6e−7/2.42

computationally convenient to compare the DAA method with the FR and DY methods. The test functions were drawn from the collection of Andrei [2].

The test was conducted on ten problems with 5000 and 10000 as the dimension (n). In each of Tables 1, 2 and 3 above, the numerical results are presented in the form $Itr/F/G/T$ where Itr is the number of iterations, F is the function evaluation, G is the gradient evaluation and T is the CPU time. Foremost, we observed from the tables that all the problems were solved by DAA and DY methods. With respect to the computing time, our hybrid method dominates the FR and DY method showing that the method is computationally efficient. The same trend was observed in the total number of iterations performed by each method for the ten problems.

5   Conclusion

Based on two previous works on the construction of hybrid CG algorithm, a new hybrid method have been proposed in this paper. An interesting feature of the method is that it contains five different terms corresponding to five different CG methods. The descent and global convergence properties of the method were established under the standard Wolfe line search. Numerical tests with the five-terms hybrid method revealed that the method can really compete with the well established methods. As part of future study, the method will be compared comprehensively with the WYL, MHS and MLS methods which are recently proposed. Doing this will afford us the opportunity to compare the performance of the method according to certain features.

Acknowledgments. The support of the DST-NRF Centre of Excellence in Mathematical and Statistical Sciences (CoE-MaSS) towards this research is hereby acknowledged. Opinions expressed and conclusions arrived at, are those of the authors and are not necessarily to be attributed to the CoE. The authors also wish to appreciate Covenant University Centre for Research, Innovation and Discovery (CUCRID) for funding the publication of this research output.

References

1. Al-Baali, A.: Descent property and global convergence of the Fletcher-Reeves method with inexact line search. IMA J. Numer. Anal. **5**, 121–124 (1985)
2. Andrei, N.: An unconstrained optimization test functions collection. Adv. Mod. Optim. **10**(1), 147–161 (2008)
3. Dai, Y.H., Yuan, Y.: A nonlinear conjugate gradient method with a strong global convergence property. SIAM J. Optim. **10**, 177–182 (1999)
4. Fletcher, R., Reeves, C.: Function minimization by conjugate gradients. Comput. J. **7**, 149–154 (1964)
5. Hestenes, M.R., Stiefel, E.L.: Method of conjugate gradients for solving linear systems. J. Res. Nat. Bur. Stand. **49**, 409–436 (1952)
6. Jian, J., Han, L., Jiang, X.: A hybrid conjugate gradient method with descent property for unconstrained optimization. Appl. Math. Model. **39**, 1281–1290 (2015)

7. Jiang, X.Z., Han, L., Jian, J.B.: A globally convergent mixed conjugate gradient method with Wolfe line search. Math. Numer. Sin. **34**, 103–112 (2012)
8. Li, X., Zhao, X.: A hybrid conjugate gradient method for optimization problems. Natural Sci. **3**(1), 85–90 (2011)
9. Liu, Y., Storey, C.: Efficient generalized conjugate gradient algorithms, Part 1: theory. J. Optim. Theory Appl. **69**, 129–137 (1991)
10. Polak, E., Ribiére, E.: Note sur la convergence de directions conjugeés, Rev. Francaise Informat Recherche Operationelle, 3e Année, 16, pp. 35–43 (1969)
11. Polyak, B.T.: The conjugate gradient method in extreme problems. USSR Comp. Math. Math. Phys. **9**, 94–112 (1969)
12. Wei, Z., Yao, S., Liu, L.: The convergence properties of some new conjugate gradient methods. Appl. Math. Comput. **183**, 1341–1350 (2006). https://doi.org/10.1016/j.amc.2006.05.150
13. Yao, S.W., Wei, Z.X., Huang, H.: Note bout WYL's conjugate gradient method and its application. Appl. Math. Comput. **191**, 381–388 (2007)
14. Zoutendijk, G.: Nonlinear programming, Computational methods. In: Abadie, J. (ed.) Integer and Nonlinear Programming, pp. 37–86. North-Holland, Amsterdam (1970)

Chemical Reaction Optimization for Traveling Salesman Problem Over a Hypercube Interconnection Network

Ameen Shaheen[✉], Azzam Sleit, and Saleh Al-Sharaeh

Computer Science Department, King Abdullah II School for Information Technology, The University of Jordan, Amman 11942, Jordan
ameen1223@gmail.com

Abstract. Traveling Salesman Problem is a well-known NP-Hard problem, which aims at finding the shortest path between numbers of cities. Chemical Reaction Optimization (CRO) is a recently established meta-heuristic algorithm for solving optimization problems which has successfully solved many optimization problems. The main goal of this paper is to investigate the possibility of parallelizing CRO for solving the TSP problem called (PCRO). PCRO is compared with Genetic Algorithm (GA), which is a well-known meta-heuristic algorithm. Experimental results show relatively better performance for PCRO in terms of execution time, Speedup, optimal cost and Error rate.

Keywords: Chemical reaction optimization · Meta-heuristics
Traveling salesman problem

1 Introduction

With the increasing number of cities, mobility has become one of the challenges of daily life, because of the existence of many dissimilar ways to reach the same city [1].

Some Algorithms could be used to guide people who use any of transportation or movement methods, like (car, walking, train, and bus), to reach its destination in the shortest path [2]. From the point of driving travel, we can discover that the driving travel problem can be classified as a kind of Travel Salesman Problem.

Traveling Salesman Problem (TSP) has received much consideration from computer researchers and mathematicians; as it's so easy to describe and so difficult to solve. The centrality of the TSP is that, it is illustrative of a bigger class of problems known as combinatorial optimization problems (COP) [3]. The TSP problem has a place in the class of such problems known as NP-complete. Specifically, if one can find an efficient (polynomial-time) algorithm for the TSP, at that point efficient algorithms could be found for all others [4].

Chemical reaction optimization (CRO), proposed in [5], is a recently established meta-heuristics for optimization, inspired by the nature of chemical reactions.

In real world, there are many complicated events which happen at the same time and in a temporal sequence like galaxy formation, weather and rush hour traffic. Hence, the concept of parallel computing appeared to increase the performance and reduce the

© Springer International Publishing AG, part of Springer Nature 2019
R. Silhavy (Ed.): CSOC 2018, AISC 765, pp. 432–442, 2019.
https://doi.org/10.1007/978-3-319-91192-2_43

computation time for solving these problems. TSP is one of those problems. Parallel computing is much better suited for simulating and modeling complex problems [6, 7]. Hypercube is a multi-dimensional mesh of processors with exactly 2 processors in each dimension; this means a d-dimensional hypercube consists of $p = 2^d$ processors. For instance, a zero-dimensional hypercube is a single processor. In general, a $(d + 1)$-dimensional hypercube is constructed by connecting the corresponding processors of 2-dimensional hypercubes [8].

We use hypercube because of its performance proved compared with static network topologies [9] in terms of diameter and cost. Also, the hypercube is implemented in many supercomputers such as Endeavour Supercomputer by NASA [10].

The main contributions of this study are outlined as follows:

- This study adapted CRO to solve TSP problem and it's executed sequentially on an instance of TSP map(lim963) taken from the World TSP and the performance metrics are measured in terms of execution time and optimal cost.
- In order to compare the result of CRO with other meta-heuristic algorithms, GA (Genetic algorithm) is chosen and adapted to solve TSP and the performance metrics in terms of execution time and optimal cost are computed.
- The development of parallel CRO. Parallel CRO is developed based on both of data and computation distribution technique over a hypercube interconnection network. The data distribution technique is designed based on dividing the data set instances (cities) to achieve load balancing among the interconnection network. The computation distribution is provided by distribution the CRO iterations over the interconnection network in order to reduce the computation time.
- A comparison between PCRO and PGA (Parallel Genetic algorithm) in terms of execution time, Speedup, optimal cost and Error rate. The PCRO shows better performance results than PGA.

The remaining of this paper is structured as follows: In Sect. 2, a work related to this paper is presented. Section 3 discusses the methodology. Section 4 presents the Traveling Salesman problem formulation. Section 5 addresses our parallel model for CRO for solving TSP. Section 6 shows the experimental results and discussions. Finally, conclusions and future work are in Sect. 7.

2 Related Works

In [11], authors parallelized the ant colony optimization algorithm for solving TSP over the Apache Hadoop Map-Reduce framework. They conducted several experiments with different parameters, in order to arrive at optimum values for each of the parameters used. In [12], authors examined a Genetic Algorithm (GA) solution, by finding near-optimal solutions within a reasonable time. Although GA is time efficient and a good approximation for TSP, sometimes it stuck to local optimal or it takes considerable time when the number of cities increases. They used Map-Reduce parallelization. At the end, they compared their sequential GA with other studies. They found that their sequential GA always gives better solutions than the others in terms of quality and time. But Map-Reduce GA finds better solutions and takes a shorter time than sequential GA when the problem size increases. Also, the work, in [13], proposed an algorithm based on Chemical Reaction Optimization and Lin-Kernighan local search for the Traveling Salesman Problem as a sequential approach. Experimental results show that the proposed algorithm, PCRO, is efficient. So in this paper, the CRO algorithm is implemented in parallel over hypercube to solve the TSP problem.

3 Methodology

The parallel CRO (PCRO) is implemented and then compared with the parallel Genetic Algorithm (PGA) on the same platform. Firstly, both algorithms were implemented sequentially. Then, they are parallelized and applied over a hypercube interconnection network. Afterwards, we test both algorithms against a dataset and the cost function evaluation is the same for all algorithms. Finally, we compare both algorithms in terms of computation time, speed up and optimal cost.

4 Traveling Salesman Problem Formulation

The TSP, is the problem of finding the shortest path between nodes or cities [4]. The problem is to create the shortest tour in aggregation manner to visit each node exactly once and then return to the starting node. TSP is an NP-Complete problem, where it requires a considerably large amount of computational time and resources for solving it, since it is a permutation problem. This class of problems usually much harder to solve than subset problems; as there are n! different permutations of n objects while there are 2^n different subsets of n objects ($n! > 2^n$) [14].

TSP is formulated as follows:

- Let $V = \{1, 2, \ldots, n\}$ be the vertices of graph G.
- Then a permutation $p = \{p_1, p_2, \ldots, p_n\}$ of vertices in V defines a unique tour consisting of edges $(p_i, p_i + 1)$, where $i = 1, 2, \ldots, n - 1$, and the edge (p_n, p_1).
- The cost of p, denoted as $C(p)$ is the sum of the cost of edges in the associated tour, the problem is to find a minimum cost permutation p, given the edge weights.

A tour is a simple cycle, which start and ends at vertex 1. Every tour consists of an edge $(1, k)$ for some k in $V - \{1\}$ and a path from vertex k to vertex 1. The path from vertex k to vertex 1 goes through each vertex in $V - \{1, k\}$ exactly once and if the tour is optimal then the path from k to 1 must be the shortest path going through all vertices in $V - \{1, k\}$. Hence, the principle of optimality holds, so if we let $g(i, S)$ be the length of the shortest path starting at vertex i going through all vertices in S and terminating at vertex 1 then $g(1, V - \{1\})$ is the length of an optimal salesman tour.

5 Parallel Chemical Reaction Optimization

CRO gets its name by the nature of chemical reactions, which there is an arrangement of chemical substances that change to another with the goal of achieving the minimum state of free energy. In chemical reactions, there is a molecule, which is the smallest indivisible unit in a chemical reaction that has unique chemical properties. A chemical reaction can break or make new chemical bonds to dissociate or form molecules respectively. This happens in a sequence of intermediate reactions and the consequent molecules head to stay at the most stable state with the lowest free energy.

A Chemical reaction is designed to occur in a closed container with a certain number of molecules and the structure of each molecule considered as a solution of the optimization problem.

There are four types of elementary reactions in CRO: (1) On-wall ineffective collision. (2) Decomposition, (3) Inter-molecular ineffective collision and (4) Synthesis as explained by [15].

The first two types of reactions occur when one of the molecules collide on the wall of the container, while the last two types of reactions take place just on molecules that interact with each other. Also, they need only one molecule to initiate the reaction while reactions 3 and 4 need more than one molecule. The number of molecules for reactions 1 and 2 are the same before and after the reactions but for reactions 3 and 4, all molecules are combined in one molecule. These reactions will produce new solutions that are different from the original one.

Since CRO is a metaheuristic algorithm, this means, it has the same structure of other metaheuristic algorithms, so, CRO can be divided into three main stages:

1. Initialization stage: contains the initial values for the parameters of the algorithm. In CRO, initialization stage contains the initial values for the population size, initial KE of molecules, MoleColl and KE loss rate, and generate random solutions to be enhanced in the iterations stage.
2. Iterations stage: when the algorithm enters this stage, it keeps working until a stopping criterion is met. In CRO, iterations stage contains the four types of elementary reactions.
3. Final output stage: contains the best solution which provided from the algorithm.

After one of elementary collision (one of the random solutions) selected randomly from the initial stage, to decide which type of reaction will apply. We use MoleColl $\in [0, 1]$. To do this, a random value t is generated in the interval $[0, 1]$. If t is less than MoleColl, it will be an intermolecular collision. Otherwise, the unimolecular collision will take place.

To decide between decomposition and synthesis reactions, α and β thresholds, are used respectively. α equals the maximum number of hits for a molecule without finding an enhanced solution. So, if a molecule cannot find a better solution with the number of hits bigger than α, decomposition will happen. Otherwise, the on-wall ineffective collision will take place. For the last two types of reactions, when both molecules do not have sufficient KE, which mean KE_{w1} and $KE_{w2} \leq \beta$, then the synthesis reaction will take place. Otherwise, the inter-molecular ineffective collision happens. Where β will be calculated in initial Population(Pop).

The main goal of this paper is to investigate the possibility of parallelizing CRO for solving the TSP problem. In our proposed model, we create the parallel version of CRO (PCRO) which can be efficiently executed on Hypercube Interconnection network. PCRO will be compared with PGA (Parallel genetic algorithm). We compared our proposed model with PGA, because GA (Genetic algorithm) is similar to CRO except that in GA only individuals (solutions) can be exchanged, while in CRO the molecules (solutions) with associated attributes that are swapped during the communication interval. Moreover, in GA new solutions are produced just by crossover and mutation operations, while in CRO new solutions are produced by on-wall ineffective collision, decomposition, inter-molecular ineffective collision and synthesis operations.

In this approach, the TSP problem is divided into sub-problems by creating districts from original map in the partition operation as Eq. (1):

$$D = m/p \qquad (1)$$

Where m equal the full map, p equal the number of processers and D contains the array of districts.

Then applying CRO and GA steps over each part of the map generated by partition operation as Fig. 1. Iterations are divided into small and equals groups. These groups are distributed to different processors as in Eq. (2):

$$pt = \text{iterations}/p \qquad (2)$$

Where iterations equal the total number of iterations, p equal the number of processors and pt value will have the number of iterations for each processor.

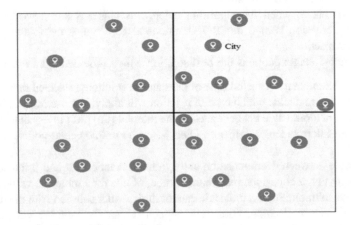

Fig. 1. Split the map into number of districts

Each processor will make pt iterations, then, it will send the best solution to the master processor. For instance, assume iterations = 16 and 2–d hypercube, this means, hypercube with 4 processors. So, by Eq. 2, pt is 4, and each processor will work four times on d sub-problems.

After each node finishes all its intended iterations, it sends the best solution to the node which contains a full tour for d (shortest path) and linked the solution with the solution that it's provide, at the end, node 00 will contain the final solution (full shortest path). Figure 2 shows the communication mechanism.

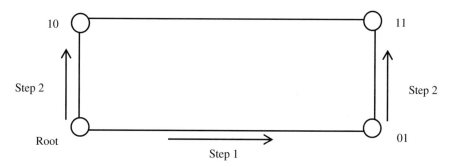

Fig. 2. Communication mechanism over hypercube.

Since sub-problems and iterations were distributed, one of the four types of CRO reactions will be implemented as follow:

1. On-wall Ineffective Collision:

When the on-wall Ineffective Collision criterion is met, then one of the molecules will be selected randomly from the population. This molecule will hit the wall and bounces back. So, we can get a new structure from the original one. In TSP, the molecule value is a random tour (solution). For example, Let city c_1 in the initial tour is the starting point: In k step, edge (c_1, c_{2k}) will be deleted, and edge (c_{2k}, c_{2k+1}) will be added. Then edge, (c_{2k+1}, c_{2k+1}) is picked. As a result, deleting edge (c_{2k+1}, c_{2k+1}), and joining edge (c_{2k+2}, c_1), will reduce the tour cost.

We implement on-wall Ineffective Collision with Mutation operator which used by genetic algorithm [16].

2. Decomposition:

When a decomposition occurs, a molecule hits the wall and then decomposes into two pieces and the result will be a new molecule different from the original one.

We implement decomposition with a Crossover operator which used by genetic algorithm [16], but in genetic algorithm crossover operator needs two solutions as input and the result will be the best one. In decomposition, when the molecule hits the wall and decomposes into two pieces, crossover is implemented over them to get the new solution.

3. Inter-Molecular Ineffective Collision:

In this phase, two molecules collide with each other and then bounce away. The change's effect from this collision is same as the effect of the on-wall ineffective collision but Inter-Molecular Ineffective Collision involves two molecules.

The inter-molecular ineffective collision was implemented with basic crossover operator which is used in genetic algorithm [16].

4. Synthesis

In Synthesis reaction, two molecules collide and combine together, and the molecule with better cost will be selected.

We implement Synthesis with the mutation operator used in genetic algorithm [16]. The difference is that in genetic algorithm mutation operator needs one solution as input and the result will be also one. But in synthesis, we apply mutation operator over two molecules and the best one is taken.

Algorithm 1 shows the pseudo code of CRO for solving TSP. Also, Algorithm 2 shows the pseudo code of data Partitioning and distribution into districts over the hypercube.

6   Experimental Results

For experiments, we used a computer with Intel Core i5-3317U CPU 1.70 GHz with 8 GB of RAM. The simulation for PCRO and PGA was written in Java JDK 8 programming language. For fairness, we also use the same operators for sequential CRO and GA. The algorithm parameters are shown in Table 1. The algorithms were tested by an instance of TSP map taken from the World TSP [17]. Since PCRO and PGA are meta-heuristic, the results obtained in different runs may be different. We repeat the simulation 50 times and we record the results as shown in Tables 2 and 3. In each run, the initial molecules (solutions) are generated randomly.

Table 1. Parameters for PCRO used in simulation

Parameter	Value
Pop size	Number of cities/2
Num of processors	1, 2, 4, 8, 16, 32
Iterations	5000
Sub iterations size	5000/number of P
MoleColl	0.8
Thresholds	$\alpha = 3$ and and $\beta = $ current cost/2
Initial KE	2000

```
Assign values to parameters PopSize, Initial KE, MoleColl, Iterations, t, α and β
Let PopSize equals the set of molecule 1, 2.....
for each molecule
    Assign a random tour.
end for
while the stopping criteria not met
    Get t randomly in interval [0, 1]
        if t > MoleColl then
            Select a molecule M from Pop randomly
                if decomposition criterion met then
                    molecules M undergo decomposition
                    if decomposition success then
                        update Pop
                    end if
                else
                    do on-wall ineffective collision on M
                    if the on-wall ineffective collision success then
                        update Pop
                    end if
                end if
        else
            select M₁ and M₂ form Pop randomly
                if synthesis criterion met then
                    do synthesis between M₁ and M₂
                    if synthesis success then
                        update Pop
                    end if
                else
                    M₁ and M₂ implemented by inter-molecular ineffective collision
                    if inter-molecular ineffective collision success then
                        update Pop
                    end if
                end if
            end if
        end if
end while
output minimum solution
```

Algorithm 1. Pseudo code of CRO for solving TSP.

```
Input:
Hypercube_Route (D, d)
where:
D: Array of districts.
d: Dimension of the hypercube.
STN: Function responsible to split the original array of districts D.
UpdateD: Function responsible to update the districts array value.
Output:
District partition and distribution over hypercube.

for (i:=0 to d) {
    part = |D| / 2.
    STN(D, 0, part).
    D = UpdateD(D, part+1, |D|)
end for
```

Algorithm 2. Hypercube routing algorithm.

Table 2. Execution time (in seconds) & speedup for PCRO & PGA

Number of Processors	PCRO			PGA		
	Avg	Best	Speedup	Avg	Best	Speedup
1	14.432	14.261	N/A	18.767	18.612	N/A
2	8.761	8.632	1.652	11.505	10.251	1.815
4	6.702	6.621	2.153	9.732	8.704	2.138
8	4.654	4.564	3.124	8.552	7.471	2.257
16	3.405	3.312	4.305	6.442	6.393	2.911
32	2.342	2.994	4.763	5.841	5.721	3.253

Table 3. The fitness value and the error rate for PCRO & PGA.

Number of Processors	PCRO			PGA		
	Best	Avg	Error rate	Best	Avg	Error rate
1	2900	2921	3.979	3010	3029	8.2825
2	2936	2938	5.270	3082	3108	10.864
4	2995	4004	7.386	3101	3127	11.545
8	3040	3056	8.999	3112	3135	11.939
16	3089	3094	10.756	3121	3147	12.262
32	3111	3119	11.545	3229	3246	16.185

In Table 2, the first column shows the number of processors and the time column shows the time it takes to run the entire program in seconds on one node (sequential time). Speedup is an important measure for a parallel algorithm, which used to calculate the ratio of the computation time of the sequential time and the parallel time as Eq. 3:

$$Speedup_m = E[T_1]/E[T_m] \tag{3}$$

In Table 3, the first column shows the best-obtained fitness value where the second column shows the average value of 50 runs, the error column shows the error value of the fitness function (minimum) of the best individual that both algorithms provide and the TSPLIB optimum. The error is calculated as in formula (4):

$$Error = \left(\frac{BestSolution - OptimalSolution}{OptimalSolution}\right) * 100 \tag{4}$$

From Fig. 3, we can observe that the speedup is almost linear for PCRO when using 2, 4, 8 and 16 processors, while it is sub-linear when there are 32 processors. This is because of the overhead of communication in 32 processors scenario which is much more than that in 16 processors or less. From comparing PCRO obtained better Speedup more than PGA.

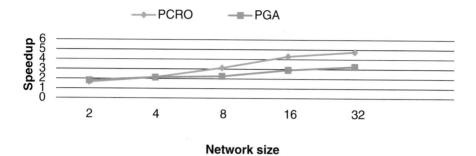

Fig. 3. Speed up for PCRO and PGA.

Comparing the Error rate of fitness values from Fig. 4, we find that in most cases the quality of solutions generated by PCRO is better than PGA. This is because in CRO there are four types of reactions which enhance the solutions more than the GA. Also, by increasing the number of processors, the Error rate increased. This is mainly due dividing the number of iterations over processors.

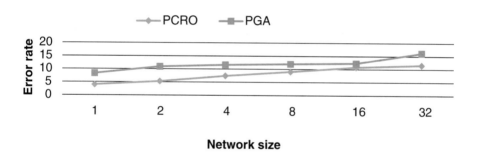

Fig. 4. Quality of solutions for PCRO and PGA

7 Conclusions and Future Work

This paper introduces a parallel model of CRO algorithm for solving the TSP problem on a hypercube interconnection network. PCRO is compared firstly with the sequential CRO and then with the PGA. Simulations are performed by one TSP instance with large problem sizes. For fairness, same stopping criteria are used in our simulations for all algorithms. The results show that PCRO for TSP can improve the fitness value as well as reduce the computation time with a higher speedup.

For future work, we plan to compare PCRO with other meta-heuristics algorithms, design and try a deferent interconnection networks.

References

1. Vukmirović, S., Pupavac, D.: The Travelling Salesman Problem in the Function of Transport Network Optimization. Fakulty of Economics, Interdisciplinary Management Research IX, University in Osijek, Osijek (2013)
2. Zhan, F., Noon, C.: Shortest path algorithms: an evaluation using real road networks. Transp. Sci. (1996)
3. Al-Shaikh, A., Khattab, H., Sharieh, A., Sleit, A.: Resource utilization in cloud computing as an optimization problem. Int. J. Adv. Comput. Sci. Appl. (IJACSA) 7(6), 336–342 (2016)
4. Hoffman, K.L., Padberg, M., Rinaldi, G.: Traveling salesman problem. In: Encyclopedia of Operations Research and Management Science, pp 1573–1578. Springer (2016)
5. Lam, A.Y.S., Li, V.O.K.: Chemical reaction optimization: a tutorial. Memet. Comput. 4, 3–17 (2012)
6. Barney, B.: Introduction to Parallel Computing. Lawrence Livermore National Laboratory (2007). https://computing.llnl.gov/tutorials/parallel_comp/
7. Sleit, A., Salah, I., Jabay, R.: Approximating images using minimum bounding rectangles. In: ICADIWT 2008, pp. 394–396 (2008) https://doi.org/10.1109/ICADIWT.2008.4664379
8. Ostrouchov, G.: Parallel computing on a hypercube: an overview of the architecture and some applications. In: Heiberger, R.M. (ed.) Proceedings of the 19th Symposium on the Interface of Computer Science and Statistics, pp. 27–32. American Statistical Association (1987)
9. Kiasari, A., Sarbazi-Azad, H.: Analytic performance comparison of hypercubes and star graphs with implementation constraints. J. Comput. Syst. Sci. 74(6), 1000–1012 (2008)
10. Cathleen, L.: "Inside a NASA Production Supercomputing Center" Concept To Reality magazines, Summer/Fall issue (2011)
11. Mohan, A., Remya, G.: A parallel implementation of ant colony optimization for TSP based on MapReduce framework. Int. J. Comput. Appl. 88(8), 9–12 (2014)
12. Er, H.R., Erdogan, N.: Parallel genetic algorithm to solve traveling salesman problem on MapReduce framework using Hadoop cluster". arXiv preprint arXiv:1401.6267 (2014)
13. Sun, J., Wang, Y., Li, J., Gao, K.: Hybrid algorithm based on chemical reaction optimization and Lin-Kernighan local search for the traveling salesman problem (2011)
14. Shaheen, A., Sleit, A.: Comparing between different approaches to solve the 0/1 Knapsack problem. Int. J. Comput. Sci. Netw. Secur. 16(7), 1–10 (2016)
15. Barham, R., Sharieh, A., Sliet, A.: Chemical reaction optimization for max flow problem. (IJACSA) Int. J. Adv. Comput. Sci. Appl. 7(8), 189–196 (2016)
16. Deb, K.: An introduction to genetic algorithms. Sadhana 24(4–5), 293–315 (1999)
17. TSP Website: A collection of worldwide benchmark datasets (2009). http://www.math.uwaterloo.ca/tsp/world/countries.html. Accessed 15 Dec 2017

The Concept of the Method for Dynamic Control of Traffic Flows on Multi-lane Roads Based on Configurable Information Systems

Sergey Kucherov, Yuri Rogozov$^{(\boxtimes)}$, Julia Lipko, and Dmitry Elkin

Institute of Computer Technology and Information Security, Southern Federal University,
Taganrog, Russian Federation
{skucherov,yrogozov,ylipko,delkin}@sfedu.ru

Abstract. The random nature of the occurrence of road congestion has a negative impact on the capacity of the road network. Such congestion occurs at the delay points-sections of the road network on which the vehicle has to stop or slow down. Effective control of the capacity of delay points, thus, can be considered an important and urgent scientific problem. Despite the variety of effective methods of assessing the road situation, traffic routing, and road network planning, the task of dynamic control of capacity is not solved in a complex manner. Also an important aspect is the lack of effective means of mass informing the participants of road traffic on ways to avoid congestion. The random character (both spatial and temporal) of the appearance of delay points, in turn, requires the creation of automated control methods based on configurable information systems. The article presents the concept of a method for managing the movement of traffic flows, based on the elimination of the delay point by means of information technology.

Keywords: Delay point · Transport stream · Decision point
Configurable information system · Control

1 Introduction

In addition to effective traffic management on sections of the road network, important for large cities and megacities, the problem of eliminating the negative consequences from the delay points is topical. Below the delay point, here and below will be understood the portion of the road network on which the vehicle has to stop or slow down. The greatest difficulty is the fight against unforeseen delay points that occur on multi-lane roads at an arbitrary time in an arbitrary location. Examples of such delay points include traffic accidents, damage to the roadway, foreign objects on the lane, etc. The retention of the capacity of a section of the road network in this case can be ensured by a dynamic redistribution of traffic flows along lanes.

Infrastructural methods involving planning and physical construction of sections of the transport network are not applicable in this case, since the place and time of the delay point are unknown a priori. Individual information methods based on online services (Yandex.Traffic, Google maps) and systems of the VANET class allow forming advisory

R. Silhavy (Ed.): CSOC 2018, AISC 765, pp. 443–449, 2019.
https://doi.org/10.1007/978-3-319-91192-2_44

actions, and the degree of their spread and the presence of inertia (due to overhead costs in determining the actual traffic situation) can lead to unpredictable consequences.

For effectively eliminating the delay points on multi-lane sections of the road, it is necessary to apply massive prescriptive control actions, which requires the complex application of various methods for assessing the road situation, making decisions and informing road users.

For this reason, the problem of ensuring harmonization of different methods used in the field of dynamic control of transport streams at the delay points and building a sustainable architecture of an intelligent transport system with the potential for multiple use and interoperability of solutions is topical. The solution to this problem can be accomplished by the use of configurable information systems [1, 2], which allow to work under conditions of constant change of functional requirements, integrate heterogeneous data sources and are oriented to adaptation by the end user.

2   Related Work

The analysis of scientific publications and open sources has shown that today the problem of automated dynamic regulation of traffic flows is solved only in the field of adaptive traffic light regulation and speed limits of traffic flows.

From the point of view of the fundamental issues of dynamic control of transport flows, groups of particular methods for estimating the congestion of sections of the transport network are known [3–5], technical solutions for redirecting the transport stream along the lanes [6], methods and algorithms for routing [7, 8], technical solutions for operational notification of road accidents [9, 10]. The quality and efficiency of solving the tasks for the particular solutions known to the applicant is suitable for practical application. At the same time, complex methodological solutions to the problem of dynamic traffic management in the field of influence of delay points do not exist [11–15]. The main problem of the complex application of different methods is in the area of their harmonization and integration into a system of methods, which today is a topical scientific problem that has not found an unambiguous solution [16–18].

From the point of view of applied problems in the construction of intelligent transport systems, one of the key tasks is the development of a basic architecture of such a system, capable of working both with different methods of one type and with sets of newly introduced methods. The first Russian standard in the field of intelligent transport systems [19] raises the question of a consistent and terminologically defined approach to building the ITS architecture for the possibility of multiple use and interoperability of solutions. Thus, the intelligent transport system, primarily in terms of software, must have the property of configurability [1]. Ensuring the creation of such systems is also an actual scientific problem.

The main breakthrough in this field on the world stage was achieved in the 70s of the XX century with the development and implementation of the 70th SCOOT (Split, Cycle and Offset Optimization Technique) system [3], which had the ability to adaptively control the traffic light regulation based on the analysis sensors transport stream.

In further studies, for the most part, the questions of improving such algorithms began to be considered.

The main directions of research in the world science can be divided into two categories: improving existing intellectual methods of managing traffic flows and involving vehicles in the analysis and decision making on behavior in road conditions.

In the first category, the study can be divided into:

- improvement of algorithms for analysis of the road network based on new types of wireless sensors [4, 5], integration of additional devices directly into vehicles [4], analysis of load on the towers of cellular communication [6].
- development of new standards for the interaction of vehicles with roadside infrastructure [7].
- embedding intelligent transport systems technologies in existing traffic management methodologies, such as PATH, Dolphin, Auto21 CDS [4].

The second category is dominated by approaches based on the concept of cloud computing and the Internet of things [8–10]. The most relevant in the world scientific literature is the question of the organization of smart cities [9], based on the VANET (Automobile ad hoc network) architecture. The VANET architecture assumes that all vehicles are connected to each other and to a single information center in which data on their location, traffic status is processed and analyzed. In response to vehicles, information on specific recommendations for traffic on the road network site is sent.

From the point of view of the authors of the first category research, it is not possible to fully solve the task of dynamic traffic flow control on lanes, since they are oriented to efficient data collection. Studies of the second category are closest to author's ideas, but they are aimed at analyzing traffic in general and do not allow to eliminate the negative consequences of the appearance of delay points on individual lanes. In addition, the inclusion of cars in the contour of the intelligent transport system requires the transition to new brands and models, or retrofitting with the necessary equipment, which in most countries, including the Russian Federation, is unacceptable.

3 The Method Concepts

To solve the above problems, we propose a method for dynamic redistribution of traffic flows across bands, depending on the arising delay points, based on an agreed set of existing methods for detecting the moment of the delay point occurrence, estimating and forecasting traffic congestion, and deciding on the redistribution of transport streams along the lanes. The essence of the proposed method can be reflected in the following:

1. Identification of an event triggering the occurrence of a delay point (road accident, damage to the roadway, repair work)
2. Estimation and forecasting of the influence region of the delay point;
3. Decision-making on the redistribution of traffic flows along the lanes;
4. Determination of the point of application of control actions on traffic flows (distances for early rebuilding);
5. Application of control actions to the transport flow by technical means.

In more detail, we describe each item by dividing them into the following steps:

Step 1. Identify the delay point. The delay points can be static and stochastic, we will focus on stochastically occurring delay points and since static delay, points arise because of an improperly constructed scheme of traffic organization and are eliminated by reorganizing it [20]. Stochastically occurring delay points, most often road accidents and repair work can be detected using transport detectors, video cameras, messages from ERA-GLONASS [9], as well as a roadwork plan. After identifying the delay point, it is necessary to describe the area of its influence. An example of the work of the first stage is shown in Fig. 1.

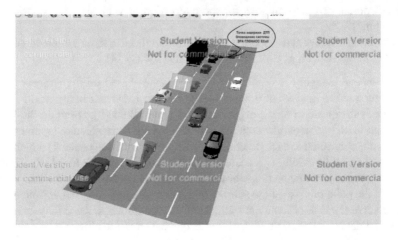

Fig. 1. Detection of the delay point

Step 2. It is necessary to determine what the situation is now on the road network section near the delay point. This can be done using a transport detector installed on the road network. The configurable information system, configured by the operator on the available set of transport detectors, evaluates through the business logic embedded in it. If the number of vehicles in this section exceeds the saturated flow boundary [21], therefore, in this section of the road network, there is a congestion we will describe a set of measures for this situation in the stage 2.1

In the other case, if the delay point has arisen, and the congestion is still not formed, then it is necessary to take measures to control the transport flows such as to prevent the formation of congestion on this site. In more detail, consider the set of measures in phase 2.2.

Stage 2.1. After determining the presence of traffic congestion, the configurable information system, based on the settings made by the operator, determines the growth rate of the influence area of the delay point. This is necessary in order to identify the nearest point of redirection of traffic flows on the border of the region (intersection, sign of variable information). Once the transport stream redirection point has been defined, the configurable information system using known traffic distribution models [8] offers the road users an alternative route with a return to the original path outside the zone of influence of the delay point. An example of the work of the second stage is shown in Fig. 2.

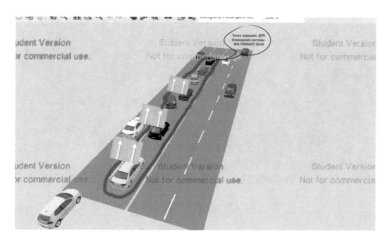

Fig. 2. Estimation of the influence region of the delay point

Stage 2.2. In the second case, when a delay point has arisen and a mash can only be formed, it is necessary to estimate the area of influence of the delay point and calculate the start time of the control of the transport stream for unhindered passage of vehicles through the area of influence of the delay point without significant losses in the speed of movement. This functionality is laid in a configurable information system and adapted by the operator to the capabilities of the monitored road network.

Stage 3. The configurable information system, based on the technical means available on the road network (signs of variable information, traffic lights, directing devices) redistributes traffic flows so that vehicles do not fall into the area of influence of the delay point or overcome it with minimal losses in the speed of movement. For example, if a delay point of the "road accident" type was formed on one of the lanes on the 2 + n lane road, a warning about the obstacle's detour is issued to the variable information signs in advance, and movement along the lane is prohibited shortly before the delay point itself. An example of the work of the third stage is shown in Fig. 3.

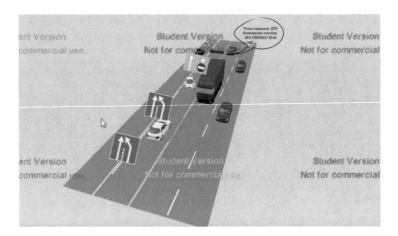

Fig. 3. Transport stream redirection

The use of this method for driving on a two-lane road in the event of a delay point caused by a traffic accident is shown in the illustrative material available by reference [22].

4   Conclusions and Future Work

The proposed method has two major advantages, such as a complex solution to the problem of managing traffic flows (through the harmonization and use of advanced methods and algorithms for estimating, forecasting, making managerial decisions, etc.). In addition, it have a massive nature of control actions, which will make the development of the road situation predictable.

Further work within the framework of the project will be devoted to the development of methods for harmonizing the methods of dynamic control of transport flows, allowing interchanging the methods of one class and the formation of the architecture of a configurable information system for controlling the movement of traffic on multi-lane roads.

Acknowledgment. The reported study was partially supported by RFBR, research project No. 18-07-00854 A.

References

1. Kucherov, S.A.: User-configurable information systems Magazine. Informatization and Communication No. 5, 135–137
2. Gercia, J., Goldszmidt, G.: Building SOA composite business services. http://www.ibm.com/developerworks/webservices/library/ws-soa-composite/
3. Kukadapwar, S.R., Parbat, D.K.: Evaluation of traffic congestion on links of major road network: a case study for Nagpur City. Int. J. Res. Eng. Sci. Technol. (2015). Deepam Publication's ISSN 2395-6453

4. Nagare, A., Bhatia, S.: International Journal of Applied Information Systems (IJAIS). Foundation of Computer Science FCS, New York, USA, vol. 1- No. 2, January 2012. ISSN 2249-0868
5. He, F., Yan, X., Liu, Y., Ma, L.: A traffic congestion assessment method for urban road networks. Procedia Eng. **137**, 425–433 (2016)
6. GOST 32865-2014. Automobile roads of general use. Variable message signs. Technical requirements
7. Popkov, V.K., Akhmediyarova, A.T., Kuandykova, D.R.: On one problem of transport route routing on urban transport networks. Natural and mathematical sciences in the modern world: Sat. Art. by mater. XXVII Intern. scientific-practical. Conf. No. 2(26), Novosibirsk, SibAK (2015)
8. Shvetsov, V.I.: Algorithms for distribution of transport flows. Avtomat. and tele-mekh., No. 10, 148–157, Autom. Remote Control, 70(10), 1728–1736 (2009)
9. GOST R 54619-2011 - Global navigation satellite system. Accident emergency response system. Protocols of data transmission from in-vehicle emergency call system/device to emergency response system infrastructure (2011)
10. GSM-Schutzengel (Automatic Emergency Call System to locate accident victims using GSM Technology). Jugend forscht (2001)
11. Feofilova, A.A.: Substantiation of conditions for the distribution of traffic on the street-road network of cities: dis. ... cand. tech. Sciences: 05.22.10/Feofilova Anastasiya Aleksandrovna; (Place of protection: Volgograd. Gos. Techn. un-t). - Rostov-on-Don, 150 p. (2013). ill. RSL OD, 61 14- 5/1784
12. Rupert, B., Wright, J., Pretorius, P., Cook, G., et al.: Traveler information systems in Europe (2003). www.international.fhwa.dot.gov
13. Lerner, N., Singer, J., Robinson, E., Huey, R., Jenness, J.: Driver Use of En Route Real-Time Travel Time Information. Final Report, 124 p. (2009)
14. Guan, J., Zheng, C., Zhu, X., et al.: VMS release of traffic guide information in Beijing olympics. J. Transp. Syst. Eng. Inf. Technol. **8**(6), 115–120 (2008)
15. Strickland, S.G., Berman, W.: Congestion Control and Demand Management. Public Roads On-Line, Winter (1995). www.tfhrc.gov/pubrds/winter95/p95wi1.htm
16. Rogozov, Y.: Approach to the construction of a systemic concept. Advances in Intelligent Systems and Computing. vol. 679, pp. 429–438 (2018)
17. Rogozov, Y.: Methodological approach to the construction of the concept of system. In: Proceedings of ISA RAS. T. 65, No. 1/2015, pp. 89–109 (2015)
18. Elkin, M.S.D., Rogozov, Y., Kucherov, S.: International multidisciplinary scientific GeoConference surveying geology and mining ecology management. SGEM 17(21), pp. 263–270
19. GOST R ISO 14813-1-2011. Intelligent transport systems. Reference model architecture (s) for the ITS sector. Part 1. ITS service domains, service groups and services, 32 p. (2012)
20. Elkin, D., Kucherov, S.: Development of the method of road regulation based on the analysis of delay points. Innovative technologies and didactics in teaching (ITDT 2017): collected papers, pp. 118–123. MVB Marketing- und Verlagservice des Buchhandels GmbH, Berlin (2017)
21. Borovskoy, A.E., Shevtsova, A.G.: The real flow of saturation depending on the class of the car. Bulletin of the Donetsk Academy of Automobile Transport, No. 2 (2012)
22. The method demonstration. https://yadi.sk/i/DzKKiZp63GYNNk

Author Index

A

Abrosimov, Mikhail, 272
Adeleke, Olawale J., 423
Ako, Churchill T., 423
Ali, Ammar Alhaj, 246
Al-Sharaeh, Saleh, 432
Antokhin, Yury, 176
Anusha, R., 290
Astorga, Gino, 1
Atallah, Rasha, 196
Aziz, Izzatdin Abdul, 13

B

Baranov, Sergey G., 186
Barot, Tomas, 53
Benes, Peter, 144
Botezatu, Nicolea, 82
Brovko, Alexander, 272
Bukovsky, Ivo, 144

C

Cauvery, N. K., 346
Chramcov, Bronislav, 233, 246
Crawford, Broderick, 1

D

Dai, Yu, 259
Dočkalová, Pavla, 223

E

Elkin, Dmitry, 443

G

Gilal, Abdul Rehman, 13

H

Hazarika, Jiten, 196
Honc, Daniel, 35
Horák, Tibor, 44
Hromada, Martin, 280
Huraj, Ladislav, 44
Hussain, Sadiq, 196

I

Indumathi, T. S., 413
Ivanov, Dmitry, 102

J

Jaafar, Jafreezal, 13
Jaafar, Nur Syakirah Mohd, 13
Jasek, Roman, 233, 246
Jaya Sankar, K., 356

K

Kadi, Mohamad, 233
Kamsin, Amirrudin, 196
Kavitha, G. R., 413
Krayem, Said, 233, 246
Kucherov, Sergey, 443

L

Lei, Jianjun, 259
Lipko, Julia, 443
Loukili, Mohamed, 337

M

Mahmood, Ahmad Kamil, 13
Mane, Preethi K., 212
Maniyath, Shima Ramesh, 63
Manuilov, Jury S., 155

© Springer International Publishing AG, part of Springer Nature 2019
R. Silhavy (Ed.): CSOC 2018, AISC 765, pp. 451–452, 2019.
https://doi.org/10.1007/978-3-319-91192-2

Matušů, Radek, 25
Maximova, Tatyana, 176
Mordane, Soumia, 337

N
Narasimha Rao, K., 212
Narsimha, G., 321
Navrátil, Pavel, 25
Neeli, Jyoti, 346

O
Oladepo, David A., 423

P
Padmaja, K. V., 364
Pakharev, Ruslan, 272
Patil, Premajyothi, 393
Patil, Shamshekhar S., 403
Patil, Shantakumar B., 393
Patra, Chiranjib, 82
Pavlov, Alexander N., 155
Pavlov, Dmitry A., 155
Pekař, Libor, 25
Peña, Alvaro, 1
Peng, Shengjie, 259
Pokorný, Pavel, 223
Pradeep Kumar, K., 311
Prasad Babu, B. R., 311
Prathiba, B., 356
Pudikov, Anton, 272

R
Raghvendra Rao, G., 300
Reddy, Chanda V., 364
Revák, Martin, 96
Reznikov, Konstantin, 272
Rogozov, Yuri, 443
Roopashree, H. R., 393
Ryzhikov, Yuri, 125

S
Seetharam, K., 374
Ševčík, Peter, 96
Shaheen, Ameen, 432
Shahina Parveen, M., 321
Sheliazhenko, Yurii, 74
Šimon, Marek, 44
Skala, Vaclav, 109
Sleit, Azzam, 432
Slin'ko, Alexey A., 155
Smolik, Michal, 109
Sokolov, Boris, 102
Sokolova, Irina, 176
Soto, Ricardo, 1
Spacek, Lubos, 133, 167
Sultana, Najiya, 382
Sumalatha, V., 356
Sunitha, N. R., 403

T
Thanikaiselvan, V., 63
Trofimova, Inna, 102
Tsypyschev, Vadim N., 116
Tukkoji, Chetana, 374

V
Vanamala, C. K., 300
Vávra, Jan, 280
Veena Devi Shastrimath, V., 290
Verzilin, Dmitrii, 176
Vojtesek, Jiri, 133, 167

Y
Yogeesh, A. C., 393

Z
Zacek, Petr, 233, 246
Žák, Samuel, 96
Zatopek, Jiri, 167

Printed in the United States
By Bookmasters